METHODS IN MOLECULAR BIOLOGY™

AF180895

Series Editor
John M. Walker
School of Life Sciences
University of Hertfordshire
Hatfield, Hertfordshire, AL10 9AB, UK

For other titles published in this series, go to
www.springer.com/series/7651

Protocols for In Vitro Cultures and Secondary Metabolite Analysis of Aromatic and Medicinal Plants

Edited by

S. Mohan Jain and Praveen K. Saxena

Department of Applied Biology, University of Helsinki, Helsinki, Finland
Department of Plant Agriculture, University of Guelph, Guelph, ON, Canada

Humana Press

Editor
S. Mohan Jain
Department of Applied Biology
University of Helsinki
Helsinki
Finland

Praveen K. Saxena
Department of Plant Agriculture
University of Guelph
Guelph, ON
Canada

ISSN: 1064-3745 e-ISSN: 1940-6029
ISBN: 978-1-61737-886-7 e-ISBN: 978-1-60327-287-2
DOI: 10.1007/978-1-60327-287-2

© Humana Press, a part of Springer Science+Business Media, LLC 2009
Softcover reprint of the hardcover 1st edition 2009
All rights reserved. This work may not be translated or copied in whole or in part without the written permission of the publisher (Humana Press, c/o Springer Science+Business Media, LLC, 233 Spring Street, New York, NY 10013 USA), except for brief excerpts in connection with reviews or scholarly analysis. Use in connection with any form of information storage and retrieval, electronic adaptation, computer software, or by similar or dissimilar methodology now known or hereafter developed is forbidden.
The use in this publication of trade names, trademarks, service marks, and similar terms, even if they are not identified as such, is not to be taken as an expression of opinion as to whether or not they are subject to proprietary rights.
While the advice and information in this book are believed to be true and accurate at the date of going to press, neither the authors nor the editors nor the publisher can accept any legal responsibility for any errors or omissions that may be made. The publisher makes no warranty, express or implied, with respect to the material contained herein.

Printed on acid-free paper

springer.com

Preface

Plants consistently synthesize, accumulate, and use a bewildering range of secondary metabolites as a part of their overall defense strategy. Many of these metabolites have been used around the world as medicines for various human health problems. In fact, more than 80% of the world's population relies on plants for principle health care. Nearly half of the medical prescriptions in the developed world are of plant origin.

In recent years the quest for quality of life and a common belief that plants are "natural and therefore safe" has paved the way for a wider acceptance of plant-based medicines worldwide. International trade in medicinal plants has become a major force in the global economy and the demand is increasing in both developed and developing countries. Thus, the continued rise in consumer demand for plant-based medicines and the expanding world population have resulted in indiscriminate harvest of wild species of medicinal plants. As well, a reduction of natural habitats for medicinal plants has placed many wild species in danger of extinction. The impact of rapid climate changes may also have an adverse effect on wild-plant species leading to the loss of useful genetic material. Most medicinal plants are harvested from the wild and the traditional agricultural and horticultural practices have not been developed even for most commonly used medicinal plant species.

The quality and consistency of the products are the most challenging issues facing the plant-based medicines. The production of medicinal metabolites in plants is affected by plant genotype, growth environment cultivation, harvesting, processing, and distribution. Medicinal plant preparations may also be contaminated with microbes and soil contaminants such as heavy metals, herbicides, pesticides, and other agricultural chemicals which can cause qualitative and quantitative changes in the levels of medicinal metabolites. The widespread occurrence of chemical variability and compromised quality of medicinal plants remain the major factors in inconsistent results of clinical trials of plant-based medicines. New regulations are currently being developed internationally to ensure consistency, safety and efficacy of plant-based medicines as well as how they are developed, manufactured, and marketed. Clearly, there is an imminent need for the development of new technologies and production approaches to improve the overall strategy of medicinal plant production to comply with up-coming legal regulations.

In vitro cell culture and controlled environment production systems offer an excellent opportunity for the selection and season-independent propagation of elite lines with specific, consistent levels of medicinal metabolites with minimum contamination. Additionally, the plant materials produced by in vitro techniques allow efficient application of the emerging analytical methods—such as metabonomics—for complete chemical profiling which has enormous potential for the discovery of new medicinal compounds.

Traditional breeding programs for medicinal plants are generally difficult to establish primarily due to lack of defined chemistry of medicinal components. Little knowledge of the genetic regulation of pathways of potential bioactive molecules further compounds this problem. In vitro techniques such as somaclonal variation, chemical mutagenesis, haploidy, protoplast fusion, and genetic transformation are applicable to create novel germplasm. The impact of these techniques is perhaps greatest in the improvement of medicinal plants since the resulting genetic diversity may open avenues for the discovery of new medicinal metabolites and treatments. Furthermore, the genetic manipulation of plant cells and organs has great advantages for producing secondary metabolites and other bioactive natural products. Together with the tools of chemical and genomic analyses, the in vitro culture methods hold the key to fundamental research on the biochemical and molecular basis of the mode of action of plant-based medicines. This book provides a detailed step-by-step description of protocols for the establishment of in vitro cultures of important medicinal plants, their mass multiplication in a controlled environment, and step-wise secondary metabolite analysis, genetic transformation, large-scale metabolite production in a bioreactor, and molecular markers. The role of altered microgravity in plant metabolite production is also described. In addition, many of these protocols will provide a basis for much needed efforts of in vitro germplasm conservation or cryopreservation of medicinal plant species at the brink of extinction as well as efforts to protect them from the adverse impact of rapid climatic changes. This book will certainly appeal to graduate and post graduate students, researchers, biotechnologists, industry, government agencies, and could be used as a text book.

This book contains 31 book chapters, divided into five sections. Section I contains 16 chapters describing step-wise protocols for micropropagation and chemical analysis of secondary compounds of different medicinal plants. Section II contains five chapters which address the transgenic approach for producing secondary metabolites. Section III contains two chapters which cover molecular markers/microsatellites. Section IV has six chapters which address biotransformation, bioreactors, and metablomics. Finally, section V contains two review chapters describing plant secondary metabolites in altered microgravity and role of biotechnology in producing anti-cancer compounds. Each chapter has been peer reviewed and revised accordingly.

We appreciate the time and effort that all reviewers have put into the development of these chapters which aided in improving the quality of the material presented herein. We extend our most sincere thanks to the staff of Humana Press for giving us the opportunity to present this book to our audience.

Helsinki, Finland ***S. Mohan Jain***
Guelph, Canada ***Praveen K. Saxena***

Contents

Contributors

ABDUL BAKRUDEEN ALI AHMED • *Department of Plant Science, School of Life Sciences, Bharathidasan University, Tiruchirappalli, Tamil Nadu, India*

A. AHUJA • *Biodiversity and Applied Botany Division, Indian Institute of Integrative Medicine, Jammu, India*

TAKASHI ASANO • *Graduate School of Pharmaceutical Sciences, Chiba University, Chiba, Japan, CREST, Japan Science and Technology Agency, Saitama, Japan*

SHASHI B. BABBAR • *Department of Botany, University of Delhi, Delhi, India*

K. NIRMAL BABU • *Indian Institute of Spices Research, Kerala, India*

INDIRA BALACHANDRAN • *Centre for Medicinal Plants Research, Arya Vaidya Sala, Kerala, India*

ROGER A. BARTHELSON • *Plant Sciences, University of Arizona, Tucson, AZ*

MARK BERHOW • *New Crops and Processing Technology Research, National Center for Agricultural Utilization Research, Agricultural Research Service-USDA, Peoria, IL*

SABHYATA BHATIA • *National Institute of Plant Genome Research, Aruna Asaf Ali Marg, New Delhi, India*

NISHRITHA BOPANA • *TERI University, New Delhi, India*

FRÉDÉRIC BOURGAUD • *UMR Nancy-Université-INRA Agronomie et Environnement Nancy-Colmar, Vandoeuvre-Lès-Nancy, France*

KATARÍNA BRUŇÁKOVÁ • *P. J. Šafárik University in Košice, Faculty of Science, Institute of Biology and Ecology, Košice, Slovakia*

ELIZABETE CATAPAN • *Departamento de Botânica, Universidade Federal de Santa Catarina, Florianopolis, SC, Brazil*

IAN B. COLE • *Chemistry, I.K. Barber School of Arts & Sciences, University of British Columbia, Okanagan, Kelowna, BC, Canada*

MÁRCIO LUÍS BUSI DA SILVA • *Departamento de Botânica, Universidade Federal de Santa Catarina, Florianopolis, SC, Brazil*

ALBERTO C.P. DIAS • *Departamento de Biologia, Universidade do Minho, Campus de Gualtar, Braga, Portugal*

PAULO FERNANDO DIAS • *Plant Morphogenesis and Biochemistry Laboratory, Federal University of Santa Catarina, Florianopolis, SC, Brazil*

MINOO DIVAKARAN • *Providence Women's College, Kerala, India*

SÉBASTIEN DOERPER • *UMR Nancy-Université-INRA Agronomie et Environnement Nancy-Colmar, Vandoeuvre-Lès-Nancy, France*

P. DUTT • *Natural Product Chemistry Division, Indian Institute of Integrative Medicine, Jammu, India*

FAISAL T. FAROOQ • *Chemistry, I.K. Barber School of Arts & Sciences, University of British Columbia, Okanagan, Kelowna, BC, Canada*

VALDIR CECHINEL FILHO • *Núcleo de Investigações Químico-Farmacêuticas, Universidade do Vale do Itajaí, Itajaí, SC, Brazil*

G. FRANKLIN • *Departamento de Biologia, Universidade do Minho, Campus de Gualtar, Braga, Portugal*

S. Pillai Geetha • *Crop Improvement and Biotechnology Division, Centre for Medicinal Plants Research, Arya Vaidya Sala, Kerala, India*
Satheesh George • *Centre for Medicinal Plants Research, Arya Vaidya Sala, Kerala, India*
M.K. Goel • *Plant Tissue Culture Division, Central Institute of Medicinal and Aromatic Plants (CIMAP), CSIR, Lucknow, India, Department of Genetics, University of Delhi, South Campus, New Delhi, India*
Eric Gontier • *EA3900 Biologie des Plantes et contrôle des Insectes ravageurs, Université de Picardie Jules Verne, Faculté des Sciences, Amiens cedex, France*
Dragoljub Grubišić • *Institute for Biological Research "Siniša Stanković", University of Belgrade, Belgrade, Serbia*
Zsuzsanna György • *Department of Process and Environmental Engineering, University of Oulu, Oulu, Finland*
Hely Häggman • *Department of Biology, University of Oulu, Oulu, Finland*
Eun Joo Hahn • *Research Center for the Development of Advanced Horticultural Technology, Chungbuk National University, Cheongju, South Korea*
Alain Hehn • *UMR Nancy-Université-INRA Agronomie et Environnement Nancy-Colmar, Vandoeuvre-Lès-Nancy, France*
Anja Hohtola • *Department of Biology, University of Oulu, Oulu, Finland*
Sung Jin Hwang • *Department of Biology, Chonnam National University, Gwang-ju, Korea*
Laura Jaakola • *Department of Biology, University of Oulu, Oulu, Finland*
Teodora Janković • *Institute for Medicinal Plants Research "Josif Pančić", Belgrade, Serbia*
Slađana Jevremović • *Institute for Biological Research "Siniša Stanković", University of Belgrade, Belgrade, Serbia*
Nebojsa Jockovic • *Institute of Pharmacy, Pharmaceutical Biology and Pharmacology, Martin-Luther University Halle-Wittenberg, Halle/Saale, Germany*
Riitta Julkunen-Tiitto • *Faculty of Biosciences, University of Joensuu, Joensuu, Finland*
M.K. Kaul • *Biodiversity and Applied Botany Division, Indian Institute of Integrative Medicine, Jammu, India*
Sushma Kaul • *Biodiversity and Applied Botany Division, Indian Institute of Integrative Medicine, Jammu, India*
Amandeep Kaur • *National Research Centre on Plant Biotechnology, Indian Agricultural Research Institute, New Delhi, India*
S.P.S. Khanuja • *Genetic Resource and Biotechnology Division, Central Institute of Medicinal and Aromatic Plants (CIMAP), CSIR, Lucknow, India*
Suk Weon Kim • *Biological Resource Center, Korea Research Institute of Bioscience and Biotechnology (KRIBB), Daejeon, South Korea*
Ján Košuth • *P. J. Šafárik University in Košice, Faculty of Science, Institute of Biology and Ecology, Košice, Slovakia*
A.K. Kukreja • *Plant Tissue Culture Division, Central Institute of Medicinal and Aromatic Plants (CIMAP), CSIR, Lucknow, India*
Paul Lacoste • *UMR Nancy-Université-INRA Agronomie et Environnement Nancy-Colmar, Vandoeuvre-Lès-Nancy, France*

HEMANT LATA • *National Center for Natural Products Research, School of Pharmacy, The University of Mississippi, University, MS, USA*
LANFANG H. LEVINE • *Dynamac Corporation, Space Life Sciences Laboratory, Kennedy Space Center, FL, USA*
KARINE LIÈVRE • *UMR Nancy-Université-INRA Agronomie et Environnement Nancy-Colmar, Vandoeuvre-Lès-Nancy, France*
CHUN-ZHAO LIU • *National Key Laboratory of Biochemical Engineering, Institute of Process Engineering, Chinese Academy of Sciences, Beijing, P.R. China*
MAZIAH MAHMOOD • *Department of Biochemistry, Faculty of Biotechnology and Biomolecular Sciences, Universiti Putra Malaysia, Selangor, Malaysia*
MARCELO MARASCHIN • *Plant Morphogenesis and Biochemistry Laboratory, Federal University of Santa Catarina, Florianopolis, SC, Brazil*
GERALD MARTIN • *Centre for Medicinal Plants Research, Arya Vaidya Sala, Kerala, India*
S. MEHROTRA • *Plant Tissue Culture Division, Central Institute of Medicinal and Aromatic Plants (CIMAP), CSIR, Lucknow, India, Department of Biotechnology, Institute of Engineering and Technology, Lucknow, India*
CASSIA S. MIZUNO • *USDA-ARS, Natural Products Utilization Research Unit, University, MS, USA*
RITA M. MORAES • *National Center for Natural Products Research, School of Pharmacy, The University of Mississippi, University, MS, USA*
HILIANA NUNES FERREIRA MORAIS • *Plant Morphogenesis and Biochemistry Laboratory, Federal University of Santa Catarina, Florianopolis, SC, Brazil*
FÁBIO NETTO MORENO • *Departamento de Botânica, Universidade Federal de Santa Catarina, Florianopolis, SC, Brazil*
SUSAN J. MURCH • *Chemistry, I.K. Barber School of Arts & Sciences, University of British Columbia, Okanagan, Kelowna, BC, Canada*
HOSAKATTE NIRANJANA MURTHY • *Research Center for the Development of Advanced Horticultural Technology, Chungbuk National University, Cheongju, South Korea, Department of Botany, Karnatak University, Dharwad, India*
MARY E. MUSGRAVE • *Department of Plant Science, University of Connecticut, Storrs, CT, USA*
KAROLIINA NIEMI • *Department of Applied Biology, University of Helsinki, Helsinki, Finland*
RIVALDO NIERO • *Departamento de Quimico, Universidade Federal de Santa Catarina, Florianopolis, SC, Brazil*
ROSLI NOORMI • *Department of Biochemistry, Faculty of Biotechnology and Biomolecular Sciences, Universiti Putra Malaysia, Selangor, Malaysia*
MYUNG JIN OH • *Biological Resource Center, Korea Research Institute of Bioscience and Biotechnology (KRIBB), Daejeon, South Korea*
MARGARIDA M. OLIVEIRA • *Instituto de Biologia Experimental e Tecnológica (IBET), Av. da República – Quinta do Marquês, Oeiras, Lisboa, Portugal*
MICHEL FLEITH OTUKI • *Departamento de Botânica, Universidade Federal de Santa Catarina, Florianopolis, SC, Brazil*
KEE YOEUP PAEK • *Research Center for the Development of Advanced Horticultural Technology, Chungbuk National University, Cheongju, South Korea*

ÊNIO LUIZ PEDROTTI • *Plant Morphogenesis and Biochemistry Laboratory, Federal University of Santa Catarina, Florianopolis, SC, Brazil*
ZHENG-SONG PENG • *Institute of Rare Animals and Plants, China West Normal University, Nanchong, Sichuan, P.R. China*
ANNA MARIA PIRTTILÄ • *Department of Biology, University of Oulu, Oulu, Finland*
A.V. RAGHU • *Centre for Medicinal Plants Research, Arya Vaidya Sala, Kerala, India*
ADHIKARLA SURYANARAYANA RAO • *Department of Biotechnology, School of Life Sciences, Bharathidasan University, Tiruchirappalli, Tamil Nadu, India*
MANDALI VENKATESWARA RAO • *Department of Plant Science, School of Life Sciences, Bharathidasan University, Tiruchirappalli, Tamil Nadu, India*
KAISU RIIHINEN • *Department of Biosciences, University of Kuopio, Kuopio, Finland*
ANETA SABOVLJEVIC • *Institute of Botany, Faculty of Biology, University of Belgrade, Belgrade, Serbia*
MARKO SABOVLJEVIC • *Institute of Botany, Faculty of Biology, University of Belgrade, Belgrade, Serbia*
KAZUKI SAITO • *Graduate School of Pharmaceutical Sciences, Chiba University, Chiba, Japan, RIKEN Plant Science Center, Yokohama, Japan*
M. SAMBYAL • *Biodiversity and Applied Botany Division, Indian Institute of Integrative Medicine, Jammu, India*
TYTTI SARJALA • *Finnish Forest Research Institute, Parkano Research Unit, Parkano, Finland*
PRAVEEN K. SAXENA • *Department of Plant Agriculture, University of Guelph, Guelph, ON, Canada*
SANJAY SAXENA • *The Energy and Resources Institute (TERI), New Delhi, India*
K. SHANKER • *Analytical Chemistry Division, Central Institute of Medicinal and Aromatic Plants (CIMAP), CSIR, Lucknow, India*
BHUMIKA SHOKEEN • *National Institute of Plant Genome Research, Aruna Asaf Ali Marg, New Delhi, India*
ANGELINA SUBOTIĆ • *Institute for Biological Research "Siniša Stanković", University of Belgrade, Belgrade, Serbia*
HIROSHI SUDO • *Graduate School of Pharmaceutical Sciences, Chiba University, Chiba, Japan*
K.A. SURI • *Natural Product Chemistry Division, Indian Institute of Integrative Medicine, Jammu, India*
BRIGITTE THOMASSET • *UMR-CNRS 6022, Génie Enzymatique et Cellulaire, UTC, Compiègne Cedex, France*
BRENT TISSERAT • *New Crops and Processing Technology Research, National Center for Agricultural Utilization Research, Agricultural Research Service-USDA, Peoria, IL, USA*
THI LÊ MINH TRAN • *Department of Biotechnology, Nong Lam University, Ho Chi Minh City, Vietnam*
LINDSEY K. TUOMINEN • *Warnell School of Forestry and Natural Resources, University of Georgia, Athens, GA, USA*
JAGAN V. VALLURI • *Cell Biology Laboratory, Division of Biological Sciences, Marshall University, Huntington, WV, USA*

STEVEN F. VAUGHN • *New Crops and Processing Technology Research, National Center for Agricultural Utilization Research, Agricultural Research Service-USDA, Peoria, IL, USA*
ANA MARIA VIANA • *Departamento de Botânica, Universidade Federal de Santa Catarina, Florianopolis, SC, Brazil*
NEETIKA WALIA • *The Energy and Resources Institute (TERI), Delhi, India*
KARL VERNON WOOD • *Department of Chemistry, Campus-Wide Mass Spectrometry Center, Purdue University, West Lafayette, IN, USA*
MAMI YAMAZAKI • *Graduate School of Pharmaceutical Sciences, Chiba University, Chiba, Japan, CREST, Japan Science and Technology Agency, Saitama, Japan*
JUN YANG • *College of Life Science, China West Normal University, Nanchong, Sichuan, P.R. China*
ROSENDO AUGUSTO YUNES • *Departamento de Química, Universidade Federal de Santa Catarina, Florianópolis, SC, Brazil*

Part I

Micropropagation and Chemical Analysis

Chapter 1

Establishment of Adventitious Root Cultures of *Echinacea purpurea* for the Production of Caffeic Acid Derivatives

Kee-Yoeup Paek, Hosakatte Niranjana Murthy, and Eun-Joo Hahn

Summary

Echinacace purpurea (purple cone flower) is an important medicinal plant, and widely used for phytochemical purposes. The roots are traditionally used in herbal medicines and dietary supplements as an immunostimulant in treating inflammatory and viral diseases. Extensive research work has been carried out on both the induction of adventitious roots from *E. purpurea* as well as established small-scale (shake flask) to large-scale (bioreactor) cultures for the production of adventitious root biomass and caffeic acid derivatives. This chapter describes the methodologies of induction of adventitious roots from explants of *E. purpurea*, propagation of adventitious roots in suspension cultures, estimation of total phenolics, flavonoids, and antioxidant activities. The detailed methodology for high-performance liquid chromatographic analysis of caffeic acid derivatives present in the adventitious roots is also discussed.

Key words: Adventitious roots, Bioreactor cultures, Caffeic acid derivatives, *Echinacea purpurea*, flavonoids, Medicinal plant, Phenolics

1. Introduction

Echinacea purpurea (L.) Moench., also known as purple cone flower, is an important medicinal plant native to North America. It is widely grown commercially in various parts of the world. The plant extracts have shown antioxidative, antibacterial, antiviral and antifungal properties, and are being used in treatment of the common cold as well as respiratory and urinary diseases *(1)*.

S. Mohan Jain and Praveen K. Saxena (eds.), *Methods in Molecular Biology, Protocols for* In Vitro *Cultures and Secondary Metabolite Analysis of Aromatic and Medicinal Plants, vol. 547*
© Humana Press, a part of Springer Science+Business Media, LLC 2009
DOI: 10.1007/978-1-60327-287-2_1

The main bioactive constituents are caffeic acid derivatives, alkylamides, polyacetylenes, and polysaccharides *(2)*. With regard to caffeic acid derivatives, several important compounds have been identified from the hydrophilic fractions of *E. purpurea*; these being are caftaric acid, cichoric acid, and chlorogenic acid.

Plant cell, tissue, and organ culture technology are ideal for the production of useful plant-specific metabolites including pharmaceuticals *(3, 5)*. In recent years, various types of bioreactor systems have been developed for plant cell and organ cultures (e.g., mechanically-agitated [stir tank] type and pneumatically agitated [airlift type] bioreactors *(4, 6)*. Bioreactors are glass or steel vessels in which cells and organs are cultured. They allow for sampling of cultures or medium, addition of fresh medium, and mixing of cultures without risk of compromising the aseptic nature of the culture. In addition, some bioreactors are capable of monitoring pH, temperature, and dissolved oxygen. Bioreactors can be installed in environmentally controlled rooms or chambers. Among pneumatically agitated bioreactors, airlift bioreactors are more convenient for cell and organ cultures *(6)*. These types of bioreactors use compressed air to aerate and mixing the contents of the reactor (*see* **Fig. 1**). Such bioreactors have been used for the production of biomass and metabolites of various medicinal plants including Korean ginseng (*Panax ginseng*) and Siberian ginseng (*Acanthopanx senticosus*). We have recently induced adventitious roots from explants of *E. purpurea* and established a suspension culture system for the production of caffeic acid derivatives (CAD) *(7–9)*.

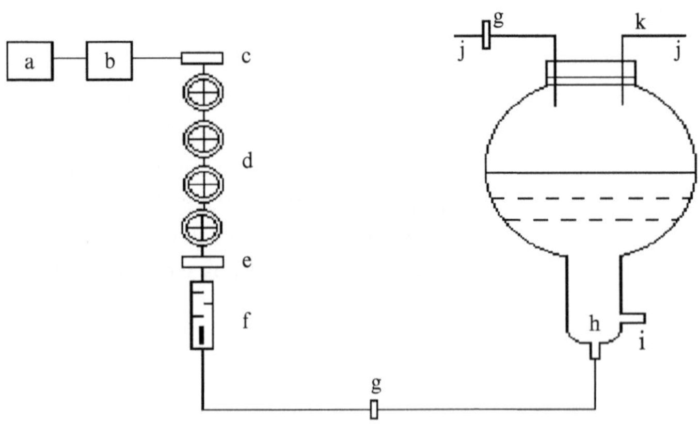

Fig. 1. Schematic diagram of an air lift bioreactor. (**a**) air compressor, (**b**) air reservoir, (**c**) air cooling device, (**d**) air filter system, (**e**) air dryer, (**f**) air flow meter, (**g**) membrane filter, (**h**) glass sparger, (**i**) medium sampling port, (**j**) vent, (**k**) pre filter.

2. Materials

2.1. Induction of Adventitious Roots from Root Explants and Establishment of Suspension Cultures

1. Roots of *Echinacea purpurea* from field- or greenhouse-grown plants.
2. Murashige and Skoog (MS) *(10)* medium stocks (MS stock I, II, III, and IV). Stock may be prepared using individual chemicals (**Table 1**). Store in the freezer or cold room at 4°C (*see* **Note 1**).

Table 1
Composition of MS medium[a]

Chemical constituents	Concentration (mg/L)	Volume per liter (mL)
Major inorganic nutrients		
NH_4NO_3	33,000	
KNO_3	38,000	
$CaCl_2{\cdot}2H_2O$	8800	
$MgSO_4{\cdot}7H_2O$	7400	
KH_2PO_4	3400	50
Minor inorganic nutrients		
KI	166	
H_3BO_3	1240	
$MnSO_4{\cdot}4H_2O$	4460	
$ZnSO_4{\cdot}7H_2O$	1720	
$Na_2MoO_4{\cdot}2H_2O$	50	
$CuSO_4{\cdot}5H_2O$	5	
$CoCl_2{\cdot}6H_2O$	5	5
Iron source		
$FeSO_4{\cdot}7H_2O$	5,560	
$Na_2EDTA_2H_2O$	7,460	5
Organic supplement		
myo-Inositol	20,000	
Nicotinic acid	100	
Pyridoxine–HCl	100	

(continued)

Table 1 (continued)

Chemical constituents	Concentration (mg/L)	Volume per liter (mL)
Thiamine HCl	100	
Glycine	400	5
Carbon source		
Sucrose	As per the experiment	

[a]After dissolving all the stock solutions in enough deionized water make it up to 1 L, adjust the pH to 6.0 (add 2 g/L gelrite if semi-solid medium) and autoclave at 25 min at 120°C

3. Indole butyric acid (IBA) (Duchefa, The Netherlands). Prepare stock solution (100 μM) and store in the freezer at −20°C (*see* **Note 2**).
4. Gelrite (Duchefa Biochemie BV; Haarlem, The Netherlands).
5. Petri dishes (15 mm × 100 mm).
6. 250-mL Erlenmeyer flasks.
7. 5-L Balloon-type bubble bioreactors (available with Samsung Scientific Company; Seoul, South Korea) (*see* **Note 3**).
8. Polytetrafluoroethylene air filters: 0.20 μm PTFE membrane filters (Midisart 2000) (Sortorius Inc.; Mississauga, Canada).

2.2. Estimation of Total Phenols

1. 2 N folin-cicalteu reagent (Sigma). Store at 4°C.
2. Gallic acid (Sigma). Generally 1,000 ppm stock is prepared and stored at 4°C in the dark (*see* **Note 4**).
3. 20% Sodium carbonate solution. Prepare a stock solution and store at 4°C.
4. 80% methanol, high-performance liquid chromatography (HPLC) grade.

2.3. Estimation of Total Flavonoids

1. (+) Catechin (Sigma). Prepare 1000 ppm stock solution and store at 4°C in the dark.
2. 10% Aluminum chloride solution (Sigma). Store stock solution at 4°C.
3. 5% Sodium nitrate solution. Store stock solution at 4°C.

2.4. Scavenging Effect on 2, 2-Diphenyl-1-picrylhydrazyl (DPPH) Radical (Scavenging Activity of Natural Antioxidants)

1. 2, 2'-diphenyl-1-picrylhydrazyl (DPPH) (*see* **Note 5**).
2. Ultraviolet (UV)-visible spectrophotometer (UV-1650PC, Shimadzu, Japan).

2.5. HPLC Analysis of Caffeic Acid Derivatives

1. HPLC-grade acetonitrile (Duksan Pure Chemicals Co.; Ansan, South Korea) (*see* **Note 6**).
2. HPLC-grade water (Duksan Pure Chemicals Co.; Ansan, South Korea) (*see* **Note 6**).
3. Standard caffeic acid derivatives: caftaric acid, chlorogenic acid and chichoric acid (ChromaDex; Laguna Hills, CA).

3. Methods

3.1. Induction of Adventitious Roots from Root Explants

1. *Echinacea purpurea* plants are grown under controlled conditions of a 16 h photoperiod 200 μmol/m^2 s, at 24 ± 2°C/21 ±12°C, day/night. Water the plants daily and supply fertilizers once in 15 d; foliar spray, N:P:K, 10:4:4. (Wyu-il Co.; Seoul, South Korea).
2. Collect the healthy roots from the plants and wash thoroughly in tap water. Immerse in 95% ethanol for 10 s, and surface sterilize in a 20% sodium hypochlorite solution containing few drops of Tween-30 detergent for 10 min. (*see* **Note 7**).
3. Rinse 3× in cold, sterile, distilled water for 5 min.
4. Cut the roots into 1-cm long segments under the laminar flow bench and culture individually on a semi-solid MS medium. Supplement with 50 g/L sucrose, 4.9 μM IBA, and 0.2% gelrite, pH 6.0.
5. Incubate the cultures at 24 ± 2°C for 4 wk in the dark. Within 4 wks adventitious roots should develop from the root explants (**Fig. 2a, b**). The explants should be subcultured to the fresh medium at 4-wk interval.
6. Either adventitious roots or calli is induced from the root explants after 4 wk of culture. Adventitious roots are induced from the calli on 0.75 strength MS medium amended with 50 g/L sucrose and 4.9 μM IBA and 0.2% gelrite.
7. The adventitious roots proliferate on the same medium at 4-wk intervals, which were used as explants for further experiments.

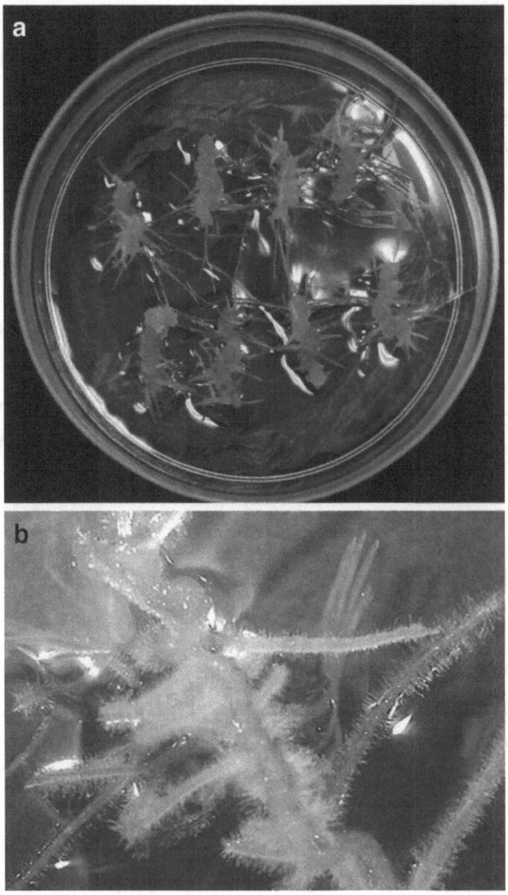

Fig. 2. Adventitious roots developing from the root explants of *Echinacea purpurea* on MS medium supplemented with supplemented with 50 g/L sucrose and 4.9 μM IBA (**a**, **b**).

3.2. Culture and Multiplication of Adventitious Roots in the Liquid Medium

1. Collect actively growing roots from the semi-solid cultures, cut them into 2-cm long segments and inoculate minimum 1 g fresh biomass into a 250-mL flask containing 100 mL half strength MS liquid medium containing 50 g/L sucrose and 4.9 μM IBA (*see* **Note 8**).
2. Incubate cultures in the dark and shake at 100 rpm. Maintain the cultures by regular sub culturing at 4-wk interval.

3.3. Suspension Cultures for the Production of CADs in Flask Scale Cultures

1. Collect the roots in the exponential phase or stationary phase and initiate suspension cultures for the production of CADs from the adventitious root.
2. Inoculate 2 g of fresh 2-cm long root biomass (inoculum density 10 g/L) in a 500-mL Erlenmeyer flask containing 200 mL half strength MS medium. Modify with nitrogen source, 5 NH_4^+: 25 NO_3^- mM, supplemented with 9.8 μM IBA and 5% sucrose.

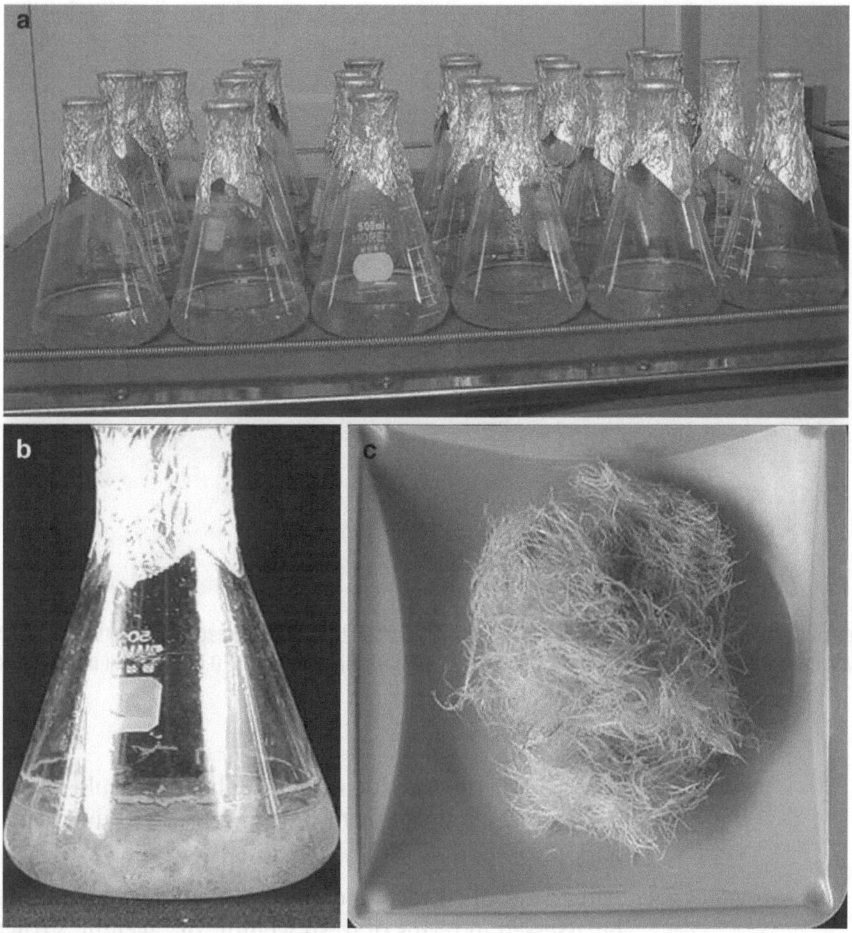

Fig. 3. Suspension culture of *Echinacea purpurea* adventitious roots in flasks (250 mL Erlenmeyer shake flasks, (**b**); Growth after 4 wk in culture (**b**); Harvested fresh roots (**c**).

3. Incubate the cultures and shake at 100 rpm in the dark at 25°C. Adventitious roots grow and multiply further (**Figs. 3a, b**).
4. After 4 wk of culture, the growth of adventitious roots should be assessed in terms of fresh weight (**Fig. 3c**), dry weight, growth ratio, and content of phenols, flavonoids and CADs.

3.4. Adventitious Root Cultures in Bioreactors for the Production of CADs

1. Culture 7 g/L adventitious roots in a 5 L capacity airlift bioreactor containing 4 L half strength modified MS medium having ammonium and nitrate ratio is 5: 25 mM and is supplemented with 9.8 μM IBA and 50 g/L sucrose.
2. Bioreactor cultures are maintained in the dark at 25°C. The cultures should be aerated with an air flow of 0.1 vvm (air volume/culture volume per min). Adventitious roots grow profusely and multiply in bioreactor cultures (**Fig. 4**) (*see* **Note 9**).

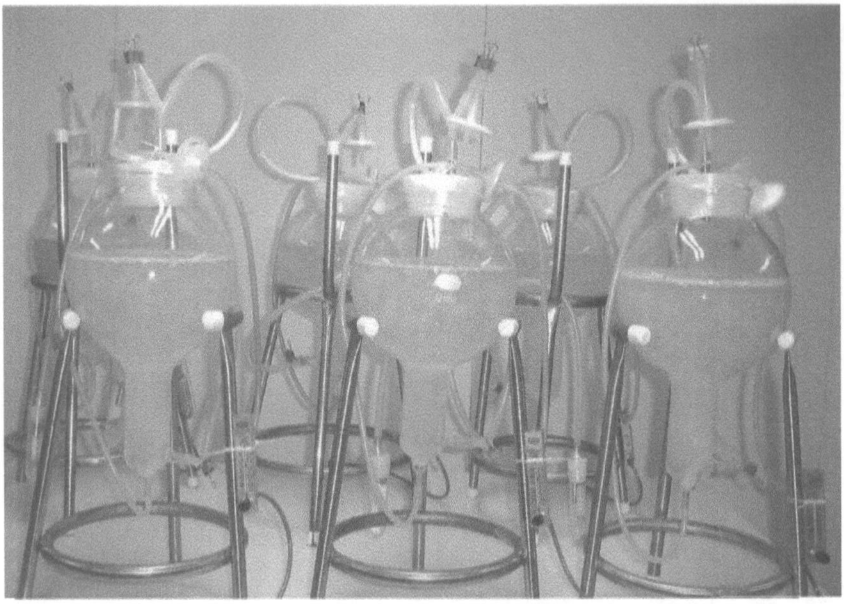

Fig. 4. Adventitious roots of *E. purpurea* growing in 5 L capacity balloon type airlift bioreactors containing 4 L of half strength modified MS medium (ammonium and nitrate ratio is 5: 25 mM) supplemented with 9.8 μM IBA and 50 g/L sucrose.

3. After 3 wk of dark incubation cultures should be irradiated under 3/21 h fluorescent light intensity 40 μmol/m^2 s and a dark photoperiod for an additional 2 wk.
4. After 5 wk of culture, the growth of adventitious roots should be assessed in terms of fresh weight, dry weight, growth ratio, and content of phenols, flavonoids and CADs.

3.5. Estimation of Root Biomass

1. The roots were separated from the media by passing them through a stainless steel sieve. Root fresh weight/fresh biomass is determined after the rinsing of roots in sterile water and blotting of excessive surface water. Biomass produced in the bioreactors is shown in **Fig. 5**.
2. Dry weight/dry biomass is recorded after drying of roots at 40°C in forced air oven to a constant weight (*see* **Note 10**).
3. The growth ratio was determined by: GR = harvested dry biomass (g) – inoculated dry biomass (g)/inoculated dry biomass (g).

3.6. Preparation of Root Extract for Estimation of Phenols and Flavonoids

1. 0.5 g Ground dried root material is put into a 50 mL polypropylene conical tube containing 10 mL 80% ethanol and stir for 15 min.
2. Cdentrifuge contents at 3000 rpm (11.76 × g) in a refrigerated centrifuge for 10 min;. The supernatant solution is filtered under vacuum into a volumetric flask and the filtrate is saved.

Fig. 5. Adventitious root biomass of *E. purpurea* spread in trays for drying.

3. Re-extract the resdiue as in **steps 1** and **2** and increase the final volume to raised to 25 mL.

3.7. Estimation of Total Phenol Contents in the Adventitious Root Extract

1. The amount of total phenols in the adventitious root extracts can be analyzed spectrophotometrically by using Folin-Ciocalteu reagent.
2. 100 μL Methanolic extract is mixed with 2.5 mL deionized water, and add 0.1 mL 2-*N* folin-ciocalteu reagent. The contents are mixed well and stand for 6 min.
3. After 6 min, 0.15 mL 20% sodium carbonate solution is added. The color will develop after incubation at room temperature for 30 min.
4. The absorbance of solutions is detected at 760 nm on a UV visible spectrophotometer. The measurements are compared to the standard curve for gallic acid. The results are expressed as the mg of gallic acid equivalent per gram of dry roots (*see* **Note 11**).

3.8. Estimation of Total Flavonoid Contents in the Adventitious Root Extract

1. The amount of total flavonoids in the adventitious root extracts should be analyzed spectrophotometrically by aluminum chloride colorimetric method (*see* **Note 12**).
2. Mix a 0.25 mL Methanolic root extract and a (+) catechin standard solution with 1.25 mL deionized water Add to 75 μL 5% sodium nitrate solution and let stand for 6 min.

3. After 6 min, add 0.15 mL 10% aluminum chloride solution and allow mixture to stand for 5 min. Add 0.5 mL 1 M sodium hydroxide and mix the contents well.
4. The absorbance is measured immediately at 510 nm on a spectrophotometer (UV-1650PC; Shimadzu, Japan). The results are expressed as mg of (+) catechin equivalents per gram of dry roots.

3.9. Scavenging Effect on 2, 2-Diphenyl-1-picrylhydrazyl (DPPH) Radical (Scavenging Activity of Natural Antioxidants)

1. For the analysis of antioxidants, extract 0.5-mL aliquots of each extract and mixed with 300 μL 1 mM methanolic solution of DPPH*. in a 4 mL cuvet. Bring to a total volume of 3.0 mL with methanol. Prepare fresh daily. Store in the dark at 4°C in a flask covered with aluminum foil.
2. After incubation in the dark at room temperature for 15 min, the reaction mixture is assayed at 517 nm using a UV-visible spectrophotometer (*see* **Note 13**).
3. In order to eliminate interference with the DPPH* reaction by extracted pigments, blanks of the extracts are assayed using 300 μL methanol instead of the DPPH* solution. A DPPH *blank sample, containing 2.7 mL methanol and 300 μL of DPPH* solution, is prepared and assayed daily. All experiments are carried out in duplicate and repeated at least 2×.
4. The percentage decrease in the absorbance at 517 nm is recorded for each concentration and the percentage of quenching of the DPPH* radical is calculated on the basis of the observed decrease of the radical. The inhibition percentage is calculated according to the formula:

$$\text{Inhibition percentage} = [(A_{\text{DPPH}} - A_{\text{Extr}})/A_{\text{DPPH}}] \times 100$$

where A_{DPPH} is the absorbance value of the DPPH* blank sample and A_{Extr} was evaluated as the difference between the absorbance value of the test solution and that of its blank. Curves showing inhibition percentage/μL of extract are used to find the concentrations at which 50% radical scavenging occurred (EC_{50}).

3.10. HPLC Analysis of CADs

1. Extraction of caffeic acid derivatives should be carried out as described in the **Subheading 3.6**. All the extracts are filtered through a 0.45 PTFE filter in an HPLC vial and cap them.
2. Chromatography is performed on any modular model 110 systems consisting of vacuum degasser, quaternary pump, auto sampler, thermo stated column compartment, and diode array detector (DAD).
3. Analyses were performed with XTerra RP 18 column, particle size 3.0 μM, 150 mm × 3 mm. The mobile was (a) aqueous phosphoric acid solution (0.1%) and (b) acetonitrile.

The caffeic acid fractions were separated by gradient elution as follows: initial 10% b for 40 min; 25% b for 11 min; and 50% b for 1 min; with recycling to the initial condition for 8 min, at a flow rate of 0.3 mL/min. The detector monitored the eluent at 330 nm. The column temperature was set at 26°C. The sample injection wass set at 5 μL. Three injections were performed for each sample.

4. Peaks were identified on the basis of their retention time values and UV spectra by comparison with those of standard solution (**Fig. 6**). Peak identity is also confirmed by spiking the extracts with pure standards (*see* **Note 14**).
5. Peak purity test are also performed using a photo diode array detector coupled to the HPLC system, comparing the UV spectra of each peak with those of authentic reference samples.
6. The stock standard solution of each caffeic acid derivative was prepared as follows: about 2.0 mg of each compound is accurately weighed and placed into a 5-mL volumetric flask. Eighty percent methanol in water was added and the solution was diluted to volume with the same solvent.

4. Notes

1. The most efficient way of preparing plant culture media is first to make up stock solutions of major inorganic nutrients, minor inorganic elements, iron source, vitamins, and individual plant growth regulators. See **Table 1** for the preparation of the MS medium. Store the vitamins at –20°C in small batches. Thaw, and mix fully before use. The other stock solutions should be kept in the refrigerator at 4°C, but frequently checked and discarded if precipitation occurs. Do not store inorganic stork solutions longer than 1 m. It is advisable to make up small stocks of growth regulators fresh for each batch of media, as small changes in concentration due to precipitation can seriously affect growth of the cultures. Alternatively, stock solutions of macro elements; micro elements, vitamins are commercially available.
2. Filter sterilized IBA should be added to the medium after autoclaving and cooling to below 40°C.
3. Silicon tubes connecting to incoming air filters, and out going air filters should be clipped air tight and sealing of the lid should be checked thoroughly before and after autoclaving. Air filters should be of high quality, 0.20 μm PTFE membrane filters, and should be changed after using 3–4 times.

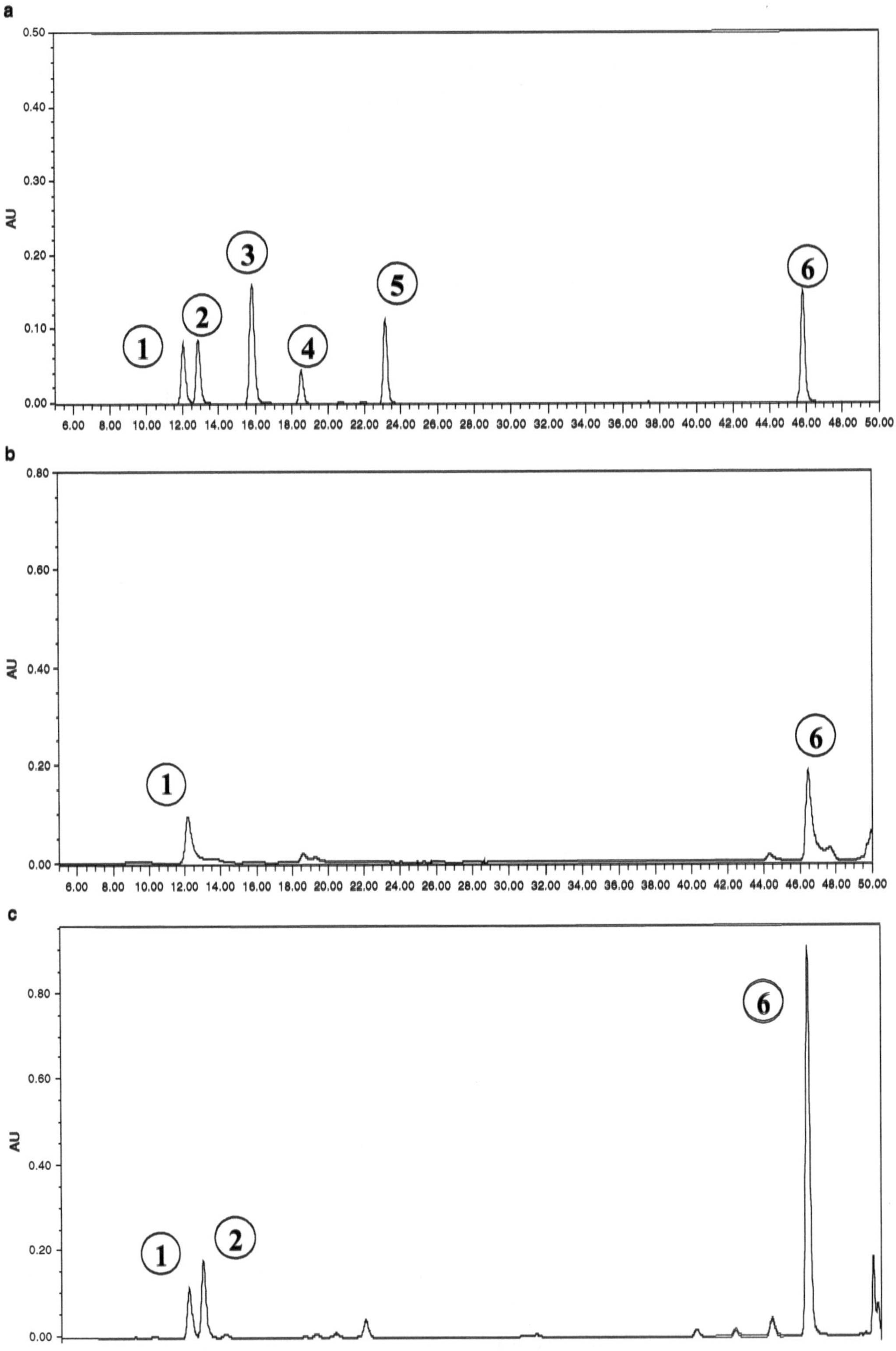

Fig. 6. HPLC profiles of standard compounds (**a**), from the extracts of dried roots of field grown plants (**b**) and adventitious cultures (**c**). Key to peak identity. (**a**) (1) caftaric acid; (2) chlorogenic acid; (3) caffeic acid; (4) cyanarin; (5) echinacoside; and (6) chichoric acid. (**b**) (1) caftaric acid; and chichoric acid. (**c**): (1) caftaric acid; (2) chlorogenic acid; and (6) chichoric acid.

4. Prepare a stock solution of gallic acid by weighing 1.0050 g gallic acid and dissolve in 1L deionised distilled water. The addition of 1 mL ethanol will help in the dissolution. Store this stock solution in an amber glass container in a refrigerator and to use for fresh working standards. Bring a portion of the stock solution to the room temperature before use.
5. Stock solution (1 mM methanolic solution) of DPPH* should be prepared fresh daily and stored in the dark at 4°C in a flask covered with aluminum foil.
6. All solvents should be of HPLC grade; solvents and solutions to be analyzed should be filtered through 0.45 μM polytetrafluoroethylene (PTFE) filters before use.
7. Sometimes explants cultured on the medium are prone to infection and in such cases stringent surface sterilization of explants is needed; use 0.1% mercuric chloride and sterilize explants for 10–15 min. Subsequently wash explants thoroughly with sterilized distilled water and culture on the nutrient medium.
8. During the initiation of first batch of suspension root cultures actively growing 2-cm long root segments with root tips should be subcultured. Such root explants grow well and proliferate quickly. After 1–2 subcultures, adventitious roots are randomly cut into 2-cm long explants, roots with or without root tips, which could proliferate successfully.
9. Adventitious root biomass accumulation in bioreactor cultures is dependent on physical factors, such as culture conditions other than chemical composition of the medium. For example, inoculum density, aeration volume, light/dark and temperature conditions have been developed over several series of experiments. Alterations in these parameters could severely affect the biomass production and accumulation of secondary products. Therefore physical conditions explained in the protocol should be strictly followed to obtain reproducible results.
10. Adventitious roots should be dried at 40°C in a forced-air oven to a moisture content of about 10%. Higher temperatures are not used that may result in the loss of phenolics *(11)*. After drying, the roots are stored at -20°C, protected from light and humidity until further use.
11. Prepare working standards of 20, 50, 100, 150, and 200 ppm by using 1000 ppm standard gallic acid solution. Careful measurements for all liquid dilutions and transfer steps should be made to assure precision of results. In addition, be very precise about times for all reactions in the samples and the standards.
12. The principle of aluminum chloride colorimetric method is as follows: Aluminum chloride forms acid stable complex

with C-4 keto group and either the C-3 or C-5 hydroxyl group of flavones and flavonols *(12)*.

13. The DPPH radical has widely been used to evaluate the free radical scavenging activity of natural anti-oxidants *(13)*. DPPH is a purple colored radical that, after being reduced by an antioxidant turns into a yellow product, DPPH* purple + antioxidant → Yellow non-radical product.
14. Use standard caffeic acid derivatives caftaric acid, chlorogenic acid and chichoric acid.

Acknowledgments

This work was supported partially by Korean Science and Engineering Foundation (KOSEF). Dr. H.N. Murthy is thankful to Korean Federation of Science and Technology Societies (KOFST) for the award of Brain Pool Fellowship.

References

1. Barrett, B. (2003). Medicinal properties of Echinacea: a critical review. *Phytomedicine* **10**, 66–68
2. Bauer, R. (1999). Chemistry, analysis and immunological investigations of Echinacea Phytochemicals, in *Immunomodulatory Agents from Plants* (Wagner, H., ed.), BirkhauserVerlag, Basel, pp. 41–88
3. Rao, S. R., and Ravishanakar, G. A. (2002). Plant cell cultures: chemical factories of secondary metabolites. *Biotechnol. Adv.* 20, 101–153
4. Sajc, L., Grubiscic, D., and Vunjak-Novakovic, G. (2000) Bioreactors for plant engineering: an outlook for further research. *Biochem. Eng. J.* **4**, 89–99
5. Zhong, J. J. (2001). Biochemical engineering of the production of plant-specific secondary metabolites by cell suspension cultures, in *Advances in Biochemical Engineering/Biotechnology*, Vol. 72 (Scheper, T., ed.), Springer-Verlag, Heidelberg, pp. 1–26
6. Paek, K. Y., Hahn, E. J., and Son, S. H. (2001). Application of bioreactors for large scale micropropagation systems of plants. *In Vitro Cell. Dev. Biol. Plant.* **37**, 149–157
7. Wu, C. H., Murthy, H. N., Hanh, E. J., and Paek, K. Y. (2007). Large scale cultivation of adventitious roots of *Echinacea purpurea* in airlift bioreactors for the production of chichoric acid, chlrogeneic acid and caftaric acid. *Biotechnol. Lett.* **29**, 1179–1182
8. Wu, C. H., Murthy, H. N., Hahn, E. J., and Paek, K. Y. (2007). Improved production of caffeic acid derivatives in suspension cultures of *Echinacea purpurea* by medium replenishment strategy. *Arch. Pharm. Res.* **30**, 945–949
9. Wu, C. H., Murthy, H. N., Hahn, E. J., and Paek, K. Y. (2007). Enhanced production of caftaric acid, chlorogenic acid and cichoric acid in suspension cultures of *Echinacea purpurea* by the manipulation of incubation temperature and photoperiod. *Biochem. Eng. J.* **36**, 301–303
10. Murashige, T., and Skoog, F. (1962). A revised medium for rapid growth and bioassays with tobacco tissue cultures. *Physiol. Plant.*, **15**, 473–497
11. Stuart, D. L., and Wills, R. B. H. (2003). Effect of drying temperature on alkylamides and cichoric acid concentrations of *Echinacea pupurea*. *J. Agric. Food Chem.* **51**, 1608–1610
12. Mabry, T. J., Markham, K. R., and Thomas, M. B. (ed.) (1970). *The Systematic Identification of Flavonoids.* Springer-Verlag, New York
13. Brand-Williams, W., Cuvelier, M. E., and Berset, C. (1995). Use of a free radical method to evaluate antioxidant activity. *Lebensmittel – Wissenschaft Technologie*, **28**, 25–30

Chapter 2

In Vitro Propagation of *Rauwolfia serpentina* Using Liquid Medium, Assessment of Genetic Fidelity of Micropropagated Plants, and Simultaneous Quantitation of Reserpine, Ajmaline, and Ajmalicine

M.K. Goel, S. Mehrotra, A.K. Kukreja, K. Shanker, and S.P.S. Khanuja

Summary

Rauwolfia serpentina holds an important position in the pharmaceutical world because of its immense anti-hypertensive properties resulting from the presence of reserpine in the oleoresin fraction of the roots. Poor seed viability, low seed germination rate, and enormous genetic variability are the major constraints for the commercial cultivation of *R. serpentina* through conventional mode. The present optimized protocol offers an impeccable end to end method from the establishment of aseptic cultures to in-vitro plantlet production employing semisolid as well liquid nutrient culture medium and assessment of their genetic fidelity using polymerase chain reaction based rapid amplification of polymorphic DNA analysis. In vitro shoots multiplied on Murashige and Skoog basal liquid nutrients supplemented with benzo[a]pyrene (1.0 mg/L) and NAA (0.1 mg/L) and in-vitro rhizogenesis was observed in modified MS basal nutrient containing NAA (1.0 mg/L) and 2% sucrose. In-vitro raised plants exhibited 90–95% survival under glass house/field condition and 85% similarity in the plants regenerated through this protocol. Field established plants were harvested and extraction of indole alkaloid particularly reserpine, ajmaline and ajmalicine and their simultaneous quantitation was performed using monolithic reverse phase high-performance liquid chromatography (HPLC).

Key words: Micropropagation, Liquid medium, *Rauwolfia serpentina,* Genetic fidelity, Reserpine, Ajmaline, Ajmalicine, HPLC, RAPD

S. Mohan Jain and Praveen K. Saxena (eds.), *Methods in Molecular Biology, Protocols for* In Vitro *Cultures and Secondary Metabolite Analysis of Aromatic and Medicinal Plants, vol. 547*
© Humana Press, a part of Springer Science+Business Media, LLC 2009
DOI: 10.1007/978-1-60327-287-2_2

1. Introduction

Rauwolfia serpentina has drawn special attention all over the world in the pharmaceutical field for quite some time and still holds an important position because of its great medicinal value. The antihypertensive properties of *Rauwolfia* roots are attributed to presence of reserpine (3,4,5-trimethyl benzoic acid ester of reserpic acid, an indole derivative of 18-hydroxy yohimbine type) *(1)*. It is a relatively weak tertiary base occurring in the oleoresin fraction of the roots and has depressant action on central nervous system (CNS) and produces sedation and lowering of blood pressure. Administration of reserpine depletes the brain and peripheral vessels of serotonin (5-hydroxy tryptamine) and catechol amines. Besides, it is also reported to influence the concentration of glycogen, acetylcholine, γ-amino butyric acid, nucleic acid, and anti-diuretic hormones. Reserpine is now being used as a tool in physiologic studies of body functions and pharmacologic studies of other drugs.

Poor seed viability, low seed germination rate, and low vegetative propagation rate through root cuttings has hampered large scale commercial cultivation of *R. serpentina* through conventional mode and over exploitation of the natural resources has led to its inclusion in Red Data Book of India as an endangered plant species. The increasing demand for *Rauwolfia* roots in national and international markets and decreasing availability have encouraged many farmers to cultivate this pharmaceutically important plant. Large scale cultivation through in vitro clonal propagation may therefore be beneficial for the germplasm conservation and commercial cultivation of elite plants of *R. serpentina* for the production of reserpine at desirable level. Use of liquid medium has certain advantages over the use of nutrients in an agar based semisolid matrix because of better availability of oxygen by agitation and aeration *(2–4)*. Keeping in view these problems and overwhelming interest in the techniques of plant tissue culture, the present protocol was developed covering the following objectives; (a) production of quality planting material, (b) mass propagation of *Rauwolfia serpentina* plantlets employing liquid medium; (c) assessment of genetic fidelity of the in vitro raised somaclones, and (d) simultaneous analysis of reserpine, ajmaline and ajmalicine using reverse phase HPLC. This protocol would be of great help for commercial industrial units and research groups who are associated with the optimization of micropropagation of economically important plants particularly medicinal plants and secondary metabolite studies.

2. Materials

2.1. Media Preparation

1. Murashige and Skoog (MS) salts.
2. Phytohormone(s) (*see* **Tables 1** and **2**).
3. 4.0, 7.0, and 9.2 pH buffers.
4. 1 N NaOH.

Table 1
Composition of Basal MS Medium

Designated Stock	Constituent of stock	Concentration (mg/L)	Stock (g/L)	Volume (mL) of the stock to be used for preparation of 1 L medium
A	NH_4NO_3	1,650	82.5	20
B	KNO_3	1,900	95	20
C	$CuSO_4 \cdot 5H_2O$	0.025	0.005	5
	$MgSO_4 \cdot 7H_2O$	370	74	
	$ZnSO_4 \cdot 7H_2O$	8.6	1.72	
	$MnSO_4 \cdot 4H_2O$	22.3	4.46	
D	$CaCl_2 \cdot 2H_2O$	440	88	5
E	$Na_2MoO_4 \cdot 2H_2O$	0.25	0.05	5
	H_3BO_3	6.2	1.24	
	$CoCl_2 \cdot 6H_2O$	0.025	0.005	
	KI	0.83	0.166	
	KH_2PO_4	170	34	
F	$FeSO_4 \cdot 7H_2O$	27.85	5.57	5
	Na_2EDTA	37.25	7.45	
G Vitamins	Thiamine–HCl	0.1	0.02	5
	Pyridoxine–HCl	0.5	0.1	
	Nicotinic acid	0.5	0.1	
	Glycine	2.0	0.4	
Other supplements	*myo*-inositol	100	To be added freshly	
	Sucrose	30,000		
	Agar	8,000		

Table 2
Details of Plant Growth Regulators

Plant growth regulators	Solubility	Sterilization	Storage
Auxin			
α-naphthelene acetic acid (NAA)	1 N NaOH	Co-autoclave	4°C
Cytokinin			
Benzyl- 6- amino purine (BAP)	1 N NaOH	Co-autoclave	4°C

5. 1 N HCl.
6. 250-mL wide-mouth Erlenmeyer flask.
7. 25 × 150 mm Culture tubes.
8. Beakers.
9. Measuring cylinders.
10. 0.1–10 mL pipets and/or 0.5-μL micropipets.
11. Glass rods (*see* **Note 1**).

2.2. Surface Sterilization and Establishment of In Vitro Cultures

1. 0.1% $HgCl_2$, Teepol.
2. 80% ethanol.
3. Savlon.
4. Sterile distilled water.
5. Culture medium.
6. 8”-12” Rust-proof stainless steel forceps.
7. Scissors.
8. Scalepl and/or razor blade.
9. Petri dishes.

2.3. Assessment of Clonal Fidelity of Micropropagated Plants

2.3.1. Isolation of Plant Genomic DNA

1. Eppendorf tubes.
2. Oakridge tubes.
3. Dessicator with vacuum pump.
4. Mortar and pestle.
5. Liquid nitrogen.
6. Buffers (*see* **Table 3**).
7. 5 mg/mL Ethidium bromide (EB) stock in water.
8. Agarose.
9. 6X Loading dye.

Table 3
Details of Buffers and Dyes

Buffers (pH 8.0)					
Extraction buffer		**High salt TE buffer**		**TAE buffer (50X)**	**Loading dye (6X)**
NaCl	1.4 M	NaCl	1 M	Tris base 242 g	30% glycerol in water
CTAB	2.5%	Tris–HCl	10 mM	EDTA (0.5 M) 100 mL	0.25% bromo phenol blue
Tris–HCl	100 mM	EDTA	1 mM	Glacial 57.1 mL acetic acid	0.25% xylene cyanol
EDTA	20 mM				Store this at –20°C
β-mercapto-ethanol	0.2%			Add water to make 1 L	
PVP	1%				

2.3.2. PCR for DNA Amplification

1. Template DNA.
2. Taq DNA polymerase.
3. Deoxyribonucleoside triphosphates (dNTPs) mix (dATP; dTTP; dCTP and dGTP).
4. 10X polymerase buffer.
5. Primer(s).
6. λ marker DNA double (HindIII + EcoRI) digested (Genei, Bangalore).
7. Polymerase chain reaction (PCR) tubes.
8. Milli-Q water.
9. Ice.

In this protocol MAP, OPA, and OPB series of primers were used (*see* **Table 4**). MAP primers were designed at CIMAP *(5)* (custom synthesized by M/s Genei; Bangalore, India). OPA and OPB primers were designed at Operon Technologies Inc. (Germany).

2.4. Quantitative Analysis of Indole Alkaloids

2.4.1. Extraction of Indole Alkaloids

1. Chloroform.
2. Methanol.
3. Distilled water.
4. Hydrochloric acid.
5. Vacuum-rotavapor.
6. Filtration assembly.
7. Micropipets.
8. Rotavapor R-144 grinding device (Buchi, Switzerzerland).

Table 4
Nucleotide Sequences of Primers Used for RAPD Analysis

Code	5′ to 3′	Code	5′ to 3′
OPA-2	TGCCGAGCTG	OPB-1	GTTTCGCTCC
OPA-3	AGTCAGCCAC	MAP-4	TGCGCGATCG
OPA-7	GAAACGGGTG	MAP-6	GCACGCCGGA
OPA-9	GGGTAACGCC	MAP-9	CGGGATCCGC
OPA-11	CAATCGCCGT	MAP-10	GCGAATTCCG
OPA-13	CAGCACCCAC	MAP-13	GTGCAATGAG
OPA-20	GTTGCGATCC		

2.4.2. High-Performance Liquid Chromatography Analysis of Indole Alkaloids

1. 100 × 4.6 ID RP-18e () Chromolith HPLC Column (Merck; Darmstadt, Germany).
2. HPLC-grade acetonitrile (Merck; Darmstadt, Germany).
3. HPLC-grade water.
4. Di-sodium-di-hydrogen orthophosphate.
5. GR-grade glacial acetic acid.
6. 0.45-μm nylon HPLC filter paper (Millipore).
7. Ajmaline (Sigma).
8. Reserpine (Sigma).
9. Ajmalicine (Hi-media).

3. Methods

3.1. Media Preparation

1. Prepare the MS *(6)* stock solutions as categorized (**Table 1**) (*see* **Notes 2** and **3**).
2. Use the stock solutions as stated in the Table 1 (*see* **Note 4**).
3. Melt agar separately.
4. Properly mix the required volume of stock solutions, sucrose, and *myo*-inositol.Add to the molten agar by stirring. Make the final volume of culture medium as per requirement.
5. Shoot multiplication semisolid medium (RS): Add BAP 1.0 mg/L and NAA 0.1 mg/L and adjust to pH 5.68 using 1 N NaOH and 1 N HCl.

6. Shoot multiplication liquid medium (RL): Prepare as RS medium but do not add agar (*see* **Note 5**).
7. Root induction medium (RI): Prepare semisolid half strength MS medium with 2% sucrose. Add 1.0 mg/L NAA and adjust the pH as above.
8. Dispense this medium in appropriate culture vessels (*see* **Note 6**). Sterilize at 121°C at 15 lb pressure for 15–20 min. Make the slant by tilting the tubes 45 and allow the medium to solidify.
9. Store the culture medium at 25°C.

3.2. Explant Preparation

1. Prepare the explants by cutting 1.5–2.5 cm long nodal segments containing 2 axillary buds (*see* **Fig. 1**) with the help of a sharp blade/razor.
2. Thoroughly wash the explants in continuous flow of tap water for 1 h and then with 1% Teepol solution followed by Savlon.

3.3. Surface Sterilization of Explants and Establishment of In Vitro Cultures

1. Sterilize the explant surface with 80% ethanol for 30 s and 0.1% $HgCl_2$ for 3 min under a laminar air flow hood (*see* **Notes 7** and **8**) followed by rinsing (3–4×) with sterilized distilled water (*see* **Notes 9** and **10**).
2. Slice off the exposed end slightly with the help of sterilized sharp blade and transfer the explants to a sterile Petri plate.
3. Inoculate single nodal explant in the culture tube (*see* **Fig. 2a**) containing RS media and incubate in the culture room under 14-h photoperiod, light intensity 54 μmol/m^2 s provided by cool white fluorescent light, 25 ± 2°C temperature and 60% relative humidity.
4. After in vitro establishment, these cultures can be multiplied and maintained indefinitely in 250-mL flasks (*see* **Fig. 2b**) by inoculating 6–8 explants per culture vessel. The cultures may be used as a source of inoculum for further routine multiplication practice.

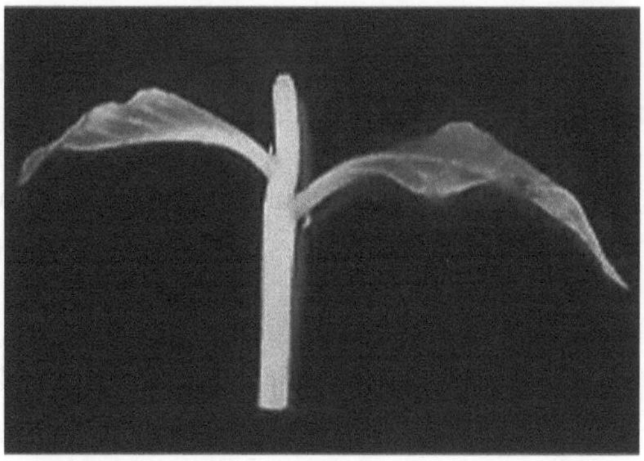

Fig. 1. Single nodal explant.

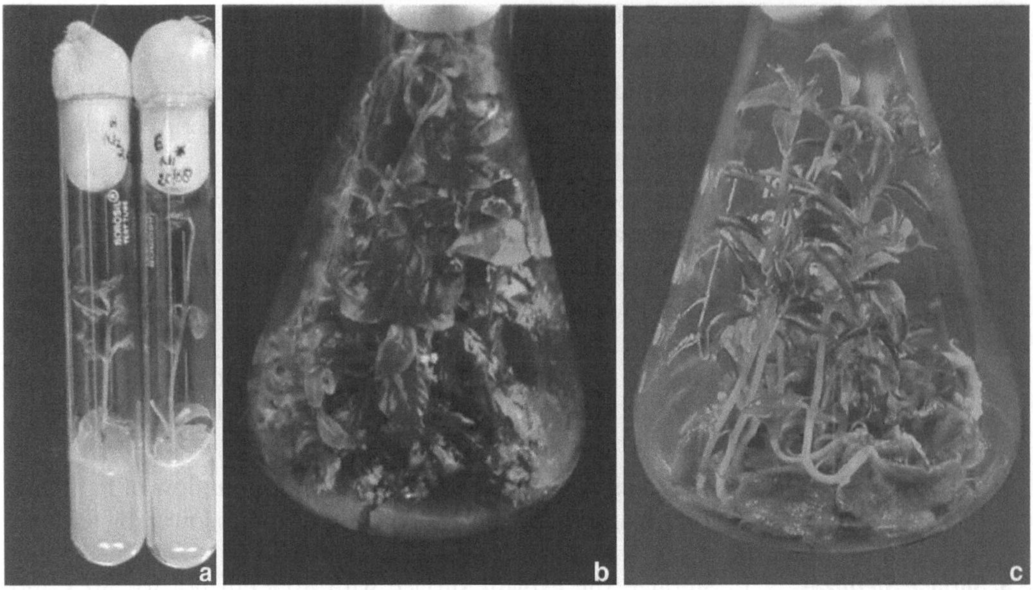

Fig. 2. In vitro establishment and multiplication of aseptic cultures of *R. serpentina* (**a–c**); semisolid medium (**a–b**) and liquid medium (**c**).

3.4. Multiplication and Maintenance of In Vitro *Cultures Using Liquid Medium*

1. Transfer 8–10 nodal explants from pre-existing aseptic stock cultures in the RL liquid medium in 250-mL flask (*see* **Fig. 2c**) containing 30 mL medium.
2. Place the culture flasks on an orbital shaker at 75 rpm in culture room.

3.5. Root Induction

1. For in vitro rhizogenesis, transfer 6 to 8 cm long microshoots to the semisolid RI medium filled in tubes or flasks. All steps should be carried in laminar air flow.
2. Remove the in vitro developed plantlets (*see* **Fig. 3a**) after 8 wk and properly clean them gently removing the agar under the tap water (*see* **Note 11**).

3.6. Acclimatization and Field Transfer

1. Transfer the plantlets to the glass culture tubes containing tap water for in vitro hardening in such a way that only their root portion should remain inside the water (*see* **Fig. 3b**). Keep these plantlets in culture room for 10 d.
2. Transfer the in vitro hardened plantlets to glass house in earthen pots (*see* **Fig. 3c**) containing sand:soil in 2:1 ratio under 70–80% humidity by covering with glass beakers. The pots may alternatively be placed in a temporary poly house.
3. After 45–60 d as new leaves starts appearing transfer the plants to the field conditions.

Fig. 3. In vitro rhizogenesis and hardening in *R. serpentina* (**a–c**); rooted plantlets (**a**); in vitro hydroponic hardening (**b**) and in vivo hardening.

3.7. Assessment of Clonal Fidelity of Micropropagated Plants

3.7.1. Preparation of Extraction Buffers

1. Add calculated amount of Cetyl Trimethyl Ammonium Bromide (CTAB), NaCl, Ethylene diamine tetra acetic acid (EDTA), Tris–HCl and Polyvinyl pyrolidone (PVP), make the final volume by distilled water and warm at 55°C in water bath for 30 min.
2. Right before the use, add β-mercaptoethanol to the extraction buffer under fume hood (*see* **Note 8**).
3. Aliquot the desired amount of extraction buffer in to the Oakridge.

3.7.2. Isolation of Plant Genomic DNA

This is an adoption to the CTAB method modified by the Khanuja group *(7)*

1. Grind 1.0 g fresh young leaves of the test samples to a fine powder in liquid nitrogen.
2. Transfer the powdered leaf tissue to sterile oakridge tubes containing 4 mL freshly prepared extraction buffer, pre-warmed to 56°C. Shake the mixture to form a slurry (*see* **Note 12**).
3. Incubate the tubes at 60–65°C in a water bath for 1–2 h for cell wall lysis.
4. Add equal volumes of 4 mL chloroform: isoamyl alcohol (24:1) and gently mix by inversion for 10 min to form an emulsion.

5. Centrifuge the tubes for 10 min at 10,000 rpm and 25°C. Three layers will be formed: (a) a lower layer containing chloroform with dissolved proteins, lipids and chlorophylls; (b) a middle layer containing cell debris and dissolved green cell wall; and (c) an upper aqueous layer containing nucleic acid along with dissolved polysaccharides.
6. Separate the upper aqueous layer (approximately 4 mL) and transfer in to a fresh sterile oakridge tube (*see* **Note 13**) and add 1.5 mL 5 M NaCl solution and 3.3 mL (i.e., 0.6 volume of the total solution) of isopropanol. Gently mix by inversion and allow this mixture to stand for 1–2 h at room temperature. At this stage nucleic acid floating in the solution could be observed. This can either be scooped or alternatively could be centrifuged for 10 min at 10,000 rpm and 25°C.
7. Discard the supernatant and wash the pellet, containing the DNA, with 1 mL 80% ethanol by centrifugation for 5 min at 10,000 rpm and 25°C.
8. Discard the supernatant and dry the pellet under vacuum for 1–2 min to remove the traces of alcohol.
9. Dissolve the pellet into 1 mL high salt tris-ethylene diamine tatraacetic acid (TE) buffer. It may take some time to dissolve. Add 1 μL RNAse and incubate at 37°C in a water bath for 30 min.
10. Extract with equal volume 1 mL chloroform to remove the remaining proteins and other impurities by gently inversion and centrifugation for 10 min at 10,000 rpm and 25°C.
11. Transfer the upper aqueous layer approximately 1 mL to two fresh sterile Eppendorf tubes and add double volume of absolute ice cold ethanol and incubate at –20°C for 1–2 h for precipitation of DNA.
12. Centrifuge this mixture at 10,000 rpm for 10 min at 4°C. Discard the supernatant and wash the pallet with 80% ethanol at 10,000 rpm for 5 min at 4°C.
13. After vacuum drying dissolve the pellet in 100 μL sterile triple distilled/Milli-Q water and store at –20°C for further use.

3.7.3. Quantification of DNA

1. The DNA can be quantified by running on 0.8% agarose gel checking the absorbance at 260 nm (*see* **Note 14**).
2. Agarose Gel: Mix 2 mL 50X TAE buffer to a final volume of 100 mL. Add 800 mg agarsoe, boil, and cool to 50–60°C. Carefully add EB. Seal the free ends of gel tray, fix the combs, and dispense the molten gel. Allow it to solidify. Remove the comb and put the gel tray in the gel reservoir containing 1X TAE buffer. Make sure the gel is fully submerged (*see* **Notes 15** and **16**).

3. Mix the DNA sample, loading 6X dye and Milli-Q water (1 + 2 + 9 μL) by repeated pipeting. Load in the wells of gel carefully (*see* **Note 17**).
4. Close the lid of the gel reservoir and turn on the power supply. The gel runs from the – pole (black) towards the + pole (red). Check after few minutes if the gel is running. Check the gel regularly to prevent the samples from running off the gel.
5. For PCR, 20 ng amount of DNA is sufficient per reaction therefore dilution of DNA should be made with sterile Milli-Q water in such a way that 1 μL should contain approximately 20–25 ng of DNA (*see* **Note 18**).

3.7.4. PCR for DNA Amplification (8)

1. Amplification reaction was performed using Bio-Rad i-cycler version 4.006.
2. PCR is carried out in a total volume of 25 μL for each reaction in 0.2 mL PCR tube.
3. Set up the PCR reaction mixture (*see* **Table 5**) (*see* **Note 19**).
4. Taq polymerase should be added at the end.
5. All this should be carried out in ice.
6. Transfer this reaction mixture in PCR tube and spin it for few seconds for uniform mixing.

Table 5
Details of Setting up the PCR Reaction Mixture

Stock	Reagents	Final concentration	Volume (μL)/reaction
	Water		19.3
10X	Polymerase buffer	1X	2.5
10 mM	dATP	100 μM	0.25
10 mM	dGTP	100 μM	0.25
10 mM	dCTP	100 μM	0.25
10 mM	dTTP	100 μM	0.25
3 U/μL	Taq DNA polymerase	0.6 U	0.2
	Primer	5 pmole	1.0
	Genomic DNA	20–25 ng	1.0

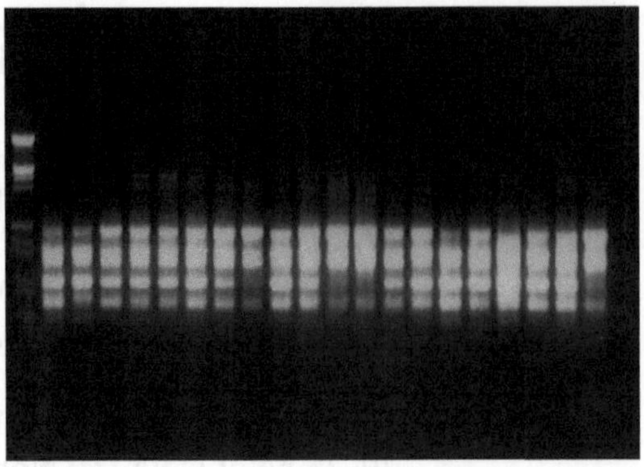

Fig. 4. RAPD profile of different samples of *R. serpentina* with primer OPA 13.

7. Carry out the PCR in the thermal cycler using the following conditions: (a) initial denaturation at 94°C for 5 min; (b) 45 cycles each consisting of a denaturation step at 94°C for 1 min, primer annealing step at 35°C for 1 min, amplification at 72°C for 2 min step; (c) final extension at 72°C for 5 min followed by arresting the reaction at 4°C for infinite period.
8. Load the amplified DNA on 1.4% agarose gel in 1X TAE buffer stained with 0.5 μg/mL EB. Photograph (*see* **Fig. 4**) on a gel documentation polaroid system.
9. Run the amplified products on the gel with molecular weight marker, λ Eco RI-Hind III digest.
10. The polymorphic bands were analyzed using image master 1-D elite software.
11. The similarity Index (SI) between two individuals = (2 × No. of common bands)/(Total No. of bands).
12. NTSys PC version 2.02j was used for analysis of the complete data set.
13. Similarity estimates were calculated by using Nei and Li coefficients *(9)* and cluster analysis was carried out by UPGMA method, Unweighted Pair Group Method Arithmetic Mean Averages.

3.8. Indole Alkaloids Analysis

3.8.1. Harvesting

1. Roots grow deeper inside the soil therefore plant is uprooted by digging out.
2. Wash the roots with tap water and allow drying under shade.

3.8.2. Extraction of Indole Alkaloids

1. Grind 1.0 g shade-dried roots of *R. serpentina* to powdered form.
2. Extract the powdered material 3× (3 × 10 mL) for 8 h with chloroform and methanol in 3:1 ratio at room temperature.

3. Pool the extracts and dry under vacuum, 417 bars at 40°C. in Rotavapor. R-144 (Buchi) to obtain a solid mass.
4. Re-dissolve the dried extract in small amount of chloroform and methanol (3:1) and transfer to small glass tube and allow the solvent to evaporate and dry in desiccators and store in refrigerator at 4°C. This extract can be used for quantitative analysis of the indole alkaloids through HPLC.

3.8.3. HPLC Analysis of Indole Alkaloids

1. Re-dissolve the dried extract in acidic methanol, methanol: HCl – 98: 2 *(v/v)* using ultra-sonication. The extract is equivalent to 1 g/mL on tissue dry weight basis.
2. Centrifuge the dissolved extract at 10,000 rpm for 30 min.
3. Prepare stock solutions of reference alkaloids i.e., ajmaline, ajmalicine and reserpine by adding 1 mg/mL of methanol separately.
4. Prepare a mixture of three alkaloids by adding the equal volume of each standard. The concentration of working solution of the standard solution is now 0.33 mg/mL.
5. Prior to injection, filter the samples and standards through 0.45 μm nylon HPLC millipore filter paper.
6. Quantitative estimation of ajmaline, ajmalicine and reserpine was carried out by reversed-phase HPLC gradient method using photodiode array (PDA) detection method *(10)*.
7. In, this protocol an analytical HPLC system consisted of a LC-20AD solvent delivery pumps, a DGU-$20A_5$ degasser, a CTO-20A column oven, 10 AF auto-sampler and a SPD-M 20A photodiode array detector was used. Data acquisition was performed on Lab Solution 3.21. The separation was achieved with a binary gradient program for pump A – acetonitrile, and pump B – 0.01 M phosphate buffer (NaH_2PO_4) containing 0.5% glacial acetic acid; pH 3.5.
8. A chromolith RP-18e HPLC column, 4.6 × 100 mm ID, is used for all the analysis.
9. Column temperature was maintained at 26 ± 2°C.
10. Analysis was performed at a flow rate of 1.0 mL/min through out the gradient run (*see* **Table 6**) and the data acquisition was performed at a wavelength of 254 nm.
11. Equilibrate the HPLC with mobile phase till the base line is established (*see* **Note 20**).
12. The identity of individual alkaloid in the sample run was confirmed by R_f comparison as well as UV-spectra match with the library of the standards maintained in the LC solution software. The area under respective peak were recorded (*see* **Fig. 5**) and used for percent content of reserpine, ajmaline and ajmalicine in the *R. serpentina* root sample.

Table 6
Gradient Program for the Separation of Test Indole Alkaloids

Time (min)	Acetonitrile concentration (%) (pump A)	Phosphate buffer concentration (%) (pump B)	Flow rate (mL/min)
0.01	15	85	1.0
9.00	15	85	1.0
9.01	25	75	1.0
10.00	25	75	1.0
10.01	30	70	1.0
12.00	30	70	1.0
12.01	35	65	1.0
30.00	35	65	1.0
50.00	15	85	1.0

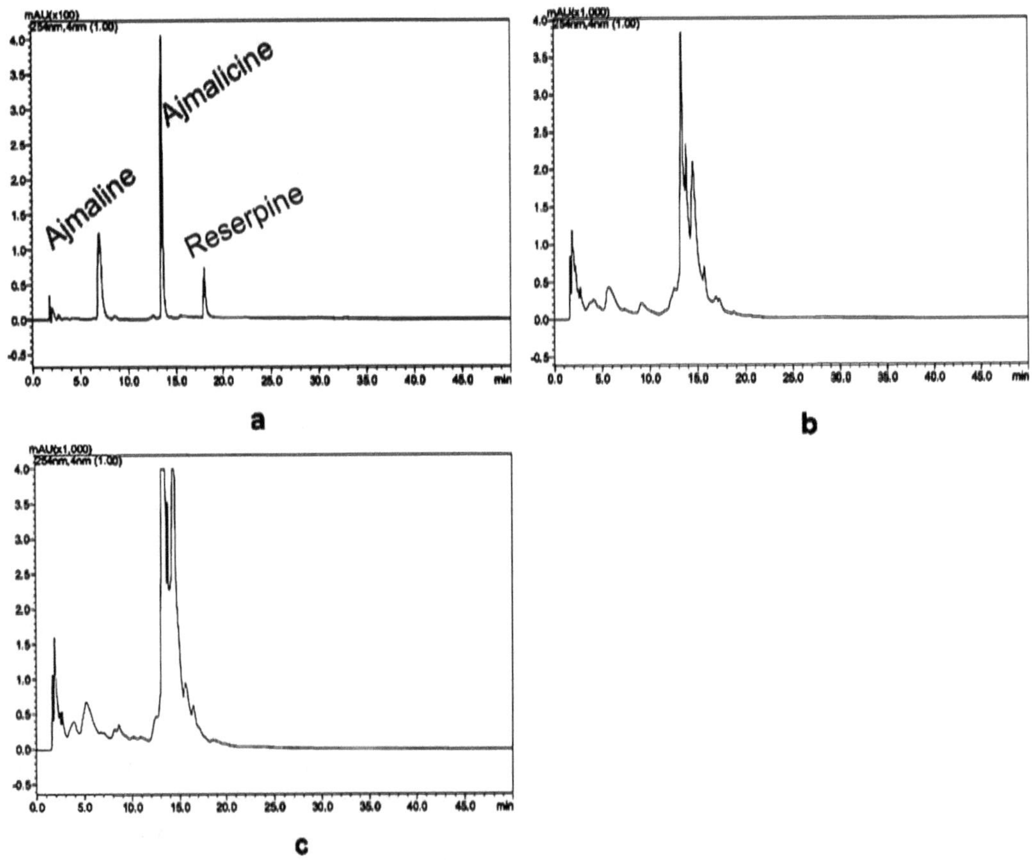

Fig. 5. Monolithic reverse phase HPLC separation of reserpine, ajmaline and ajmalicine in mixture of reference compounds (**a**) *R. serpentina* root sample of micropropagated plants showing ajmaline (**b**) and reserpine (**c**).

13. The percent content on the basis of dry weight was calculated by using the area under respective peak in sample run and standard run.
14. The results were calculated using the following formula:

$$\%\text{content} = \frac{\text{Peak area (sample)}}{\text{Peak area (standard)}} \times \frac{\text{Concentration (standard)}}{\text{Concentration (sample)}} \times 100$$

4. Notes

1. Before using, soak all the glass wares overnight in chromic acid solution. Then clean with a liquid detergent and thoroughly wash with tap water to remove the last trace of detergent. Finally, rinse glass ware with double distilled demineralized water and dry in hot air oven at 150°C for 2 h.
2. An iron stock solution is prepared separately because of the problem of iron solubility. Usually the iron stock is prepared in a chelated form as Na Fe–ethylene-diamine tetraacetic acid (EDTA).
3. Dissolve $FeSO_4$ and Na EDTA separately in warm water, mix both solutions and make the final volume.
4. The MS basal salt mixture is also commercially available in powdered form.
5. Method for preparing liquid and solid medium is almost same; the only difference is presence or absence of agar.
6. For initial establishment culture tubes are preferred to avoid cross contamination.
7. Regularly check the air flow gauge of laminar air flow chamber. After regular intervals clean the pre-filters of the laminar air flow by washing followed by vacuum cleaning. Switch on the UV light for 30 min. Before starting any activity in the laminar air flow bench, wipe the surface of the laminar air bench with alcohol frequently during any aseptic operation. Fumigate the entire transfer room.
8. Avoid contact of body parts with $HgCl_2$, β-merceptoethanol, and EB as all these chemicals are highly mutagenic, carcinogenic and hazardous chemicals.
9. Possible source of danger exists if a person, after flaming an instrument reinserts the hot instrument into the alcohol dip. **CAUTION:** Ethanol is inflammable! One should be very careful.
10. UV irradiation may also cause some serious health risks.
11. In vitro roots are delicate do not let them break.

12. Mix extraction buffer and samples quickly; do not let the samples thaw.
13. Cut the tip to widening the orifice of pipette enables the easy suction and separation of upper fraction without causing any disturbance to the lower layer.
14. Alternatively DNA yield may also be also checked spectrophotometrically.
15. While preparing the agarose gel, take care that final volume should never be reduced due to evaporation during boiling. Take care do not entrap any air bubble.
16. Clean the gel tray, gel reservoir, combs and other materials with ethanol properly before and after use.
17. While loading the sample, do not let the sample spill out. Wear gloves during the entire operation and prevent contamination.
18. Pay attention to the quality of DNA-bad when there are many small fragments; and loading the amount of RNA – high when there is much smear above sample on the gel.
19. To minimize the error and for convenience mix all the dNTPs in equal amount and make a stock in advance, and then take 1 μL for each PCR reaction.
20. The method used for the analysis is precise, accurate, robust and validated. The general steps of chromatography should be followed. First of all the loop of the auto-sampler should be cleaned with at least 100 μL methanol three times and the system should be allowed to run at least for 30 min to achieve minimum noise signals.

Acknowledgments

Financial assistance from Council of Scientific and Industrial Research (CSIR) Government of India to MKG is gratefully acknowledged. Authors are also thankful to Director, CIMAP, for providing the facilities.

References

1. CSIR, (1969). Wealth of India: raw materials. *Council of Scientific and Industrial Research*. N. Delhi, pp. 376–390
2. Mehrota, S., Goel, M.K., Kukreja, A.K., and Mishra, B.N. (2007). Efficiency of liquid culture systems over conventional micropropagation: a progress towards commercialization. *Afr. J. Biotechnol.* **6(13)**: 1484–1492
3. Goel, M.K., Kukreja, A.K., and Khanuja, S.P.S. (2007). Cost-effective approaches for *in vitro* mass propagation of Rauwolfia serpentina Benth. *Ex. Kurz. Asian J. Plant Sci.* **6**: 957–961
4. Goel, M.K. (2007). Applications of bioreactors for micropropagation and secondary metabolite production in *Rauwolfia serpentina L.* Benth. Ex. Kurz. Ph.D Thesis submitted

to the H.N.B. Garhwal University, Srinagar, Garhwal (India)

5. Khanuja, S.P.S., Shasany, A.K., Srivastava, A., and Kumar, S. (1998). Assessment of genetic relationship in *Mentha* species. *Euphytica* **111**: 121–125
6. Murashige, T., and Skoog, F. (1962). A revised medium for rapid growth and bioassays with tobacco tissue cultures. *Physiol. Plant.* **15**: 473–497
7. Khanuja, S.P.S., Shasany, A.K., Darokar, M.P., and Kumar, S. (1999). Rapid isolation of DNA from dry and fresh samples of plants producing large amounts of secondary metabolites and essential oils. *Plant Mol. Biol. Rep.* **17**: 74–80
8. Goel, M.K., Kukreja, A.K., and Khanuja, S.P.S. (2006). PCR based RAPD analysis of *in vitro* raised plants of *Rauwolfia serpentina Benth. Ex. Kurz. In: National Symposium on Plant Biotechnology*. Dehradun, India
9. Nei, M., and Li W. (1979). Mathematical model for studying genetic variation in terms of restriction endonucleases. *Proc. Natl. Acad. Sci. U.S.A.* **76**: 5269–5273
10. Srivastava, A., Tripathi, A.K., Pandey, R., Verma, R.K., and Gupta, M.M. (2006). Quantitative determination of reserpine, ajmaline, and ajmalicine in *Rauwolfia serpentina* by reversed-phase high-performance liquid chromatography. *J. Chromatogr. Sci.* **44**: 557–560

Chapter 3

Medicinal Properties, In Vitro Protocols and Secondary Metabolite Analyses of Scots Pine

Hely Häggman, Anna Maria Pirttilä, Karoliina Niemi, Tytti Sarjala, and Riitta Julkunen-Tiitto

Summary

Scots pine (*Pinus sylvestris* L.) is known as an economically important forest tree with a wide distribution throughout the Northern hemisphere. Recently, the species has also become recognized as a novel source of functional food and bioactive compounds with medicinal properties. The present paper provides up-to-date information on protocols for somatic embryogenesis (i.e., the most promising in vitro method for vegetative propagation of Scots pine). Endophyte protocols cover the topics of endophyte isolation, identification and elimination from in vitro cultures. Moreover, the protocols for secondary metabolite analyses are described in order to emphasize the emerging role of Scots pine as a medicinal plant.

Key words: *Pinus sylvestris*, Somatic embryogenesis, Endophytes, Phenolic compounds

1. Introduction

Scots pine (*Pinus sylvestris* L.) belongs to the family Pinaceae and has a wide distribution in Northern hemisphere. Especially in Nordic countries the species is economically important and used as a raw material for sawn timber as well as pulp and paper production. Traditionally Scots pine tissues, such as bark and needles, have been used as supplements in food and folk medicine and lately Scots pine has proved to be a novel source of functional food and bioactive compounds with medicinal properties.

The most promising in vitro propagation method for Scots pine is somatic embryogenesis (i.e., development of embryos from somatic cells) *(1–3)*. The success of somatic embryogenesis

S. Mohan Jain and Praveen K. Saxena (eds.), *Methods in Molecular Biology, Protocols for* In Vitro *Cultures and Secondary Metabolite Analysis of Aromatic and Medicinal Plants, vol. 547*
© Humana Press, a part of Springer Science+Business Media, LLC 2009
DOI: 10.1007/978-1-60327-287-2_3

depends on several factors such as genotype and family, and also the fact that tissues of Scots pine contain high numbers of endophytic microbes, which may create problems during in vitro culture. For example, endophytic fungi *Alternaria alternata* (Fr.) Keissl., *Epicoccum purpurascens* Ehrenb., and *Ulocladium atrum* (Preuss) Sacc are found in the seeds *(4)*, whereas *Hormonema dematioides* Lagerberg & Melin and *Rhodotorula minuta* (Saito) FC Harrison and many bacterial species, such as *Methylobacterium extorquens*, *Mycobacterium* sp. and *Pseudomonas synxantha* (Ehrenb.) Holl. colonize in the shoot tips of Scots pine *(5–7)*. Therefore, isolation, identification and elimination of endophytic microbes may be necessary to manage Scots pine in vitro culture. On the other hand, endophytes are not only harmful during in vitro cultures, as they can produce compounds beneficial for plant tissues *(8)*.

Generally, Scots pine tissues are rich in different phenolic and terpenoid secondary compounds, whereas lysine-derived piperine alkaloids are also found in small amounts *(9)*. The main phenolic compounds in pine extracts are phenolic acid glucosides, (+)-catechin and its derivatives, taxifolin and taxifolin derivatives, acylated flavonols, lignans, and proanthocyanidins *(10, 11)*. Most of the flavonoids are glycosylated and localized in the epidermal tissue of needles and young shoots *(11)*. Moreover, more than 15 different mono- and diterpenes are found in needles of pine, and the most abundant are alpha-pinene, abietic acid and neoabietic acid *(12)*.

The phenolic extracts of Scots pine bark *(10)* and knots *(13)* have been reported to possess high antioxidant activities and anti-inflammatory properties. Especially ferulic acid, pinoresinol, and matairesinol extracted from bark have proved to inhibit production of nitric oxide (NO) and prostaglandine E2 *(14)*. Main groups of components in Scots pine hydrophilic knotwood extracts are lignans, nortrachelogenin, oligolignans, pinosylvins, and pinosylvin monomethyl ether *(15)*; yet matairesinol, secoisolariciresinol, and liovil are present in small amounts *(13)*. Lignans are interesting compounds showing antioxidant and antitumor effects and, thus, have potential for cancer chemoprevention *(16)*. Pinosylvin is also known to have antioxidant *(17)* as well as fungicide and bactericide properties. Furthermore, phloem tissue extracts of Scots pine have shown antibacterial effect against *Staphylococcus aureus* Rosenb *(18)*.

The aim of this paper is to provide up-to-date information on protocols for Scots pine somatic embryogenesis and isolation, identification, and elimination of endophytic fungi and bacteria from in vitro cultures. Moreover, in order to emphasize the emerging role of Scots pine as a medicinal plant protocols for secondary metabolite analyses are described.

2. Materials

2.1. In Vitro Protocol for Somatic Embryogenesis

2.1.1. Initiation, Induction and Proliferation

1. One-year-old seed cones containing immature seeds.
2. Laminar flow hood.
3. Scalpels.
4. Tweezers.
5. 70% ethanol.
6. Sterile distilled water.
7. 9-cm Petri dishes.
8. 250-mL Erlenmeyer flasks.
9. Sterile filter papers (Whatman No.2, 5.5 cm diameter).
10. Büchner funnel attached to low vacuum pump.
11. Growth chamber.

Initiate and proliferate on either solid DCR *(19)* or Litvay (LM) *(20)* medium with modifications:

1. DCR medium containing 3 mg/L, 13.6 μM, 2, 4-D (2, 4-dichlorophenoxyacetic acid), 0.5 mg/L, 2.2 μM, BA (benzylaminopurine), and 30 g/L, 88 mM, sucrose *(21)*.
2. ½ LM medium (half-strength macro and full-strength micro nutrients, and 1.0 g/L casein hydrolysate) supplemented with 2.2 mg/L, 9.9 μM, 2, 4-D, 1.0 mg/L, 4.5 μM, BA, and 30 g/L, 88 mM, sucrose *(3, 22)*, (**Tables 1** and **2**) (*see* **Notes 1–3**).
3. Arrest of proliferation on either solid hormone-free DCR medium *(23)* or liquid hormone-free DCR *(24)* or ½ LM medium *(3)*. All media include 30 g/L, 88 mM, sucrose.

2.1.2. Maturation

1. Laminar flow hood.
2. Tweezers.
3. 9-cm Petri dishes.
4. Growth chamber.
5. (a) Solid DCR media for maturation: DCR with 23.8 mg/L, 90 μM, abscisic acid (ABA), 7% *(w/v)* polyethylene glycol (PEG) 4000, and 60 g/L, 175 mM, sucrose; DCR with 23.8 mg/L, 90 μM, ABA and 175 mM, 60 g/L, sucrose; hormone-free DCR with 30 g/L, 88 mM, sucrose *(2, 23)* (**Table 2**).

   (b) Solid LM medium for maturation: ½ LM with 15.9 mg/L, 60 μM, ABA and 30 g/L, 88 mM, sucrose *(3)*.

2.1.3. Conversion and Acclimatization to Ex Vitro

1. Laminar flow hood.
2. 150-mL Tissue culture jars.

Table 1
Concentrations of Ingredients in Basic DCR *(22)* and ½ LM media *(3, 23)* Used for Somatic Embryogenesis of Scots Pine

	DCR		½ LM	
Ingredient	**mg/L**	**mM**	**mg/L**	**mM**
Inorganic				
NH_4NO_3	400	5	825	10.3
KNO_3	334	3.3	950	9.4
$Ca(NO_3)_2{\cdot}4H_2O$	543	2.3	–	–
$MgSO_4{\cdot}7H_2O$	370	1.5	925	3.75
$CaCl_2{\cdot}2H_2O$	84	0.57	11	0.07
KH_2PO_4	163	1.2	170	1.25
KI	0.8	0.005	4.15	0.025
H_3BO_3	6.2	0.1	31	0.5
$MnSO_4{\cdot}H_2O$	22	0.13	21	0.12
$ZnSO_4{\cdot}7H_2O$	8.6	0.03	43	0.15
$Na_2MoO_4{\cdot}2H_2O$	0.24	0.001	1.25	0.005
$CuSO_4{\cdot}5H_2O$	0.25	0.001	0.50	0.002
$CoCl_2{\cdot}6H_2O$	0.024	0.0001	0.125	0.0005
$NiCl_2{\cdot}6H_2O$	0.024	0.0001	–	–
$FeSO_4{\cdot}7H_2O$	27.8	0.1	27.8	0.1
Na_2–EDTA	34	0.1	34	0.1
Organic				
Nicotinic acid	0.5	0.0041	0.5	0.0041
Pyridoxine–HCl	0.5	0.0024	0.1	0.0048
Thiamine–HCl	1	0.003	0.1	0.0003
Myo-inositol	198	1.1	100	0.56
Casein hydrolysate	500		1000	
L-Glutamine	248	1.7	500	3.4
Glycine	2	0.0266	–	–
Sucrose	30,000	88	30,000	88
Agar	600			
Gelrite (*see* **Note 8**)	1,900			4,000

Table 2
Composition of the DCR Culture Media Used During Initiation and Proliferation, Maturation as well as Conversion of Embryogenic Cultures of Scots Pine *(2, 22, 24)*

Medium composition	Initiation and proliferation	Maturation				Conversion
Inorganics and organics	DCR		DCR			½ DCR
Sucrose (g/L)	30	30	60	60	30	20
PHs (mg/L)						
2,4-D	3.0					
BA	0.5					
ABA			23.8	23.8		
PEG 4000 (g/L)			70			

PHs phytohormones

3. Tweezers.
4. Growth chamber.
5. (a) Solid DCR medium for conversion: hormone-free ½ DCR with 20 g/L, 58 mM, sucrose *(2)* (**Table 2**).
6. Solid LM medium for conversion: hormone-free ½ LM with 30 g/L, 88 mM, sucrose *(3)*.
7. Plastic table greenhouses containing non-fertilized horticultural peat and perlite (2:1).
4. 8 Plastic containers, commercial fertilized peat (VAPO, Finland) with 1 kg/L basic fertilizer: 9.7% N, 7.5% P, 14.4% K, 5.0% Ca, 6.6% S, 3.8% Mg, 0.27% Fe, 0.13% Mn, 0.04% B, 0.05% Zn, 0.25% Cu, and 0.09% Mo and 3 kg/L limestone dust with Mg), commercial 0.2% 5-Superex fertilizer (Kekkilä, Finland).

2.2. Isolation, Identification and Elimination of Endophytes

2.2.1. Isolation of Bacterial and Fungal Endophytes

1. Scots pine tissues from various stages of somatic embryogenesis.
2. Laminar flow hood.
3. Forceps.
4. 10% methanol.
5. Mortar and pestle.
6. Luria Bertani (LB) or King's B (KB) medium with 1.5% *(w/v)* agar in 9-cm Petri dishes (**Table 3**).
7. Potato dextrose agar (PDA; Fluka, Buchs, Germany), 2% *(w/v)* or malt extract agar (MEA; Fluka, Buchs, Germany), 2% *(w/v)* in 9-cm Petri dishes.

Table 3
Concentrations of Ingredients in King's B and Luria Bertani Medium Used for Cultivation of Bacteria

	g/100 mL	
Ingredient	**KB**	**LB**
Peptone	2	0.5
Tryptone	–	1
NaCl	–	0.5
Glycerol	1	–
K_2HPO_4	0.15	–
$MgSO_4 \cdot 7H_2O$	0.15	–
Agar	1.5	1.5
pH	7.2	7.0

2.2.2. Identification of Bacterial and Fungal Endophytes

1. DNA extraction kit (e.g., DNeasy Plant Mini Kit) (Qiagen, Valencia, CA).
2. Primers to amplify fungal 18S rDNA: NS1 (5''-GTAGTCATATGCTTGTCTC-3'', *Saccharomyces cerevisiae* positions 20–38) and NS8 (5''-TCCGCAGGTTCACCTACGGA-3'', positions 1788–1769) *(25)*. Primers for amplifying bacterial 16S rDNA: UNI-1 (5′-AGAGTTTGATC(T/C)TGG(T/C)T(T/C)AG-3′; *Escherichia coli* positions 8–27) and UNI-2 (5′-ACGGCTACCTTGTTACGACTT-3′, positions 1512–1492) *(26)*.
3. DNA Polymerase enzyme (Dynazyme; Finnzymes, Espoo, Finland).
4. 10 × Buffer (Finnzymes, Espoo, Finland).
5. Nucleotides (MBI Fermentas, Vilnius, Lithuania).
6. Sterile distilled water.
7. Thermal cycler.
8. 0.2-mL PCR tubes.
9. Agarose (Fluka; Buchs, Germany).
10. SybrSafe (Invitrogen; Carlsbad, CA).
11. DNA size marker (e.g., GeneRuler DNA Ladder Mix) (MBI Fermentas).
12. Loading buffer (MBI Fermentas; Vilnius, Lithuania).
13. Subcloning kit (pGEM T-easy vector cloning kit; Promega, Madison, WI).

14. LB medium with 100 μg/mL ampicillin, both liquid and solidified with 1.5% *(w/v)* agar in 9-cm Petri dishes.
15. X-Gal (isopropyl-ß-D-1-thiogalactopyranoside): X-Gal and IPTG (5-bromo-4-chloro-3-indolyl-ß-D-galactopyranoside) and competent *Escherichia coli* cells (Promega; cat. no. JM109 L2001).
16. pUC19-specific primers SP6 Promoter Primer (Promega; cat. no. Q5011) and T7 Promoter Primer (Promega; cat.no. Q5021,).
17. Plasmid DNA isolation kit (Qiagen Plasmid Mini Kit; Qiagen, Valencia, CA).
18. Sequencing kit (Abi Prism BigDye Terminator Cycle Sequencing Kit; Applied Biosystems, Warrington, UK).

2.2.3. Elimination of Endophytes from Tissue Culture

1. LB or KB medium containing 10, 20, and 50 μg/mL of various antibiotics (e.g., tetracycline, ampicillin, erythromycin, doxycycline, kanamycin, streptomycin) and solidified with 1.5% *(w/v)* agar in 9-cm Petri dishes, inoculation loops.
2. PDA or MEA containing 10, 20, and 50 μg/mL of various fungicides e.g., amphotericin B, miconazole, and nystatin
3. Tissue culture medium DCR **(Table 1)**.

2.3. Protocols for Analyses of Secondary Compounds

2.3.1. Extraction of Soluble Phenolics

1. Air or freeze-dried needles, young stems, or embryogenic cultures of Scots pine.
2. Methanol (flavonoids) (p.a. or HPLC-grade).
3. 2-mL Eppendorf tubes.
4. Stainless steel balls.
5. Acidic butanol (50 mL/L conc. HCl in *n*-butanol).
6. Iron-reagent (20 g/L $FeNH_4(SO_4)_2 \times 12H_2O$ in 2 N HCl).
7. 60 μg/100 μL, 2.19 mM, internal standard salicin, Sephadex® LH-20 (Sigma-Aldrich).
8. Ball mill tissue homogenizer (e.g., PrecellysR24).
9. Ultra-Turrax homogenizer T8.
10. Roughened class rod.
11. Table centrifuge.
12. 20-mL vials.
13. Gaseous nitrogen or a Speed-Vac equipment for evaporation of extraction solvent, syringe filters.

2.3.2. Identification and Quantification of Phenolics with HPLC–DAD and Colorimetry (Fig. 1)

1. Standards: Quercetin 3-galactoside (Apin Chemicals Ltd., Oxon, UK), quercetin 3-glucoside (Apin Chemicals Ltd., Oxon, UK), quercetin 3-rhamnoside (Apin Chemicals Ltd., Oxon, UK), myricetin 3-rhamnoside (Apin Chemicals Ltd., Oxon, UK), kaempferol 3-glucoside (Extrasynthese, Genay,

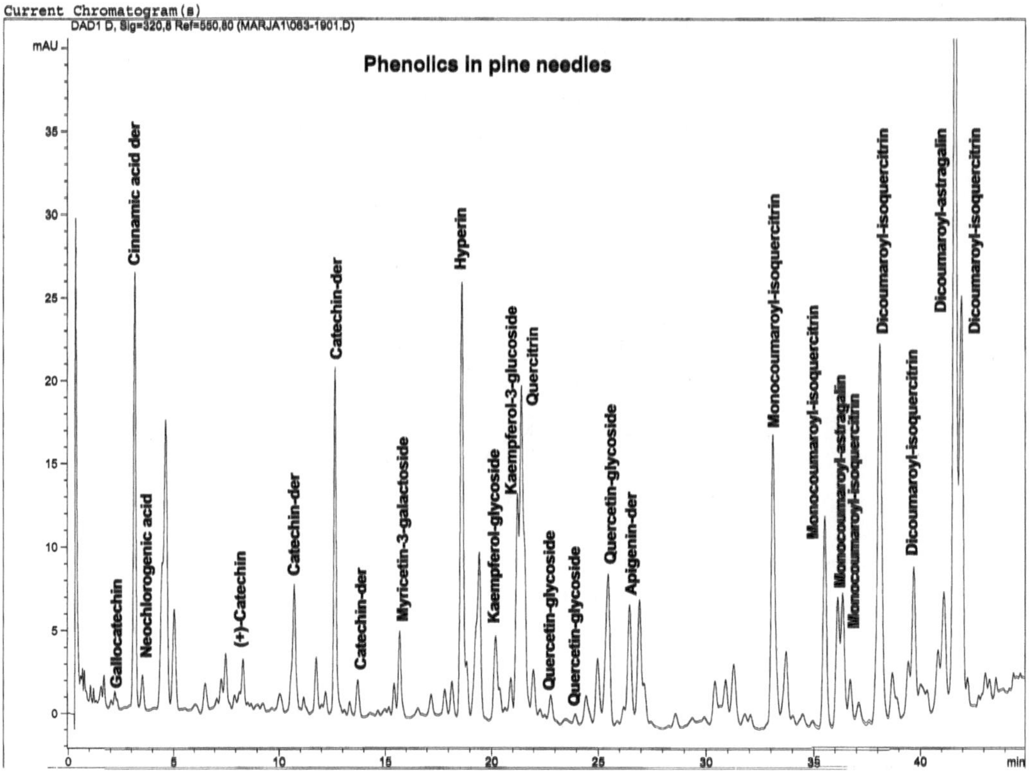

Fig. 1. Polyphenolic compounds in Scots pine needles.

France), (+)-catechin (Aldrich, Steinheim, Germany), taxifolin (Aldrich, Steinheim, Germany), piceatannol (Sigma-Aldrich, Steinheim, Germany) dissolved in methanol:MilliQ water (1:1) at a concentration of 0.1 mg/L. Purity of standards is noted in exact concentrations. Purified proanthocyanidins from pine needles and stems *(27)*.

2. Elution solvents for HPLC are 1.5% tetrahydrofurane-0.25% *o*-phosphoric acid and methanol (*see* **Note 1**).
3. 4.6 × 60 mm, 3 μm-Hypersil ODS column (Agilent Technologies; Germany).
4. 1701 capillary GC-column (Perkin-Elmer Elite).
5. Helium as a carrier gas.
6. A boiling water bath and colorimetric equipment.
7. HPLC-diode array detector.
8. GC-MS equipment.

2.3.3. Extraction of Monoterpenes

1. Freeze-dried needles or stems or embryogenic cultures of Scots pine.
2. 2-mL Eppendorf tubes.

3. Na_2SO_4.
4. Celite.
5. Liquid nitrogen.
6. Delta-2-carene as internal standard.
7. Pentane (p.a. grade).
8. Mortar and pestle.
9. Ball mill.
10. Liquid nitrogen.
11. Table centrifuge.
12. Cold room at 5°C.
13. 2-mL GC vials.

3. Methods

3.1. Somatic Embryogenesis

3.1.1. Culture Media

1. Solid DCR and LM culture media (**Tables 1** and **2**) for initiation, induction, proliferation, and maturation are prepared into 9-cm Petri dishes and liquid media to arrest proliferation into 250-mL Erlenmeyer flasks. Media for conversion are prepared into tissue culture jars. Aqueous stock solutions of L-glutamine and phytohormones 2, 4-D, BA, and ABA are filter sterilized and added to the medium after autoclaving.

3.1.2. Explant Preparation

Explants

1. Immature zygotic embryos surrounded by the megagametophyte (called immature zygotic embryos) and isolated from 1-yr-old immature seed cones (**Fig. 2a**) are most favourable material for starting somatic embryogenic cultures of Scots pine. The optimum developmental stage of immature zygotic embryos for initiation is the precotyledonary stage (i.e., after fertilization but before the formation of cotyledons) *(1, 2)*. Generally, this developmental phase occurs when the effective temperature (d.d. the heat sum unit based on the daily mean temperatures minus the adapted +5°C temperature) is between 440 and 650 d.d. (*see* **Note 4**).

Explant Excision and Sterilization

1. Immature seed cones are sprayed with 70% ethanol, after which immature seeds are removed from the cones using scalpels and tweezers and placed into sterile water in Petri dishes.
2. Seeds are surface sterilized for 1 min in 70% ethanol in Petri dishes and rinsed twice in sterile water.

Culture Initiation and Induction

1. Seed coats are opened and removed with tweezers and immature zygotic embryos and their suspensor tissues surrounded by megagametophytes are excised and placed onto DCR or LM medium for initiation.

Fig. 2. Somatic embryogenesis of Scots pine. (**a**) Developing cones. (**b**) Proliferating embryogenic culture. (**c**) A somatic cotyledonary embryo. (**d**) Germinating somatic embryo plants in vitro. (**e**) A somatic embryo plant in the greenhouse.

2. Immature zygotic embryos are first cultured for 4 wk, and thereafter transferred to new positions in the same Petri dish for an additional 2 wk.
3. Both initiation and induction are performed in the dark at 25 ± 2C°.

Proliferation of Embryogenic Cultures

1. Embryogenic cell mass starts to protrude from the micropylar end of the responsive explants (immature zygotic embryos surrounded by megagametophytes) 4–6 wk after initiation (**Fig. 2b**). Embryogenic tissues are excised from explants and transferred onto a new DCR or LM medium for proliferation in the dark at 25 ± 2C° (*see* **Note 5**).
2. To maintain embryogenic tissues, the proliferating upper and outer tissues from embryogenic cell masses are transferred onto fresh medium every 4 wk. Embryogenic cell masses can be subcultured for several months (*see* **Note 6**).

3. Embryogenic cell masses can be used as such for maturation. To arrest proliferation before maturation phase, the embryogenic cell masses are transferred onto hormone-free DCR-medium with 30 g/L, 88mM, sucrose for 2 *(23)* or 4 wk *(28)*.
4. Suspension made from embryogenic cell masses can also be used for maturation *(3, 24, 29)*. Two grams (FW) of embryogenic cell mass is transferred from proliferation medium into 50 mL of liquid hormone-free DCR or hormone-free ½LM medium with 30 g/L, 88 mM, sucrose in the 250-mL Erlenmeyer flasks. The suspensions are cultivated on a shaker in the dark. After 2–4 wk, the suspension is decanted aseptically on the moist sterile filter paper in the Büchner funnel, after which the embryogenic cell masses are let to filter out of the liquid phase.
5. Proliferation occurs under the same conditions as initiation and induction.

Maturation of Embryogenic Cultures

1. (a) Maturation on DCR-media: Embryogenic cell masses or filter paper covered by a thin layer of embryogenic cell mass is transferred for 2 wk onto DCR-medium with 23.8 mg/L, 90 μM, ABA, 7% *(w/v)* PEG, and 60 g/L, 175 mM, sucrose, and thereafter onto the same medium without PEG for further 2 wk. Maturation is continued for 2–4 wk on hormone-free DCR-medium with 30 g/L, 88 mM, sucrose (*see* **Note 7**).
   (b) Maturation on LM-medium: Filter paper covered by a thin layer of embryogenic cell mass is transferred for 9–12 wk onto ½ LM medium with 15.9 mg/L, 60 μM, ABA and 30 g/L, 88 mM, sucrose.
2. Maturation is performed under a 16 h photoperiod about 20–30 μmol/m^2 s *(3, 23)* at 25 ± 2C°.

Conversion of Somatic Embryos

1. Mature somatic embryos **(Fig. 2c)** are carefully detached from the embryogenic cell masses and transferred onto hormone-free ½ DCR medium with 20 g/L, 58 mM, sucrose or onto hormone-free ½ LM medium with 30 g/L, 88 mM, sucrose in the tissue culture jars.
2. The base of somatic embryos is gently inserted into the medium.
3. Somatic embryos are converted into plantlets under 16-h photoperiod 140–150 μmol/m^2 s **(Fig. 2d)**.

Acclimatization to Ex Vitro

1. Somatic embryo-derived plantlets are carefully detached from the medium and roots are washed. Thereafter plantlets are planted into plastic table greenhouses containing non-fertilized horticultural peat and perlite *(v:v)* (2:1). For the first 2 wk the plantlets are kept under mist in order to keep relative humidity at approximately 90% after which the humidity is gradually decreased.

2. After 1 month the plantlets are transplanted into bigger containers containing commercial fertilized peat (VAPO, Finland) in a greenhouse. During the season, plantlets are fertilized monthly with commercial 0.2% 5-Superex fertilizer (Kekkilä, Finland) (**Fig. 2e**).

3.2. Isolation, Identification and Elimination of Endophytes

3.2.1. Isolation of Endophytes

Bacterial Endophytes

1. In a laminar flow hood, somatic embryogenic tissues are transferred into a mortar (*see* **Note 1**). One milliliter of 10% methanol is added and the tissue is grinded with the pestle. The grinded plant material is transferred into an Eppendorf tube and centrifuged at 1000*g* for 1 min.
2. Serial dilutions of 1:1, 1:10 and 1:100 of the supernatant are prepared to evaluate the number of colony forming units (CFU).
3. Of each dilution, 100 μL is spread onto solid LB or KB medium, dried, and grown at room temperature for up to 4 wk.

Fungal Endophytes

1. For detection of fungi, plant material is grinded in sterile water.
2. The grinded plant material is transferred onto PDA or MEA medium and air-dried.
3. The cultures are grown at room temperature for up to 12 wk.

3.2.2. Identification of Endophytes

Polymerase Chain Reaction

1. The fungal DNA is isolated using the DNA isolation kit. For bacteria, a colony is picked up and used as the template in the PCR.
2. The PCR reaction is pipetted on ice to consist of 100 ng of fungal DNA or the bacterial colony, 0.15 μM primers, 200 μM nucleotides, 1X buffer, and 1U of PCR enzyme (Dynazyme; Finnzymes, Espoo, Finland) in a 0.2-mL PCR tube.
3. The tube is placed in a thermal cycler along with a tube containing negative control (no template).
4. The following PCR program is used for bacteria: 94°C 5 min, 3 cycles of 1 min at 94°C; 1 min at 62°C, and 3 min at 72°C, followed by 3 cycles with 60°C; and 30 cycles with 58°C as annealing temperatures; and extension at 72°C for 5 min.
5. The following PCR program is used for fungi: 94°C 5 min, 3 cycles of 1 min at 94°C; 1 min at 53°C; and 3 min at 72°C; followed by 3 cycles with 51°C, and 30 cycles with 49°C as annealing temperatures; and extension at 72°C for 5 min.

Agarose Gel Electrophoresis

1. Loading buffer is added at a final concentration of 1X to the PCR products.
2. The PCR products are separated on 1% agarose gel with 1X SybrSafe at 80 V for 30 min. The DNA size marker is included to identify bands of correct size.
3. The gel is visualized and photographed under UV light. A band of approximately 1.5 kb and approximately 1.8 kb is expected for bacteria and fungi, respectively.

Cloning of PCR Products

1. The products are cloned into pGEM T-easy vector according to the manufacturer's instructions.
2. The cloned PCR products (5 μL) are mixed with 200 μL competent *Escherichia coli* JM109 cells and kept on ice for 40 min.
3. The transformation mixture is quickly transferred to 42°C for 2 min and immediately back on ice. The mix is kept on ice for 2 min.
4. To the transformation mix, 250 μL of liquid LB medium is added and shaken at 37°C for 45 min.
5. Various aliquots (50, 100 and 200 μL) are plated along with 50 μL of both 2% X-gal and 100 mM IPTG and incubated at 37°C overnight.
6. White colonies are taken and checked for insert size by PCR as described (**Subheadings 3.2.2, steps 1** and **2**) with primers SP6 and T7.

Isolation of Plasmid DNA

The transformants containing a vector with correct insert size are grown in liquid Luria bertani (LB) medium with 100 μg/mL ampicillin overnight.

1. Plasmid DNA is isolated according to manufacturer's instructions.

Sequencing of PCR Products

1. The inserts are sequenced with the primers SP6 and T7 according to the manufacturer's instructions.

3.2.3. Elimination of Endophytes

1. Bacteria are inoculated onto the solid LB or KB media containing various antibiotics at concentrations of 10, 20, and 50 μg/mL.
2. Fungal mycelium is transferred onto PDA or MEA medium containing antifungal compounds nystatin, amphotericin B, and miconazole at concentrations 10, 20, and 50 μg/mL, respectively.
3. Bacteria and fungi are grown for up to 1 and 2 mo, respectively, and checked for viability. If growth is observed, the antibiotic compound is not effective at the corresponding concentration. Antibiotics at the lowest effective concentration should be used in DCR medium for somatic embryogenesis.

3.3. Protocols for Analyses of Secondary Compounds

3.3.1. Extraction of Soluble Phenolics

1. 5–10 mg dried pine tissue material is placed in 2-mL Eppendorf tube with 2 stainless steel balls. Thereafter 0.6-mL cold methanol (*see* **Notes 2–4**) and an internal standard salicin are added.
2. Samples are homogenized for 15 s using a ball mill, allowed to stand in an ice bath for 15 min (*see* **Note 5**) and centrifuged for 3 min at 11,000*g*, after which the supernatant is separated (*see* **Note 6**).

3. Homogenization process is repeated for three times, for 5 s each, while standing in an ice bath for a total of 5 min (*see* **Note 7**).
4. Methanol is evaporated from combined supernatants with gentle flow of gaseous nitrogen or with a Speed Vac, after which the samples are re-dissolved into 0.3 mL -methanol and 0.3-mL milliQ water and filtered trough a syringe filter to HPLC vials (*see* **Note 8**).

3.3.2. Extraction of Monoterpenes

1. 0.1-g frozen needles, 0.035 g Na_2SO_4 and 0.1 g celite are weighed and grinded quickly to powder with a pestle and mortar in liquid nitrogen.
2. The powder is transferred into an Eppendorf tube, and internal standard (delta-2-carene), and 1-mL pentane are added (*see* **Note 9**).
3. Tubes are shaken for 3 h and allowed to stand overnight at 5°C.
4. The samples are centrifuged, 5°C, 11,000*g*, for 10 min and the supernatant is transferred to GC-vials.

3.3.3. Analysis of Soluble Phenolics by HPLC–DAD

1. Separation of phenolics is processed using a gradient elution: in solvent A for 5 min, linear gradient from 0 to 50% of B during 40 min, rising of B to 100% and equilibration by A to 100%. Injection volume is 20 μL. The injector and column oven temperature is set to 23 and 30°C, respectively.
2. Identification of the phenolics is based on their retention times and spectral characteristics compared with commercial standards.
3. Quantification of the pine phenolics is based on external commercial standards and is expressed as mg/g dry weight.

3.3.4. Analysis of Proantocyanidins by Colorimetric Determination (see Note 10)

1. 2- to 4-mg freeze dried and milled needles or embryogenic cultures is weighed into a 20 mL vial, and 6 mL acidic butanol, 1 mL methanol, 0.2 mL ferri-reagent are added and mixed.
2. The mixture is hydrolyzed in sealed tubes for 45 min in a boiling water bath and thereafter the vials are cooled.
3. Absorbance is measured at A_{550} nm.
4. Values are expressed as tannin equivalents purified from pine needles or stems *(27)*.

3.3.5. Analysis of Monoterpenes by GC–MS

1. Separation of pine monoterpenes is processed by 1701 capillary column (Perkin-Elmer Elite) using 1-μL injection volume. The initial temperature in the oven is 35°C for 1 min followed by increases at a rate of 4°C/min to 140°C and then 45°C/min to 200°C, which is maintained for 5 min. The injector temperature is 220°C and interface temperature 180°C. The flow rate of the helium carrier gas is 1.0 mL/min.

2. Qualitative analyses are based on ions produced (SCAN) from the components by comparing them with the MS-library.
3. Selected-ion monitoring (SIM) is used for quantification of each component based on commercial monoterpenes.

4. Notes

4.1. In Vitro Protocol for Somatic Embryogenesis

1. DCR medium was originally used for tissue cultures of Douglas-fir [*Pseudotsuga menziesii* (Mirb.) Franco] by Gupta and Durzan *(19)* and Litvay medium for those of loblolly pine (*Pinus taeda* L.) *(20)*.
2. In DCR medium sucrose can be replaced with 30 g/L, 88 mM, *(1)* or 20 g/L, 58 mM, maltose *(30)*.
3. Initiation and proliferation can also be performed on DCR-medium with 819 mg/L, 8.1 mM, KNO_3; 171 mg/L, 1.16 mM, $CaCl_2 \cdot 2H_2O$; 2.0 mg/L, 9.0 μM; 2,4-D; and 1.0 mg/L, 4.4 μM, BA;and from which $Ca(NO_3)_2 \cdot 4H_2O$ is omitted *(28)*.
4. The cones can be collected and stored at 4°C for at least 2 mo without losing the ability for somatic embryogenesis *(2)*.
5. Induction of somatic embryogenic cultures in Scots pine takes place slowly, 4–6 wk after initiation. Initiation frequencies are usually low (under 10%) and genotype-dependent *(1–3, 30, 22)*.
6. During prolonged proliferation the regeneration ability of embryogenic cell masses decreases.
7. Maturation can also be performed on DCR medium with 8% *(w/v)* PEG 3500, 4 mg/L, 15.1 μM, ABA; and 20 g/L, 58 mM, maltose *(30)*.
8. DCR medium can also be solidified using only 0.25% (2.5 g/L) *(2)*, 0.325% (3.25 g/L) *(30)*, or 0.35% (3.5 g/L) *(28)* gelrite. ½ LM medium for proliferation and germination is solidified with 0.4% (4.0 g/L) gelrite and that for maturation with 1% (10 g/L) gelrite *(3)*.

4.2. Protocols for Isolation, Identification and Elimination of Endophytes

1. In case of suspension cultures, the culture medium can be sampled and plated directly onto solid media for bacterial or fungal growth.

4.3. Protocol for Analyses of Secondary Compounds

1. Ammonium formate in formic acid (A) and methanol (B) or water/formic acid and acetonitrile may be used *(14, 31, 32)*.
2. Prehandling of the plant material before extraction is of utmost importance to obtain the real image of the chemical composition *(33)*. Phenolic compounds can easily be enzymatically degraded when the collected plant material is fresh *(34)*.

3. For the extraction of pine phenolics, the solvent is generally chosen based on the function and the type of the component required. Due to the high number of components of varying polarity, the choice of solvent is important. Weakly polar phenolics (such as isoflavonoids, flavanones, methylated flavones and flavonols) can easily be extracted with chloroform, dichloromethane, diethyl ether, or ethyl acetate; whereas more polar phenolics (such as flavonoid glycosides) should preferably be extracted with alcohols or alcohol–water mixtures *(35)*. Soluble tannins are quantitatively extracted with aqueous acetone whereas insoluble condensed tannins can be determined by acid hydrolysis of dried plant material and by quantification of the resulting anthocyanidins *(36)*. Pine monoterpenes can easily be extracted by hexane *(12)* or pentane *(11)*, while petroleum ether-diethyl ether is used for resin acids *(37)*.
4. Aqueous methanol, aqueous acetone or ethanol can also be used for extraction of phenolics *(38, 39)*.
5. The extraction time in methanol may be continued overnight at 4°C *(32)*.
6. Depending on the tissue, the homogenization can be processed further with a roughened class rod, in addition to the ball mill or the Ultra-Turrax. Pre-freezing of the material is needed when the ball mill is used for fresh stem tissues. The first homogenization should result in a fine powder. Homogenization may be faster using liquid nitrogen *(32)*.
7. The residue should be colourless after the last extraction phase, otherwise additional washing steps are needed.
8. Instead of a syringe filter, the dissolved sample can be clarified by centrifugation 11,000*g* for 3 min.
9. *n*-Hexane as solvent and 1-chloro-octane as an internal standard can also be used for monoterpene analyses.
10. Soluble proanthocyanidins from pine needles can be extracted by 70% aqueous acetone, purified by Sephadex LH-20, and analyzed by RP- and NP-HPLC as described by Karonen et al. *(39)*.

Acknowledgments

This work was supported by the Academy of Finland (project numbers 126488 to KN, 121994 to TS, 118569 to AMP, 123826 to HH and 45591 and 64308 to RJT).

References

1. Keinonen-Mettälä, K., Jalonen, P., Eurola, P., von Arnold, S., and von Weissenberg, K. (1996). Somatic embryogenesis of *Pinus sylvestris*. *Scand. J. For. Res.* **11**, 242–250
2. Häggman, H., Jokela, A., Krajnakova, J., Kauppi, A., Niemi K., and Aronen, T. (1999). Somatic embryogenesis of Scots pine: cold treatment and characteristics of explants affecting induction. *J. Exp. Bot.* **50**, 1769–1778
3. Lelu, M.-A., Bastien, C., Drugeault, A., Gouez, M.-L., and Klimaszewska, K. (1999). Somatic embryogenesis and plantlet development in *Pinus sylvestris* and *Pinus pinaster* on medium with and without growth regulators. *Physiol. Plant.* **105**, 719–728
4. Lilja, A., Hallaksela, A.M., and Heinonen, R. (1995). Fungi colonizing Scots pine cone scales and seeds and their pathogenicity. *For. Pathol.* **25**, 38–46
5. Pirttilä, A.M., Laukkanen, H., Pospiech, H., Myllylä, R., and Hohtola, A. (2000). Detection of intracellular bacteria in the buds of Scots pine (*Pinus sylvestris* L.) by in situ hybridization. *Appl. Environ. Microbiol.* **66**, 3073–3077
6. Pirttilä, A.M., Pospiech, H., Laukkanen, H., Myllylä, R., and Hohtola, A. (2003). Two endophytic fungi in different tissues of Scots pine buds (*Pinus sylvestris* L.). *Microb. Ecol.* **45**, 53–62
7. Pirttilä, A.M., Pospiech, H., Laukkanen, H., Myllylä, R., and Hohtola, A. (2005). Seasonal variation in location and population structure of endophytes in buds of Scots pine. *Tree Physiol.* **25**, 289–297
8. Pirttilä, A.M., Joensuu, P., Pospiech, H., Jalonen, J., and Hohtola, A. (2004). Bud endophytes of Scots pine produce adenine derivatives and other compounds that affect morphology and mitigate browning of callus cultures. *Physiol. Plant.* **121**, 305–312
9. Seigler, D.S. (1998). *Plant secondary metabolism*. Kluwer Academic Publishers, Boston, MA
10. Vuorela, S., Kreander, K., Karonen, M., Nieminen, R., Hämäläinen, M., Galkin, A., Laitinen, J., Salminen, J.-P., Moilanen, E., Pihlaja, K., Vuorela, H., Vuorela, P., and Heinonen, M. (2005). Preclinical evaluation of rapeseed, raspberry, and pine bark phenolics for health related effects. *J. Agric. Food Chem.* **53**, 5922–5931
11. Räisänen, T., Ryyppö, A., Julkunen-Tiitto, R., Peltola, H., and Kellomäki, S. (2008). Effects of long-term elevation of CO2 and temperature on secondary compounds in the needles of Scots pine. *Trees* **22**, 121–135
12. Turtola, S., Sallas, L., Holopainen, J., Julkunen-Tiitto, R., and Kainulainen, P. (2006). Long-term exposure to enhanced UV-B radiations has no significant effect on the growth or secondary compounds outdoor-grown Scots pine and Norway spruce seedlings. *Environ. Pollut.* **144**, 166–171
13. Willför, S., Hemming, J., Reunanen, M., and Holmbom, B. (2003). Phenolic and lipophilic extractives in Scots pine knots and stemwood. *Holzforschung* **57**, 359–372
14. Karonen, M., Hämäläinen, M., Nieminen, R., Klika, K.D., Loponen, J., Ovcharenko, V.V., Moilanen, E., and Pihlaja, K. (2004). Phenolic extractives from bark of *Pinus sylvestris* L. and their effects on inflammatory mediators nitric oxide and prostaglandin E2. *J. Agric. Food Chem.* **52**, 7532–7540
15. Pietarinen, S., Willför, S.M., Ahotupa, M.O., Hemming, J.E., and Holmbom, B.R. (2006). Knotwood and bark extracts: strong antioxidants from waste materials. *J. Wood Sci.* **52**, 436–444
16. Saarinen, N.M., Wärri, A., Mäkelä, S.I., Eckerman, C., Reunanen, M., Ahotupa, M., Salmi, S.M., Franke, A.A., Kangas, L., and Santti, R. (2000). Hydroxymatairesinol, a novel enterolactone precursor with antitumor properties from coniferous tree (*Picea abies*). *Nutr. Cancer* **36**, 207–216
17. Lindberg, L.E., Willför, S.M., and Holmbom, B.R. (2004). Antibacterial effects of knotwood extractives on paper mill bacteria. *J. Ind. Microbiol. Biotechnol.* **31**, 137–147
18. Rauha, J.-P., Remes, S., Heinonen, M., Hopia, A., Kähkönen, M., Kujala, T., Pihlaja, K., Vuorela, H., and Vuorela, P. (2000). Antimicrobial effects of Finnish plant extracts containing flavonoids and other phenolic compounds. *Int. J. Food Microbiol.* **56**, 3–12
19. Gupta, P.K., and Durzan, D.J. (1985). Shoot multiplication from mature trees of Douglas-fir (*Pseudotsuga menziesii*) and sugar pine (*Pinus lambertiana*). *Plant Cell Rep.* **4**, 643–645
20. Litvay, J.D., Verma, D.C., and Johnson, M.A. (1985). Influence of a loblolly pine (*Pinus taeda* L.) culture medium and its components on growth and somatic embryogenesis of the wild carrot (*Daucus carota* L.). *Plant Cell Rep.* **4**, 325–328
21. Sarjala, T., Häggman, H., and Aronen, T. (1997). Effect of exogenous polyamines and inhibitors of polyamine biosynthesis on growth and free polyamine contents of embryogenic Scots pine callus. *J. Plant Physiol.* **150**, 597–602

22. Park, Y.S., Lelu-Walter, M.A., Harvengt, L., Trontin, J.F., MacEacheron, I., Klimaszewska, K., and Bonga, J.M. (2006). Initiation of somatic embryogenesis in *Pinus banksiana*, *P. strobus*, *P. pinaster*, and *P. sylvestris* at three laboratories in Canada and France. *Plant Cell Tissue Organ Cult.* **86**, 87–101

23. Niemi, K., Sarjala, T., Chen, X., and Häggman, H. (2002). Spermidine and methylglyoxal bis (guanylhydrazone) affect maturation and endogenous polyamine content of Scots pine embryogenic cultures. *J. Plant Physiol.* **159**, 1155–1158

24. Niemi, K., and Häggman, H. (2002). *Pisolithus tinctorius* promotes germination and forms mycorrhizal structures in Scots pine somatic embryos *in vitro*. *Mycorrhiza* **12**, 263–267

25. White, T.J., Burns, T., Lee, S., and Taylor, J. (1990). Amplification and direct sequencing of fungal ribosomal RNA genes for phylogenetics. In: M.A. Innis, D.H. Gelfand, J.J. Sninsky, T.J. White (eds.) PCR Protocols. Academic Press, San Diego, CA, USA, 315–322

26. Jalava, J., Kotilainen, P., Nikkari, S., Skurnik, M., Vanttinen, E., Lehtonen, O.P., Eerola, E., and Toivanen, P. (1995). Use of the polymerase chain reaction and DNA sequencing for detection of *Bartonella quintana* in the aortic valve of a patient with culture-negative infective endocarditis. *Clin. Infect. Dis.* **21**, 891–896

27. Hagerman, A.E. (1995). Tannin analysis. Department of Chemistry, Miami University, Oxford, OH

28. Rodríguez, M., Suárez, M., Heredia, R., Ávila, C., Breton, D., Trontin, J.-F., Filonova, L., Bozhkov, P., von Arnold, S., and Cánovas, M. (2006). Expression patterns of two glutamine synthetase genes in zygotic and somatic embryos support specific roles in nitrogen metabolism during embryogenesis. *New Phytol.* **169**, 35–44

29. Klimaszewska, K., and Smith, R. (1997). Maturation of somatic embryos of *Pinus strobus* is promoted by a high concentration of gellan gum. *Physiol. Plant.* **100**, 949–957

30. Niskanen, A.-M., Lu, J., Seitz, S., Keinonen, K., von Weissenberg, K., and Pappinen, A. (2004). Effect of parent genotype on somatic embryogenesis in Scots pine (*Pinus sylvestris*). *Tree Physiol.* **24**, 1259–1265

31. Lange, B.R., Trost, M., Heller, W., Langebartels, C., and Sanderman, H., Jr. (1994). Elicitor-induced formation of free and cell-wall-bound stilbenes in cell-suspension cultures of Scots pine (*Pinus sylvestris* L.). *Planta* **194**, 143–148

32. Turunen, M., Heller, W., Stich, S., Sanderman, H., Sutinen, M.L., and Norokorpi, Y. (1999). The effects of UV exclusion on soluble phenolics of young Scots pine seedlings in subartic. *Environ. Pollut.* **106**, 219–228

33. Waterman, P.G., and Mole, S. (1994). Analysis of phenolic plant metabolites. Blackwell Scientific Publications, Oxford

34. Julkunen-Tiitto, R., and Sorsa, S. (2001). Testing the drying methods for willow flavonoids, tannins and salicylates. *J. Chem. Ecol.* **27**, 779–789

35. Naczk, M., and Shahidi, F. (2004). Extraction and analysis of phenolics in food. *J. Chromatogr. A.* **1054**, 95–111

36. Hagerman, A.E., and Butler, L.G. (1994). Assay of condensed tannins or flavonoid oligomers and related flavonoids in plants. *Methods Enzymol.* **234**, 429–437

37. Gref, R., and Ericsson, A. (1985). Wound-induced changes of resin acid concentration in living bark of Scots pine seedlings. *Can. J. For. Res.* **15**, 92–96

38. Slimestad, R., Nerdal, W., Francis, G.W., and Andersen, O.M. (1993). Myricetin 3,4 -diglucoside and kaempferol derivatives from needles of Norway spruce, *Picea abies*. *Phytochemistry* **32**,179–181

39. Karonen, M., Loponen, J., Ossipov, V., and Pihlaja, K. (2004). Analysis of proanthocyanidins in pine bark with reversed-phase and normal-phase high-performance liquid chromtaography-electrospray ionization mass spectrometry. *Anal. Chim. Acta* **522**, 105–112

Chapter 4

Saussurea medusa Cell Suspension Cultures for Flavonoid Production

Chun-Zhao Liu and Praveen K. Saxena

Summary

Saussurea medusa Maxim. is a valuable traditional Chinese herb. The flavonoids are the main active pharmaceutical compounds in this medicinal plant species and have effective anti-tumor and anti-inflammation properties. This species is now almost extinct in China because of over-exploitation. The establishment of plant cell cultures would be a promising alternative to avoid extinction of this species and establish cultivation for the production of bioactive flavonoids. The callus is induced from leaf explants of *S. medusa* on Murashiage and Skoog medium supplemented with 0.5 mg/L 6-BA, 2 mg/L NAA, 30 g/L sucrose, and 5 g/L agar. A fine cell suspension is established from the induced light-yellow calluses in the MS liquid medium with 30 g/L sucrose, 0.5 mg/L BA, and 2.0 mg/L NAA for biosynthesis of flavonoids. The kinetics of cell growth and flavonoid accumulation in the cell suspension cultures are investigated. The highest dry weight and flavonoid production reach 17.2 g/L and 607.8 mg/L respectively after 15 d. Significantly high antioxidant activity and flavonoids accumulate in the cell suspension cultures of *S. medusa*.

Key words: *Saussurea medusa*, Cell suspension cultures, Flavonoid biosynthesis, Antioxidant activity

1. Introduction

Saussurea medusa Maxim. (**Fig. 1**) is a valuable traditional Chinese herb, which belongs to the Composite family. It is used for the treatment of various ailments such as rheumatoid arthritis, mountain sickness, cough, lumbago, kidney asthenia, and menoxenia *(1)*. Flavonoids are the main active medicinal components of

S. Mohan Jain and Praveen K. Saxena (eds.), *Methods in Molecular Biology, Protocols for* In Vitro *Cultures and Secondary Metabolite Analysis of Aromatic and Medicinal Plants, vol. 547*
© Humana Press, a part of Springer Science+Business Media, LLC 2009
DOI: 10.1007/978-1-60327-287-2_4

Fig. 1. Induced red calluses from leaf explants of *Saussurea medusa* Maxim.

S. medusa, which are highly effective in relieving pain and inflammation, and also aid in widening of blood vessels and extension of human life *(2, 3)*. The wild *S. medusa* plants only grow at rocky beaches, 3900–5100 m above sea level in Qinghai province, China. *S. medusa* is nearly extinct due to commercialization and difficulties in propogation of the species via traditional plant breeding methods.

Current cultivation protocols of *S. medusa* produce bioactive flavonoids via cell suspension *(4, 5)*. This chapter describes a protocol to establish *S. medusa* cell suspension cultures for efficient production of flavonoids under controlled environmental conditions.

2. Materials

2.1. Establishment of S. Medusa In Vitro *Plantlets*

1. Seed collection of *Saussurea medusa* Maxim.: Seeds of *S. medusa* are collected from the Qinghai province of China. Their botanical identity has already been confirmed by comparing them with the reference standards maintained at the Institute of Botany, Chinese Academy of Sciences, P. R. China.
2. Surface sterilization of *S. medusa* seeds: 50 mL, 70% ethanol; 50 mL, 20% aqueous solution of 5.4% sodium hypochlorite; and 1000 mL sterile distilled water.
3. Surface-disinfested seed germination: Solid Murashige and Skoog (MS) medium *(6)* supplemented with 20 g/L sucrose and 5 g/L agar, pH 5.8.

2.2. Callus Induction and Suspension Culture Establishment of *S. medusa*

1. Callus induction: Use leaf explants (~ 0.5 × 0.5 cm size) from 30-d-old in vitro plantlets. The callus induction medium is MS medium containing 30 g/L sucrose, 0.5 mg/L BA, and 2.0 mg/L NAA.
2. Establishment of cell suspension cultures: Liquid MS culture medium with similar composition as of callus induction medium, but devoid of agar.
3. Incubation conditions: Growth cabinet.

2.3. Analysis

1. Dry weight determination: Filter papers to remove excessive water and a dry oven.
2. Conductivity of liquid medium: Measure with digital conductivity meter.
3. Determination of flavonoid content: 20 mg dry cells, 5 mL 80% *(v/v)* ethanol, 1 mL filtrate, 10-mL test tubes, 0.3 mL 5% *(w/w)* $NaNO_2$, 0.3 mL 10% *(w/w)* $AlCl_3$, and 4 mL 4% *(w/w)* NaOH, rutin as a standard *(7)*.
4. Assessment of antioxidant potential: 1, 1-diphenyl-2-picryl-hydrazyl (DPPH); 500 mg dry tissue; 50 mL, 80% ethanol; crude extract and a series of its dilutions (2 mL diluted test sample and 2 mL DPPH (200 μM) solution); spectrophotometer.

3. Methods

3.1. Induction and Maintenance of Callus Cultures

1. Surface sterilize seeds in 70% ethanol for 30 s followed by treatment with 20% aqueous solution of 5.4% sodium hypochlorite for 20 min. Rinse 3× with sterile distilled water.
2. Germinate sterile seeds and maintain on solid MS medium supplemented with 20 g/L sucrose and 5.0 g/L agar for 30 d in a growth cabinet under a 16-h photoperiod from cool-white fluorescent tubes, 60 μmol/m^2 s PFD, at 25 ± 1°C. Adjust pH of the solid medium is to 5.8 with 1 N NaOH before autoclaving at 121°C for 20 min.
3. Use leaf explants, approximately 0.5 × 0.5 cm in size, from the geminated plantlets for callus induction on MS medium containing 30 g/L sucrose, 0.5 mg/L BA, and 2.0 mg/L NAA.
4. Callus is induced in *S. medusa* leaf explants after 3 mo of culture (**Fig. 1**), which is sub-cultured onto the fresh solid callus induction medium at every 18-d interval.

5. The callus growth is relatively slow on the solid culture medium. 15.3 g/L dry weight and 490.7 mg/L flavonoids are obtained after culturing for 24 d **(Fig. 2)**.

3.2. Establishment of Cell Suspension Cultures

1. Transfer 2.5 g fresh weight of 18-d-old callus into 250-mL Erlenmeyer flasks containing 50 mL MS liquid culture medium supplemented with 30 g/L sucrose, 0.5 mg/L BA, and 2.0 mg/L NAA (*see* **Note 1**).
2. Cover flasks with hydrophobic fluoropore (PTFE) membranes, and incubate them on a rotary shaker, 110 rpm, 0.2 × *g* at 25°C under 16-h light, 60 μmol/m^2 s PFD per day.
3. The cell suspension cultures are established by regular subculturing at least 10× in the same medium at 14-d intervals **(Fig. 3)**.
4. A rapidly growing cell suspension cultures of *S. medusa* is shown in **Fig. 4**. After a 3-d lag phase, the suspension cells begin to grow fast, and the highest dry weight and flavonoid production reach up to 17.2 g/L and 607.8 mg/L respectively on day 15.
5 The conductivity of medium decreases in proportion to the increase of dry weight from day 0 to day 15, and then increases in the stationary phase because of decaying of the suspension cells.
6. A linear relationship between the increase in dry weight and the decrease in medium conductivity is established for on-line growth measurement, the linear equation is shown as follows: $X = K\,C + \beta$, where X = dry weight (g/L), K = empirical coefficient (g/cm ms), C = conductivity (ms/cm) and β = empirical constant. The parameters K and β are determined to be 4.27 g/cm ms and −0.85 g/L with the correlation coefficient R^2 being 0.998.

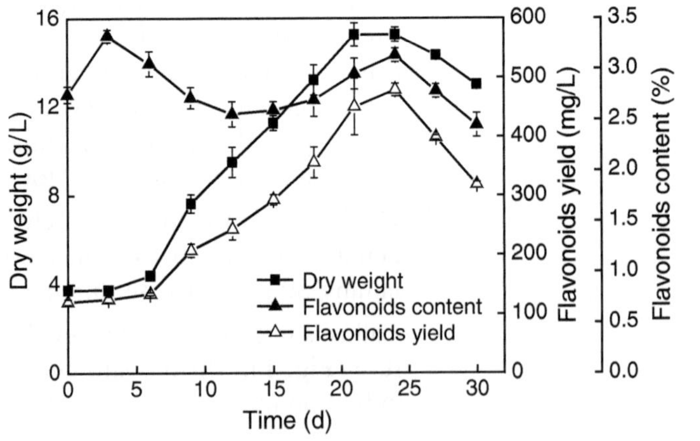

Fig. 2. Growth and flavonoid production of *S. medusa* callus cultures on solid medium. Values are means of triplicate ± standard deviation.

Fig. 3. Suspension callus cultures of *Saussurea medusa* Maxim.

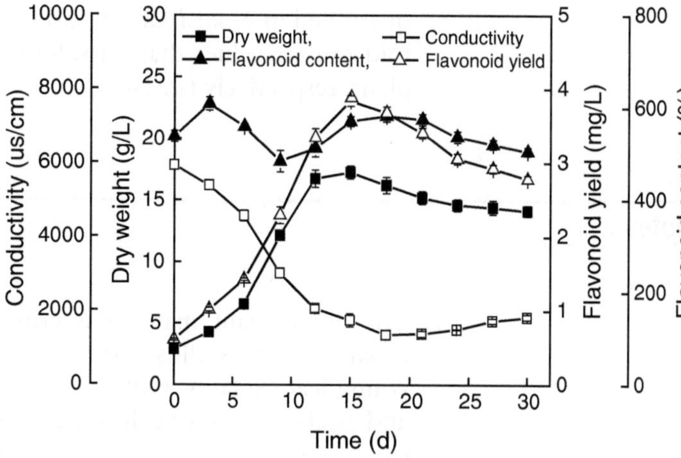

Fig. 4. Cell growth and flavonoid production of *S. medusa* cell suspension cultures in liquid medium. Values are means of triplicate ± standard deviation.

3.3. Antioxidant Activity of S. medusa Cell Suspension Cultures

1. DPPH has been widely used to evaluate the antioxidant activity of plant and microbial extracts. The DPPH radical-scavenging activity is expressed as tested tissue dry weight required for a 50% reduction in radical production and is calculated from the added volume of the crude extract solution *(8)*(*see* **Note 2**).
2. Measure antioxidant activities of *S. medusa* cell suspension cultures, callus from solid culture, and root tissue from wild plants. **Figure 5** shows cell suspension cultures detoxify

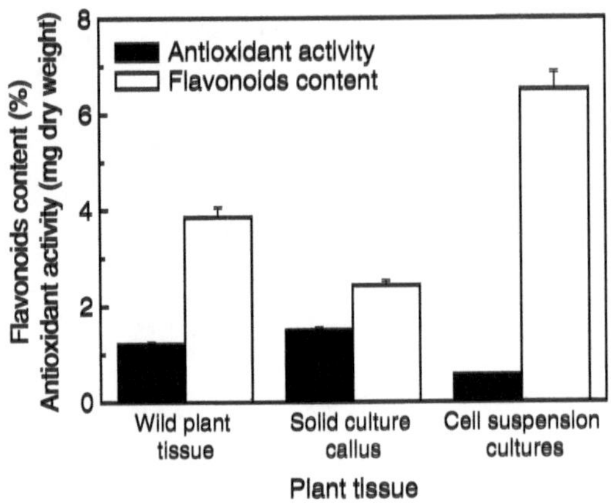

Fig. 5. Relationship between antioxidant activity and flavonoid content of *S. medusa* tissues. A lower value of tissue dry weight indicates a greater antioxidant activity. Values are means of triplicate ± standard deviation.

oxygen radicals, 2.2-fold and 2.7-fold higher than the antioxidant activity of wild plant and solid culture callus (*see* **Note 3**).

3. A lower value of tissue dry weight indicates a greater antioxidant activity. However, significantly more flavonoids are quantified in *S. medusa* cell suspension cultures, 2.7-and 1.8-fold greater values than of solid culture callus and wild-type plant, respectively (*see* **Note 4**).

4. Notes

1. The initial plant material is critical for callus induction in *S. medusa (9)*. Callus with various colors are obtained from young leaf explants. Only red callus is selected for subculture and establishment of cell suspension cultures.
2. There are increasing suggestions by considerable evidences that the free radicals induce oxidative damage to biomolecules (i.e., lipids, proteins, and nucleic acids), which eventually cause atherosclerosis, ageing, cancer, diabetes mellitus, inflammation, AIDS, and several degenerative diseases in human *(10)*.
3. Flavonoids are a new class of natural free radical scavenger, besides α-tocopherol, ascorbic acid, and ß-carotene, and many flavonoids, which are better antioxidants than -tocopherol assessed by in vitro oxygen radical absorbency capacity method *(11)*.
4. These findings provide a foundation for future investigation into medicinally active flavonoids in *S. medusa* cell cultures, and also show that it is an attractive alternative for flavonoids production by large-scale plant cell culture.

References

1. Li, G.H., Liu, F., and Xin, Z. (1979). Studies on pharmacodynamic actions in *Saussurea medusa*. *Chin. Bull. Pharma.* **14**, 86–87
2. Liu, L.S., Xiao, X.H., and Zhang, L.D. (1985). Effect of the flavonoids from *Saussurea involucrata* on DNA synthesis of cancer cells. *J. Lanzhou Univ. Nat. Sci.* **21**, 80–83
3. Schinella, G.R., Giner, R.M., Recio, M.C., Mordujovich, D., Rios, J.L., and Manez, S. (1998). Anti-inflammatory effects of southern American *Tanacetum vulgare*. *J. Pharm. Pharmacol.* **50**, 1069–1074
4. Zhao, D., Huang, Y., Jin, Z., Qu, W., and Lu, D. (2003). Effect of aggregate size in cell cultures of *Saussurea medusa* on cell growth and jaceosidin production. *Plant Cell Rep.* **21**, 1129–1133
5. Yuan, X.F., Zhao, B., and Wang, Y.C. (2004). Cell culture of *Saussurea medusa* in a periodically submerged air-lift bioreactor. *Biochem. Eng. J.* **21**, 235–239
6. Murashige, T., and Skoog, F. (1962). A revised medium for rapid growth and bioassays with tobacco tissue cultures. *Physiol. Plant.* **15**, 473–497
7. Luximon, R.A., Bahorum, T., and Soobra, M.A. (2002). Antioxidant activities of phenolic, proanthocyanidin, and flavonoid components in extracts of *Cassia firtula*. *J. Agric. Food Chem.* **50**, 5042–5047
8. Takahata, Y., Ohnishi-Kameyama, M., Furuta, S., Takahashi, M., and Suda, I. (2001). Highly polymerized procyanidins in brown soybean seed coat with a high radical-scavenging activity. *J. Agric. Food Chem.* **49**, 5843–5847
9. Zhao, D.X. (1997). Initial Studies on Tissue Culture of Snow Lotus (Saussurea involucrata). *Chin. Traditional Herbal Drugs,* **11**, 682–684.
10. Halliwell, B. (1994). Free radicals, antioxidants, and human disease: curiosity, cause, or consequence. *Lancet.* **34**, 721–724.
11. Cao, G., Sofic, E., Prior, R.L. (1997). Antioxidant and proxidant behavior of flavonoids: structure-activity relationships. *Free Radical Biology Med.* **22**, 749–761.

Chapter 5

Large-scale In Vitro Multiplication of *Crataeva nurvala**

Shashi B. Babbar, Neetika Walia, and Amandeep Kaur

Summary

Conservation and propagation of species using biotechnologic tools—such as plant tissue culture—are relevant when natural propagation is hampered for various reasons. In vitro techniques allow mass multiplication and propagation under pathogen-free conditions but also override dependence on season for availability of plant material. Moreover, in vitro genetic manipulation of a species, invariably, requires a prestandardized tissue culture protocol for its multiplication.

To fulfill these requirements, efficient, cyclic, two-step protocols for micropropagation of the medicinal tree—*Crataeva nurvala*—employing juvenile explants and those from mature trees, were developed. Both protocols can be employed at commercial scale. The seedling-derived explants (e.g., cotyledonary nodes, epicotyl nodes, hypocotyl segments, first pair of leaves, cotyledons, and root segments) developed shoots on Murashige and Skoog's (MS) or the same supplemented with different concentrations of 6-benzylaminopurine (BAP). The epicotyl and cotyledonary nodal explants developed shoots on MS basal medium. Other explants exhibited caulogenesis on BAP (0–2.0 mg/L) adjuvated media. The explants from in vitro regenerated shoots too exhibited a similar caulogenic capability. Nodal explants from a 30-yr-old-tree, when cultured on MS medium supplemented with 0.5 mg/L BAP, produced multiple shoots which elongated satisfactorily on the same medium. Similar to the microshoots developed from the seedling derived explants, nodal and leaf explants from the microshoots regenerated from the mature explants too developed shoots, thus making the process recurrent. Due to the recurrent nature of the protocol, over 5400 shoots may be produced from a single nodal explant of an adult tree over a period of six months. The addition of casein hydrolysate significantly increased the average number of shoots per explant. The regenerated shoots could be rooted on the medium supplemented with 0.02 mg/L or 0.1 mg/L NAA (α-naphthalene acetic acid). Regenerated plantlets were acclimatized and successfully transplanted to soil.

Key words: *Crataeva nurvala,* Juvenile and mature explants, Microshoots, Recurrent production, Rhizogenesis, Acclimatization

---

*The protocols described here are based on two earlier published papers listed at **ref.** *(6, 7)*.

S. Mohan Jain and Praveen K. Saxena (eds.), *Methods in Molecular Biology, Protocols for* In Vitro *Cultures and Secondary Metabolite Analysis of Aromatic and Medicinal Plants, vol. 547*
© Humana Press, a part of Springer Science+Business Media, LLC 2009
DOI: 10.1007/978-1-60327-287-2_5

1. Introduction

Micropropagation plays a distinctive role in conservation of species, particularly those having pharmacologic value *(1)*. Most of the medicinal plants are collected from the wild and very few come from the cultivated stands. In most cases, the entire plant is severed during harvesting. To counter for this harvesting methods and to safeguard the species from extinction, it is imperative that cost-effective protocols be developed in order to scale-up propogation *(2)*.

As a step in this direction, standardization of a protocol for micropropagation of an economically important medicinal tree, *Crataeva nurvala,* was undertaken. The tree produces creamy-white flowers during summer and is grown as a promenade tree *(3)*. Besides its elegance, the tree is bred for its medicinal value. Over-exploitation of the plant for its bark,—which is used to treat urinary tract infections; promotes appetite; increases bile secretions; and acts as a diuretic, laxative and anti-pyretic *(4)*—might lead to its extinction in the near future. Its natural propagation is highly impeded because of poor seed set and germination and heavy infestation by insects *(4)*. This necessitates its propagation through an alternative technique, such as tissue culture. Besides affording multiplication in limited time and space, the technique circumvents the limitation posed by the long regeneration cycle of ligneous taxa and obviates the dependence on flowering and seed set *(5)*.

We have reported in vitro regeneration of *Crataeva nurvala* from both seedling-derived *(6)* as well as adult tree-derived explants *(7)*. Because seeds are units of recombination, plants developed from them differ genotypically. Though this adds to the gene pool, the advantage of an elite genotype can be retained only through regeneration from explants derived from an adult plant *(8)*. Thus, for clonal propagation, regeneration from adult tree-derived explants has also been optimized *(7)*.

There are few earlier reports on regeneration of *Crataeva nurvala* through axillary bud break *(9)* and somatic embryogenesis *(10)*. In addition, there is also a report on micropropagation of a closely related species—*Crataeva adansonii*—through axillary buds of mature trees *(11)*. This chapter details standardized protocols developed by our team for regeneration from both seedling- and adult tree-derived explants *(6, 7)*.

2. Materials

The materials required for micropropagation using juvenile and mature explants are described below.

2.1. For Juvenile Explants

2.1.1. Plant Material

1. Seeds from mature fruits.

2.1.2. Treatment of Seeds

1. 5 N HCl.
2. Concentrated H_2SO_4 for scarification of seeds.
3. Bavistin (carbendezine) at 10 g/L (BASF India Ltd.; Mumbai).
4. 5–6 Drops of Tween 20 (HiMedia Laboratories Pvt. Ltd.; Mumbai, India) in 25 mL distilled water.
5. Mercuric chloride at 2 g/L, prepared in distilled water.

2.1.3. Propogation of Seedlings

1. Sterilized (*see* **Note 1**) Knop's medium *(12)* supplemented with 10 g/L sucrose and 8 g/L agar (Qualigens Glaxo Fine Chemicals; India), pH 5.8 (*see* **Notes 2** and **3**).
2. Rimless test tubes (25 mm × 150 mm; Borosil India).
3. Cotton plugs (non-adsorbent cotton wrapped in two layers of cheese cloth).

2.1.4. Culture of Explants

1. Sterilized (*see* **Note 1**) Murashige and Skoog's (MS) medium *(13)* supplemented with 0.5 or 1.0 mg/L BAP (6-benzylaminopurine; Sigma Chemicals) *see* **Note 4**), 30 g/L sucrose (Daurala Sugar Works; Daurala, U. P., India) and 8 g/L agar, pH 5.8 (*see* **Notes 2** and **3**).

2.2. From Adult Trees

2.2.1. Collection of Explants

1. 100 mg Ascorbic acid in 1 L distilled water.

2.2.2. Surface Sterilization

1. 10 mL/L Tween 20 in distilled water.
2. 10 g/L Bavistin (Carbendazine, BASF India Ltd.; Mumbai), in distilled water.
3. 2 g/L mercuric chloride (Qualigens Glaxo Fine Chemicals; India), in distilled water.

2.2.3. Initiation, Multiplication of Shoots

1. Sterilized (*see* **Note 1**) MS medium *(13)* supplemented with 0.5 mg/L BAP (*see* **Note 4**), 30 g/L sucrose and gelled with 8 g/L agar, pH 5.8 (*see* **Notes 2** and **3**).

2.3. Rooting of In Vitro Regenerated Shoots

1. Sterilized (*see* **Note 1**) 50% MS medium *(13)* supplemented with 0.02 mg/L NAA (α-Naphthalene acetic acid, *see* **Note 4**), 2% *(w/v)* sucrose and 8 g/L agar, pH 5.8 (*see* **Notes 2** and **3**).

2.4. Acclimatization and Field Transfer of Rooted Plants

1. Sterilized (*see* **Note 1**) garden soil, paper.
2. 20× diluted MS (inorganic salts).
3. Transparent polythene bags/plastic tents.

3. Methods

Methods involved in in vitro regeneration from juvenile as well as mature explants are described below.

3.1. From Juvenile Explants

3.1.1. Collection of Fruits and Isolation of Seeds

1. Collect fully ripe fruits **(Fig. 1a)**, with crimson colored skin from identified trees of *Crataeva* (*see* **Note 5**).
2. Open fruits and scoop out seeds. For de-pulping the seeds, treat them with 5 M HCl for 2 h.
3. Wash under running tap water in a sieve.
4. Rub the seeds against sieve to remove any adhering pulp to the seeds.
5. Air-dry de-pulped seeds **(Fig. 1b)** and store in a clean container (*see* **Note 6**).

Fig. 1. *Crataeva nurvala*: (**a**) Ripe fruits (**b**) Air-dried seeds.

3.1.2. Treatment of Seeds

1. Treat seeds with concentrated sulphuric acid for 2 h to soften seed coat. Subsequently, wash them in running tap water to remove traces of acid and remove charred outer seed coat (*see* **Note** 7).
2. Treat seeds with Bavistin for 10 min, followed by 2–3 rinses with sterile distilled water.
3. Wash seeds in Tween-20 for 5 min, followed by treatment with $HgCl_2$ for 10 min. Rinse seeds 4–5 times with sterile distilled water (*see* **Note 8**).
4. Leave seeds in sterile water for 2 h for imbibition.

3.1.3. Raising of Seedlings

1. Decant water in which seeds are soaked and inoculate seeds on Knop's basal medium. While inoculating place the hilum end in contact with the medium. Radicle will emerge from the hilum end when the seeds germinate.
2. Incubate the cultured seeds at 25 ± 2°C in continuous light of 17.76 μmol/m^2 s provided by cool daylight fluorescent tubes.

3.1.4. Culture of Explants

1. Excise cotyledonary nodal, cotyledonary, root segments, hypocotyls and leaves from 4–5-wk-old seedlings, which have attained height of 5–6 cm **(Fig. 2a)**.

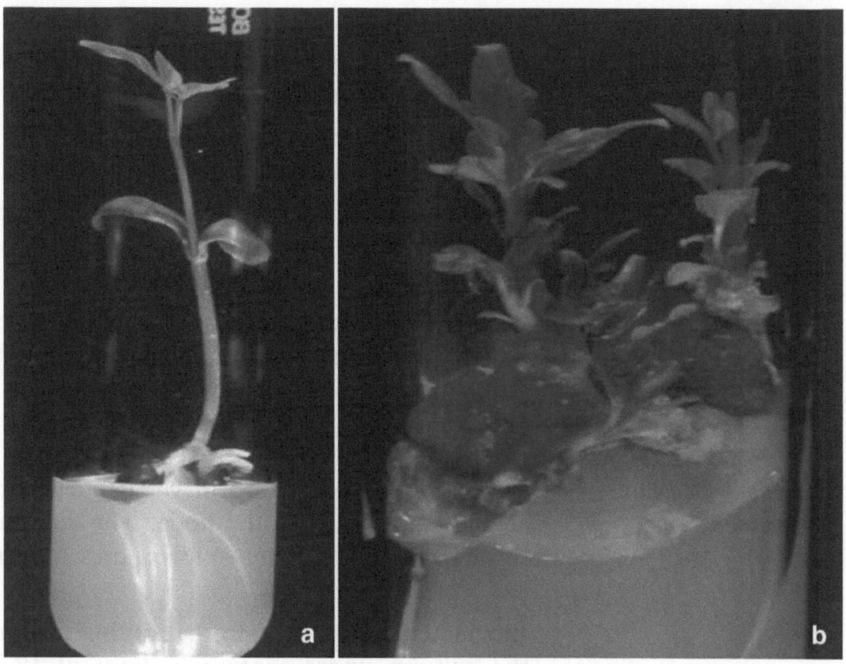

Fig. 2. *Crataeva nurvala*: (**a**) An in vitro grown seedling (**b**) Multiple shoots from the cotyledonary explants.

2. Place explants with distal end facing base of the slant of the medium. Incubate cultures in continuous light of 17.76 μmol/m^2 s provided by cool daylight fluorescent tubes, at 25 ± 2°C.
3. After 4 wk, excise nodal segments from in vitro regenerated shoots (**Fig. 2b**) from various explants and culture them onto fresh medium. Also transfer the residual mother explant to fresh medium for further development of shoots.

3.2. From Adult Trees

3.2.1. Collection of Explants

1. Explants should be collected in the month of May–June, after the flowering has ended and when the new sprouting starts. Explants should be collected essentially during morning.
2. Collect twigs with 5–6 nodes in a beaker containing ascorbic acid. The cut end should be immediately submerged in the solution.

3.2.2. Surface Sterilization

1. After bringing the collected material to the laboratory, remove leaves and wash the material under running tap water for 30 min. This is to be followed by washing with Tween-20 for 10 min.
2. Rinse the explants thoroughly with sterile distilled water for 3–4 times. Thereafter, treat the explants with Bavistin for 10–15 min.
3. After thorough rinsing with sterile water, treat the explants with $HgCl_2$ for 10–15 min. Followed by treatments with 150 mg/L ampicillin for 10 min.

3.2.3. Initiation, Multiplication of Shoots

1. After 3–4 rinses with water, excise 1-cm long nodal segments (**Fig. 3b**) starting fbetween the 6th and 9th nodes from shoot tip. Culture them onto the initiation medium.
2. Incubate cultures in continuous light of 17.76 μmol/m^2 s provided by cool daylight fluorescent tubes, at 25 ± 2°C.
3. After 3–4 wk of incubation, sub-culture micro shoots (**Fig. 3b**) regenerated from these nodal explants on the fresh initiation medium.
4. Further, leaf and nodal explants, from these in vitro regenerated micro shoots can be cultured on the same medium in subsequent passages. These explants too develop multiple shoots (**Fig. 3c**).

3.3. Rooting of In Vitro Regenerated Shoots

1. Excise 3- to -4-cm long in vitro regenerated shoots. Give an oblique cut below the node. The cut should pass partly through the node. Remove leaves present at the lower nodes (*see* **Note 9**).
2. Inoculate the shoots on to the rooting medium and incubate the cultures under continuous light of 17.76 μmol/m^2 s^1 provided by cool daylight fluorescent tubes, at 25 ± 2°C.

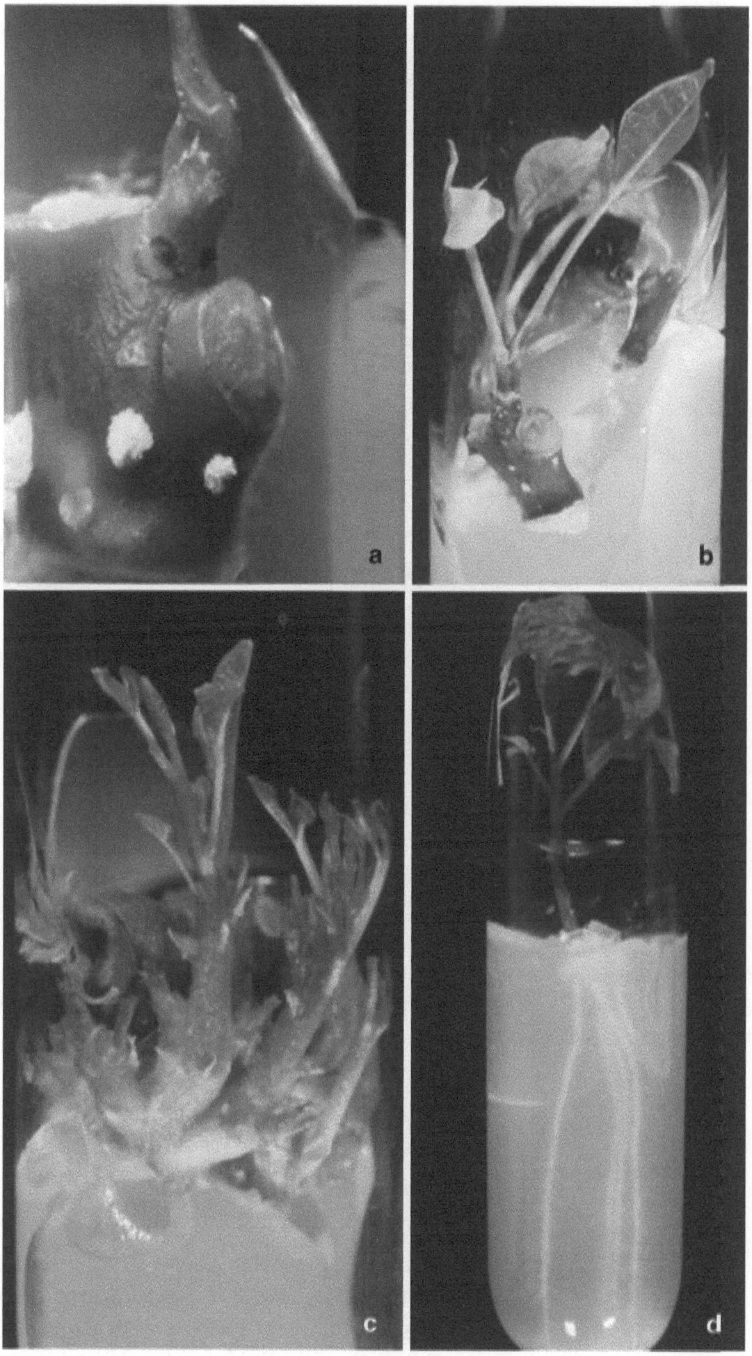

Fig. 3. *Crataeva nurvala*: (**a**) A nodal explant with an axillary bud, (**b**) An explant with elongated shoots, (**c**) Multiple shoots from the explants excised from the microshoots, (**d**) A rooted shoot.

3.4. Acclimatization and Field Transfer of Rooted Plants

1. After 3–4 wk of incubation, remove the rooted shoots/plantlets (**Fig. 3d**) from culture tubes (*see* **Note 10**). Wash the roots under running tap water to remove any adhering agar.

Fig. 4. *Crataeva nurvala*: (**a**) Plants 3 wk post-transfer from culture tubes to soil in paper cups, (**b**) Plants approximately 5 month post-transfer from paper cups to the earthen pots, (**c**) A plant, 6 month post-transfer to the field, (**d**) 5-yr-old tissue culture raised plant.

2. Put the plantlets in the paper cups containing sterilized garden soil (**Fig. 4a**, *see* **Note** 11).
3. Cover the plantlets either with transparent poly bags or put the plants in plastic tents.
4. Initially spray the poly bag/plastic tent with water. If using poly bags, make 2–3 perforations to begin with, to allow gaseous exchange. Decrease the humidity gradually after week 2 of transfer. This can be done by reducing the frequency of spraying the walls of humidity tent followed by incremental lifting of the poly tent and finally removing it completely. If poly bags are used, increase the perforations in poly bags gradually and finally remove the poly bags.
5. Fertigate the plants with diluted MS inorganic ingredients on alternate days for initial 3 wk. Thereafter, transfer plants to earthen pots (**Fig. 4b**) and irrigate them with water.

6. Once completely acclimatized (about 3 months from test tubes) transfer the plants to field (**Fig. 4c, d**), preferably in the month of March–April.

4. Notes

1. Sterilization of glassware, culture media and instruments should be carried out by autoclaving at 103 kPa at 121°C.
2. The pH of the medium is adjusted using 1 N NaOH or 1 N HCl. Before taking pH of the medium, it is advisable to calibrate pH meter.
3. Dispense 20 mL medium in 25 mm × 150-mm test tubes, which can be closed with cotton plugs, double-layered muslin cloth stuffed with non-adsorbent cotton.
4. For making the stock solution of BAP, dissolve the weighed amount in 1 mL of 1 N NaOH and then raise the volume to desired level with distilled water. Likewise, NAA is first dissolved in a little amount of ethanol and the volume is raised to the desired level with distilled water.
5. In Delhi conditions, fruits of *Crataeva* mature in August. During this month, the average maximum and minimum temperatures are 34 and 26°C, respectively and average rainfall is 173 mm.
6. The dried seeds of *Crataeva* can be stored for 6–7 m. However, during this period, their viability decreases progressively.
7. Seeds after treatment with H_2SO_4 are washed under running tap water. For better germination, seed coat should be chipped off at few places to allow imbibition of water.
8. After treating seeds with Tween-20, treatment with surface-sterilant can be followed immediately, without an intermittent step of rinsing with water. Residual Tween-20 in the surface sterilization step helps in better cleaning of the explants.
9. While putting shoots on rooting medium, leaf surface area should be reduced by clipping the leaves. It helps in faster rhizogenic response. Leaves from lower nodes should be removed completely. Leaves should not be touching the medium.
10. While taking out rooted shoots/plantlets from the culture tube, care should be taken that the roots are not damaged. Agar adhering to roots should be removed very carefully under water.

Roots from the plantlets should not coil when being placed in soil.

References

1. Afolayan, A. J. and Adebola, P. O. (2004). *In vitro* propagation: a biotechnological tool capable of solving the problem of medicinal plants decimation in South Africa. *Afr. J. Biotechnol.* **3**, 683–687
2. Bajaj, Y. P. S., Furmanowa, M., and Olszowska, O. (1988). Biotechnology of micropropagation of medicinal and aromatic plants. In: *Biotechnology in Agriculture and Forestry, vol. 4, Medicinal and Aromatic Plants I.* (Bajaj Y.P.S., ed.), Springer, Berlin, Heidelberg, New York, pp. 60–103
3. Warman, C. K. (1999). *Trees of India: Medicinal, Commercial, Religious & Ornamental, A Colour Atlas.* CBS Publishers, New Delhi
4. Joshi, S. G. (2000). *Medicinal Plants.* Oxford and IBH Publishers, Co. Pvt. Ltd., Calcutta
5. Bonga, J. M. (1982). Vegetative propagation in relation to juvenility, maturity and rejuvenation. In: *Tissue Culture in Forestry* (Bonga J.M. and Durzan D.J. eds.), Martinus Nijhoff/Dr. W Junk Publishers, Dordrecht, pp. 287–412
6. Walia, N., Sinha, S., and Babbar, S. B. (2003). Micropropagation of *Crataeva nurvala* Buch. Ham. *Biol. Plant.* **46**, 181–185
7. Walia, N., Kaur, A. and Babbar S. B. (2007). An efficient, *in vitro* cyclic production of shoots from adult trees of *Crataeva nurvala* Buch. Ham. *Plant Cell Rep.* **26**, 277–284
8. Sharma, S. K. and Ramamurthy, V. (2000). Micropropagation of 4-year-old elite *Eucalyptus tereticornis* trees. *Plant Cell Rep.* **19**, 511–518
9. Sharma, V. and Padhya, M. A. (1996). *In vitro* rapid multiplication and propagation of *Crataeva nurvala*. *Ind. J. Exp. Biol.* **34**, 243–246
10. Inamdar, J. A., Nataraj, M., Mohan, J. S. S., and Subramanian, R. B. (1990). Somatic embryogenesis from callus cultures of *Crataeva nurvala* Buch. Ham. *Phytomorphology* **40**, 319–322
11. Sharma, P. K., Tyagi, P., Sharma, K. C. and Kothari, S. L. (2003). Clonal propagation of *Crataeva adansonii* (DC.) Prodr.: a multipurpose tree. *In Vitro Cell Dev. Biol. – Plant* **39**, 156–160
12. Knop, W. (1865). Quantitative Untersuchungen über die Ernährungsprozesse der Pflanzen. *Landwirtsch. Vers. Stn.* **7**, 93–107
13. Murashige, T. and Skoog, F. (1962). A revised medium for rapid growth and bio assays with tobacco tissue cultures. *Physiol. Plant.* **15**, 473–497

Chapter 6

Bilberry In Vitro Protocols and Analyses of Phenolic Compounds

Laura Jaakola, Kaisu Riihinen, Hely Häggman, and Anja Hohtola

Summary

Bilberry or European blueberry (*Vaccinium myrtillus* L.) belongs to the most significant family of wild berries in Northern Europe and is recognized for its bioactive properties. Its fruits and leaves are rich with phenolic compounds, in particular flavonoids. Their health benefits have received notable attention in recent years. Bilberry is one of the richest sources of the anthocyanins, a subclass of flavonoids. The interest in bilberry cultivation is growing because of the high value of the fruit in global food markets. Tissue culture provides an efficient propagation method for the selected bilberry genotypes both for the breeding and cultivation purposes. This chapter describes the protocol of bilberry in vitro propagation and for the analyses of phenolic compounds.

Key words: Anthocyanins, Bilberry, Flavonols, Tissue culture, *Vaccinium myrtillus*

1. Introduction

Bilberry or European blueberry (*Vaccinium myrtillus* L.) belongs to the genus *Vaccinium*, which includes many economically important cultivated small fruit species, like blueberries (e.g., *V. corymbosum*, *V. angustifolium*), cranberry (*V. macrocarpon*), and lingonberry (*V. vitis-idaea*). Bilberry grows wild in Europe and Asia, most abundantly in Northern and Eastern Europe, and at higher elevations in Southern Europe *(1, 2)*. Bilberries are recognized as one of the best sources of flavonoids, in particular anthocyanins, whose antioxidative properties have received notable attention in recent years *(3–5)*. Among flavonoids, anthocyanin pigments are predominantly present in bilberry. Flavonols and anthocyanidins mostly occur as soluble glycosides. The reverse-phase high-performance liquid

S. Mohan Jain and Praveen K. Saxena (eds.), *Methods in Molecular Biology, Protocols for* In Vitro *Cultures and Secondary Metabolite Analysis of Aromatic and Medicinal Plants, vol. 547*
© Humana Press, a part of Springer Science+Business Media, LLC 2009
DOI: 10.1007/978-1-60327-287-2_6

Fig. 1. Different phases during the bilberry in vitro propagation. (**a**) Growth induction of bud explants. (**b**) Multiplication. (**c**) Rooting phase. (**d**) Mature bilberry plant with fruits.

chromatography (RP–HPLC) combined with diode array detection (DAD) analytical device is used frequently for the identification and quantification of flavonoids. Further research has demonstrated the vascular protective and anticancer activity of these berries *(3, 6)*. Moreover, the aroma of bilberry is regarded as delicious and special.

Propagation of bilberry by cuttings is inefficient because of their poor rooting. Also, when grown from seed, new plants take several years to reach maturity. Therefore, in vitro propagation of bilberry has proven to be the most efficient method for producing plant material for cultivation and breeding (**Fig. 1**) *(7)*. This chapter describes protocols for in vitro propagation and a quick and repeatable method to determine anthocyanins and flavonol glycosides as aglycons *(8–10)*.

2. Materials

2.1. In Vitro *Protocol*

2.1.1. Growth Induction and Multiplication

1. Dormant bilberry buds, shoot tips, leaves, or seeds.
2. Laminar flow hood, growth chamber, forceps, scalpel, 300-mL culture vessels, and 70% ethanol.

3. Calcium hypochlorite, 6%, *(w/v)*; sodium hypochlorite, 3% *(v/v)*; Tween-20; and sterile distilled water.
4. Growth medium: E-MS *(11)*, (**Tables 1** and **2**), with 10 mg/L, 49.2 μM, or 5 mg/L, 24.6 μM, 2iP (N6-(2-isopentyl) adenine) or 0.6 mg/L, 2.27 μM, thidiazuron (TDZ) in 9-cm Petri dishes.

2.1.2. Rooting and Hardening

1. Rooting medium: E-MS *(11)* (**Tables 1** and **2**) with 0.01 mg/L 0.49 μM indole-3-butyric acid (IBA) with 15 g/L sucrose in 300-mL culture vessels.
2. Plastic boxes (22 cm × 28 cm) containing *Sphagnum* peat, vermiculite, and sand (2:1:1).

2.2. Extraction of Flavonoids

1. Frozen or freeze-dried bilberry fruits.
2. Acidified methanol: concentrated 12 M HCl diluted to 0.6 M with methanol (p.a or high-performace liquid chromatography [HPLC]-grade). Store at room temperature.

Table 1
Concentrations of Inorganic and Organic Salts of E-MS Media and of Half-strength MS Media[a]

	E–MS		½ MS	
Ingredient	**mg/L**	**mM**	**mg/L**	**mM**
Inorganic				
NH_4NO_3	400	5	825	10.3
KNO_3	202	2	950	9.4
$(NH_4)_2SO_4$	132	1	99	0.75
$CaCl_2 \cdot 2H_2O$	440	3	220	1.5
$MgSO_4 \cdot 7H_2O$	370	1.5	185	0.75
KH_2PO_4	408	3	85	0.62
Na_2-Fe-Edta	37	0.1	37	0.1
H_3BO_3	6.2	0.1	3.1	0.05
$MnSO_4 \cdot H_2O$	16.9	0.1	11.5	0.068
$ZnSO_4 \cdot 7H_2O$	8.6	0.03	4.3	0.015
$NaMoO_4 \cdot 2H_2O$	0.25	0.001	0.125	0.0005
$CuSO_4 \cdot 5H_2O$	0.025	0.0001	0.013	0.00005
$CoCl_2 \cdot 6H_2O$	0.025	0.0001	0.013	0.00005
Organic				
Myoinositol	100	0.56	100	0.56
Thiamine–HCl	0.4	0.0012	0.4	0.0012
Sucrose	20,000	58.2	20,000	58.2
Agar	6,000		6,000	

[a]Adapted from references *11* and *15*.
Note: A pH of 4.8 should be maintained for optimal propagation.

Table 2
Composition of the Culture Media used During Induction, Multiplication, and Rooting of Bilberry Explants

Medium composition	Induction			Multiplication	Rooting
	Shoot tips	Seeds	Leaves		
Macro- and microelements	E-MS	E-MS	E-MS	E-MS	E-MS
Vitamins (mM)					
Thiamine	1.2	1.2	1.2	1.2	1.2
Sucrose (%)Vitamins	2.0	2.0	2.0	2.0	1.5
PGRs (mM)					
2ip	49.2	24.6		24.6	
TDZ			2.27		
IBA					0.49
Agar (g/L)	6	6	6	6	6

Abbr: PGRs, plant growth regulators.

3. Regen cellulose syringe filter (TITAN®®; Gloucester, UK) or equivalent filters.

2.3. Identification and Quantification of Flavonoids with HPLC–DAD

1. Standards for the flavonoid analyses: 3-*o*-β-glucosides of delphinidin, cyanidin, petunidin, pelargonidin, peonidin, and malvidin (Polyphenols AS; Sandnes, Norway), in addition to myricetin (Fluka; Buchs, Switzerland) and quercetin (Sigma Chemical Co.; St Louis, MO).
2. LiChroCART Purospher RP-18e column 125 × 3 mm ID, 5 μm (Merck; Darmstadt, Germany) protected with a guard column of the same material (4 × 4 mm).

3. Methods

3.1. In Vitro Protocol

3.1.1. Culture Media

1. The culture media E-MS *(11)*, (**Tables 1** and **2**) are prepared with 10 mg/L, 49.2 μM, 2iP (N6-(2-isopentyl) adenine) for bud explants, 0.6 mg/L, 2.27 μM, TDZ for leaf explants *(12)*, and with 5 mg/L, 24.6 μM, 2iP for multiplication phase. The media for the explants is prepared in 9-cm Petri dishes and in culture vessels for the multiplication phase.

3.1.2. Explant Preparation

1. The bilberry plants grow wild on acidic soils, from spruce and pine dominated forests to mountains and peat bogs. When choosing the adult starting material for the in vitro culture, either dormant buds or shoot tips or leaves can be used. Bilberry seeds are the favorable and easy material for starting cultures. Typically, 90% seed germination has been detected in bilberry in vitro cultures *(7)*.
2. The bilberry shoots are sectioned into 2- to 3-cm segments with scalpel and surface-disinfected in 100 mL 6% *(w/v)* calcium hypochlorite with a drop of Tween-20 for 10 min. When leaves are used as a starting material, shorter sterilization time (5–8 min) prevents the browning of the leaves. Bilberry seeds are sterilized in 3% *(v/v)* sodium hypochlorite for 10 min.
3. Stem sections, leaves or seeds are rinsed three times in sterile distilled water.

3.1.3. Culture Initiation

1. The sterilized and rinsed stem sections are cut into 1-cm single node pieces and placed on growth culture medium containing 10 mg/L 2 iP (*see* **Note 1**). Seeds are placed on to the growth media containing 5 mg/L 2iP. Leaves are trimmed and placed on to the media containing 0.6 mg/L TDZ.
2. Stem section explants and seeds are cultured and maintained at +22 ± 2C° under 16-h photoperiod with 20–50 μmol/m^2 s light intensity (*see* **Note 2**). Leaf explants are cultured in the dark at the same temperature for the first 2 wk, and thereafter transferred to the light.

3.1.4. Shoot Regeneration and Maintenance

1. The shoot regeneration begins normally in 2 wk on growth media for bud, shoot tip, leaf, and seed explants.
2. The multiplication begins after 5–8 wk from the initiation of the cultures (*see* **Note 2**). Subsequently, micro shoots are subcultured every 3 wk onto the same growth media, containing 2 iP at 5 mg/L (*see* **Notes 3–6**).

3.1.5. Rooting

1. 3- to 5-cm long micro shoots are inserted vertically into a depth of 1.0–1.5 cm into culture vessels containing rooting media (**Table 2**).
2. Micro shoots are cultured on the rooting media for 3–4 wk. The root formation usually begins during the first 10 d.

3.1.6. Hardening

1. The rooted micro shoots are washed carefully in water and planted into plastic boxes containing Sphagnum peat, vermiculite, and sand (*v:v:v*; 2:1:1). The boxes are sealed and kept in the humidity chamber with the relative humidity approximately 90% for the first 4–8 wk.

2. After the acclimatization phase bilberry shoots can be grown in normal greenhouse conditions before planting outside (*see* **Note 7**).

3.2. Extraction and Analyses of Flavonoids

The high content of anthocyanins can easily be analyzed directly from bilberry-solvent suspension as 15 signals in HPLC–DAD chromatogram **(Fig. 2)**. However, the lower amounts of diverse flavonol glycosides are acid hydrolyzed to the two aglycons, which simplifies the quantification **(Fig. 3)**. The important step in the sample

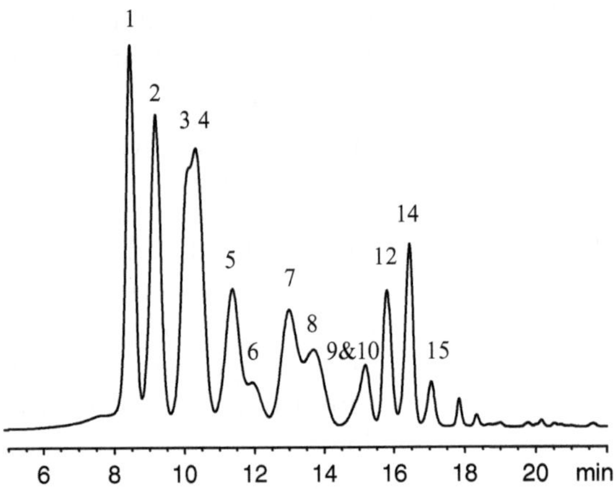

Fig. 2. HPLC-DAD profile of anthocyanins found in mature bilberry fruits at 520 nm. Peak identification in the order of the retention time: (1) delphinidin 3-galactoside, (2) delphinidin 3-glucoside, (3) cyanidin 3-galactoside, (4) delphinidin 3-arabinoside, (5) cyanidin 3-glucoside, (6) petunidin 3-galactoside, (7) cyanidin 3-arabinoside, (8) petunidin 3-glucoside, (9) peonidin 3-galactoside, (10) petunidin 3-arabinoside, (11) peonidin 3-glucoside, (12) malvidin 3-galactoside, (13) peonidin 3-arabinoside, (14) malvidin 3-glucoside, and (15) malvidin 3-arabinoside.

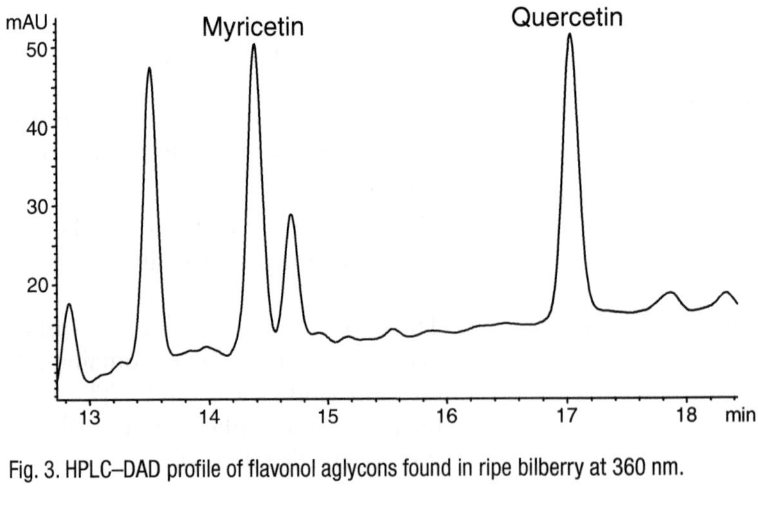

Fig. 3. HPLC–DAD profile of flavonol aglycons found in ripe bilberry at 360 nm.

handling is the solubilization of phenolic compounds in a solvent. The sample matrix is milled or crushed to a fine powder to increase the surface area and to facilitate maximum release of flavonoids during solvent extraction and chemical hydrolysis of the sample. The bilberry matrix is macerated in the extraction solvent for uniform bilberry-solvent suspension. The extraction of anthocyanins should be carried out under acidic conditions in order to obtain the highest yields of flavylium forms, which are red in color and stable in a acid medium *(13)*. Mineral acid such as HCl is commonly used, but in the case of acylated anthocyanins (e.g., in some blueberry species), an organic acid is favored *(13)*. Traditional chemical hydrolysis conducted with acid (e.g., hydrochloric acid) has been used to release flavonol aglycons from glycosides.

3.2.1. Extraction of Flavonoids

1. 10-g frozen or 2-g freeze-dried berries are crushed and powdered using a mortar and pestle in liquid nitrogen. Water content of freeze-dried berries is normalized by adding 8 mL of water. Samples of 1–2 g are weighed in round bottomed extraction flasks and add 10 mL acidified methanol. Samples are macerated in solvent by mixing with magnetic stirrer for 1–3 min until the sample is suspended.
2. Aliquots of solvent, 1 mL, are removed for the analysis of anthocyanins (*see* **Note 8**).
3. Samples are refluxed in the boiling point with electric flask heater for 2 h. Upon heating in acidic methanol flavonol glycosides are released to aglycons.
4. The refluxed samples are cooled and volume of solvent is measured. Volume of the berry-solvent suspension is adjusted to 9 mL.
5. Aliquots of samples are filtered through a syringe filter to HPLC vials (*see* **Note 8**).

3.2.2. Analysis of Anthocyanins and Flavonols with HPLC–DAD

1. These instructions assume the use of a HPLC combined to diode array detector (DAD). The quick HPLC separation of the anthocyanins is achieved on a (125 × 3 mm ID, 5 μm) LiChroCART Purospher RP-18e column protected with a guard column of the same material (4 × 4 mm). A 20-min linear gradient from 5 –30% acetonitrile in 1% formic acid is used to separate flavonols. A step gradient of acetonitrile in 5% formic acid is used to separate anthocyanins as follows: 5–10% acetonitrile for 0–5 min, 10% acetonitrile for 5–10 min, 10–40% acetonitrile for 10–25 min, and finally 40–90% acetonitrile for 25–35 min (*see* **Note 9**). Both gradients are followed by raising acetonitrile to 90% in 10 min, a return to initial conditions in 5 min and then re-equilibration of the column for 10 min. The flow rate for all separations is

0.5 mL/min. The chromatographic performance is followed by analysis of a standard mixture at the beginning of every sample series.

2. Retention times and spectra of the peaks at maximum absorption wavelengths (λ_{max}) in DAD are matched with authentic standards for identification of anthocyanins as 3-*o*-β-glucosides of delphinidin, cyanidin, petunidin, peonidin, and malvidin (λ_{max} = 520 nm) and of flavonol aglycons as myricetin and quercetin (λ_{max} = 360 nm). Prior the analysis, mixture (5 μmol of each) of 3-*o*-β-glucosides of delphinidin, cyanidin, petunidin, pelargonidin, peonidin, and malvidin (Polyphenols AS; Sandnes, Norway) is dissolved in 20 mL methanol for preparing a stock solution. Stock solutions are stored in single use aliquots at −20°C. Myricetin (Fluka; Buchs, Switzerland) and quercetin (Sigma Chemical Co.; St Louis, MO) are dissolved in methanol to a concentration of 1 mg per mL and stored as stock solutions in single use aliquots at −20°C. Purity of standards is noted in exact concentrations. Other bilberry anthocyanins (galactosides and arabinosides) may be identified according to their elution order in reversed phase system (**Fig. 1**) (*see* **Note 9**).
3. External standard curves for quantification are prepared using authentic standards as follows: flavonols 2–110 μg mg/L, and anthocyanins 1.5–70 μg mg/L in methanol (*see* **Note 10**). The concentration of 0.6 M HCl in methanolic solutions of anthocyanins is adjusted to the same level as of the samples.
4. Peaks in the chromatograms are quantified using the equivalent standard compounds listed above (*see* **Note 10**) and counted for the fresh or dry weight of the berry. Standard curve of anthocyanidin glucosides is used also in quantification of respective arabinosides and galactosides (*see* **Note 10**).

4. Notes

1. 2iP concentration of 20 mg/L induced more shoot growth at the start of the cultures, but also results in browning of the explants after 6 wk.
2. A light intensity of more than 150 μmol/m^2 s causes reddish leaves and reduction of growth of bilberry cultures.
3. The initiation of the growth using bud explants was 20% higher in the cultures started in spring compared to autumn cultures *(7)*.

4. We have also successfully used ½ MS inorganic and organic salts for the multiplication of bilberry cultures (**Table 1**), whereas basic MS media turned out to be too strong for bilberry explants due to high nitrate concentration.
5. A low pH of 4.8 is critical for the successful bilberry cultures.
6. In cold storage at +4°C, in continuous crepuscular light bilberry shoots remain green and vital on the same basal media, without subcultures for prolonged period of time even after 5 years.
7. The bilberry plants produced with this method have flowered and produced first fruits already 2 yr after planting.
8. A wide bore tip is recommended for pipetting the suspended berry matrix.
9. Separation of anthocyanins may be enhanced by changing the column length and/or pore size, manufacture or by changing the composition of mobile phases. Lifetime of the column is approximately 1000 analyses.
10. The best possible choice is to use separate standard for all the 15 individual anthocyanins in bilberry, because the intensity of absorption on line in DAD is affected by the linked sugar moiety *(14)*. A spectral maximum varies between anthocyanins aglycons, but attached sugar will not change the maxima. Therefore, the second best choice is to use presentative anthocyanidin glycosides similarly as in the presented method.

Acknowledgments

We thank The Nordic Innovation Centre, Emil Aaltonen foundation and Kone foundation for the financial support.

References

1. Salo, K. (1995). Non-timber forest products and their utilization. In: Hytönen, M. (ed.) Multiple-use Forestry in the Nordic Countries. Gummerus Press, Jyväskylä, pp. 117–155
2. Sjörs, H. (1989). Blåbär, *Vaccinium myrtillus*: ett växtporträtt. *Svensk. Bot. Tidskr.* **83**, 411–428
3. Prior, R. L., Cao, G., Martin, A., Sofic, E., McEwen, J., O'Brien, C., Lischner, N., Ehlenfeldt, M., Kalt, W., Krewer, G., and Mainland, M. (1998). Antioxidant capacity as influenced by total phenolic and anthocyanin content, maturity, and variety of *Vaccinium* species. *J. Agric. Food Chem.* **46**, 2686–2693

4. Smith, M. A. L., Marley, K. A., Seigler, D., Singletary, K. W., and Meline, B. (2000). Bioactive properties of wild blueberry fruits. *J. Food Sci.* **65**, 352–356
5. Zitnanova, I., Ranostajova, S., Sobotova, H., Demelova, D., Pechan, I., and Durackova, Z. (2006). Antioxidative activity of selected fruits and vegetables. *Biologia* **61**, 279–284
6. Katsube, N., Iwashita, K., Tsushida, T., Yamaki, K., and Kobori, M. (2003). Induction of apoptosis in cancer cells by bilberry (*Vaccinium myrtillus*) and anthocyanins. *J. Agric. Food Chem.* **50**, 519–525
7. Jaakola, L., Tolvanen, A., Laine, K., and Hohtola, A. (2001). Effect of 2iP concentration on growth initiation *in vitro* and rooting of bilberry and lingonberry microshoots. *Plant Cell Tissue. Org. Cult.* **66**, 73–77
8. Jaakola, L., Määttä, K., Pirttilä, A. M., Törrönen, R., Kärenlampi, S., and Hohtola, A. (2002). Expression of structural genes involved in anthocyanin biosynthesis in relation to anthocyanin and flavonoid levels during bilberry (*Vaccinium myrtillus* L.) fruit development. *Plant Physiol.* **130**, 729–739
9. Jaakola, L., Määttä, K., Kärenlampi, S., and Hohtola, A. (2004). Activation of flavonoid biosynthesis by solar radiation in bilberry (*Vaccinium myrtillus* L.) leaves. *Planta* **218**, 721–728
10. Määttä-Riihinen, K. R., Kamal-Eldin, A., Mattila, P. H., González-Paramás, A. M., and Törrönen, A. R. (2004). Distribution and contents of phenolic compounds in eighteen Scandinavian berry species. *J. Agric. Food Chem.* **52**, 4477–4486
11. Econonomou, A. S. and Read, P. E. (1984). *In vitro* shoot proliferation of Minnesota deciduous azaleas. *Hortic. Sci.* 19, 60–61
12. Shibli, R., and Smith M. A. L. (1996). Direct shoot regeneration from *Vaccinium phalae* (Ohelo) and *V. myrtillus* (Bilberry) leaf explants. *Hortic. Sci.* **31**, 1225–1228
13. Strack, D., and Wray, V. (1989). Anthocyanins. In: Dey, P. M., and Harborne, J. B., (eds.) Methods in Plant Biochemistry. London, UK: Academic Press, pp. 325–356
14. Mazza, G., Cacace, J. E., and Kay, C. D. (2004). Methods of analysis for anthocyanins in plants and biological fluids. *J. AOAC Int.* **87**, 129–145
15. Murashige, T. and Skoog, F. (1962). A revised medium for rapid growth and bioassay with tobacco tissue cultures. *Physiol. Plant.* **15**, 473–497

Chapter 7

In Vitro Propagation of Two Tuberous Medicinal Plants: *Holostemma ada-kodien* and *Ipomoea mauritiana*

S. Pillai Geetha, A.V. Raghu, Gerald Martin, Satheesh George, and Indira Balachandran

Summary

Medicinal plants are sources of important therapeutic aid for healing human diseases. The depletion of the wild resources has prompted conservation, propagation, and enhancement of resources for medicinal plants. Micropropagation offers an alternate method to propagate and improve medicinal plants through selection of high-yield lines and their efficient cloning. This chapter describes cost effective and efficient protocols that have been successfully applied for the micropropagation and large-scale production of quality planting material in two important tuberous medicinal plants viz., *Holostemma ada-kodien* Schult. and *Ipomoea mauritiana* Jacq.

Key words: *Holostemma ada-kodien*, *Ipomoea mauritiana*, Micropropagation, Tuberous medicinal plants

1. Introduction

Forests and natural habitats have been the supply source of raw material medicinal plants for pharmaceuticals since the start of the industry. The resurgence public concern in plant-based medicine coupled with the growth of pharmaceutical industries has resulted in an increased demand for raw material medicinal plants, which has led to over-exploitation that threatens the existence of many rare species. As the industrial use of medicinal plants is dramatically increasing, many medicinal plant species are disappearing at an alarming rate as a result of urban development, uncontrolled deforestation, and unscientific, destructive, and indiscriminate collection.

S. Mohan Jain and Praveen K. Saxena (eds.), *Methods in Molecular Biology, Protocols for* In Vitro *Cultures and Secondary Metabolite Analysis of Aromatic and Medicinal Plants, vol. 547*
© Humana Press, a part of Springer Science+Business Media, LLC 2009
DOI: 10.1007/978-1-60327-287-2_7

Many tuberous medicinal plants are used in traditional Indian system of medicine, Ayurveda. In spite of their medicinal uses, most of them have not been genetically improved or scientifically validated and there are diverse reasons for this. Hence, it has become highly essential to take urgent measures to conserve, propagate, and improve the valuable genetic wealth for posterity. The tuberous medicinal plants are harvested and destroyed to the point of elimination at a rapid rate. Tuberous medicinal plants such as *Holostemma ada-kodien* Schult., and *Ipomoea mauritiana* Jacq are regularly used in Ayurveda.

1.1. Holostemma ada-kodien Schult

Holostemma ada-kodien Schult. (Syn. *H. annulare* Roxb. K. Scumn.) is an endangered medicinal plant in the state of Kerala, India, and belongs to *Asclepiadaceae* family. It is a large, perennial, laticiferous, twining shrub grown in parts of India such as the tropical Himalayas, Konkan, Deccan, Kerala, and Kanyakumari and in hedges and open forests in Sri Lanka, Myanmar, and Western China. Tuberous roots of the plant are widely used as a drug *Jivanti* and prescribed in 40 Ayurvedic formulations. The roots are sweet, and used as an aphrodisiac, in rejuvenative medicine, and as both laxative and stimulant. It promotes health and vigor, improves the voice, and is useful in ophthalmopathy and for treating cough, fever, and burns *(1–3)*.

The major ingredients of the root powder are – starch (35.4%) and sugar (24.0%). The ash contains protein (4.07%), fiber (12.2%), calcium (5.6%), and phosphate (2.5%). The bark fiber is suitable for cordage and papermaking. The plant exudes latex and on drying yields an elastic residue. Compounds such as α-amyrin, lupeol, β-sitosterol, alanine, aspartic acid, glycine, serine, threonine, and valine have been isolated from the roots *(1, 4)*.

1.2. Ipomoea mauritiana Jacq

Ipomoea mauritiana Jacq. (syn. *I. paniculata* R. Br.) is another medicinal plant that belongs to the *Convolvulaceae* family. It is a large, glabrous, twining shrub with large tuberous roots and a source of the Ayurvedic raw drug '*Vidari*' which is used in more than 50 Ayurvedic formulations. The species is distributed throughout India in deciduous and evergreen forests and coastal tracts. The roots are sweet, have cooling properties, and are used to increase appetite, as a galactagogue, in rejuvenative medicine, and as a stimulant, carminative, and tonic *(3)*.

Alcohol extract of tubers is stimulant as well as depressant and has convalescent effect on central nervous system. β-sitosterol and fixed oil containing palmitic, stearic, oleic, linolic, and linolenic acid have been isolated from the tubers *(5, 6)*.

1.3. Advantages of Micropropagation in Tuberous Plants

The destructive harvesting of tuberous roots poses a threat to these species, resulting in the shortage of raw material. These medicinal plants are cultivated to meet the increasing demand. *H. ada-kodien* is propagated predominantly through seeds. The root buds and stem cuttings also serve as propagules, but the propagation process is extremely slow. The seedling progeny is highly heterozygous and consequently the tuber yield is not uniform. *I. mauritiana* is vegetatively propagated. Non-availability of improved varieties and good quality planting material is a great impediment in the cultivation of these species. Plant tissue culture is a suitable option for propagation and is being used extensively for the commercial propagation of many plant species, including several medicinal plants. Micropropagation has proved to be a viable technology in both the species described in the present report. Micropropagated plants of both these species are superior in tuber production when compared to the conventionally propagated plants.

In vitro propagation of *H. ada-kodien* and *I. mauritiana* was reported *(7–12)*. This chapter describes the various steps of a reliable micropropagation protocol for large-scale production of high-quality planting material of *H. ada-kodien* and *I. mauritiana*.

2. Materials

2.1. Micropropagation of Holostemma ada-kodien

2.1.1. Selection of Mother Plants

Selection of mother plants based on their performance is the first step toward the establishment of in vitro culture. Healthy and vigorous plants growing in the nursery or experimental field of Centre for Medicinal Plants Research (CMPR) in Kerala, India are the primary source of explants (*see* **Note 1**).

2.1.2. Sources of Explants

1. Excise freshly emerged actively growing shoots with approximately five nodes from the shoot apex (*see* **Note 2**).

2.1.3. Pre-conditioning of Donor Plants

1. Cut and remove the shoot tips of the selected shoots still attached to the mother plant 2–3 d prior to the collection of explants for inoculation (*see* **Note 3**).

2.1.4. Culture Media and Culture Conditions

The composition of basal medium is provided in **Table 1.**

1. Culture initiation: Murashige and Skoog basal (MS-B) medium *(13)* without any growth regulators containing 20 g/L sucrose and solidified with 7 g/L agar.

Table 1
Composition of Murashige and Skoog Medium

Components	Standard mg/L	Actual weight (gm)	Final volume of stock (mL)	Volume of the stock per litre of medium (mL)
Macroelements				
NH_4NO_3	1650	33	1000	50
KNO_3	1900	38		
$CaCl_2 \cdot 2H_2O$[a]	330	6.6		
$MgSO_4 \cdot 7H_2O$	370	7.4		
KH_2PO_4	170	3.4		
$FeSO_4 \cdot 7H_2O$[b]	27.90	1.395	500	10
Na_2EDTA[b]	37.30	1.865		
Microelements				
H_3BO_3	6.20	0.31	500	10
$MnSO_4 \cdot 4H_2O$	22.30	1.115		
KI	0.83	0.0415		
$ZnSO_4 \cdot 7H_2O$	8.60	0.43		
$Na_2MoO_4 \cdot 2H_2O$	0.25	0.0125		
$CoCl_2 \cdot 6H_2O$	0.025	0.00125		
$CuSO_4\ 5H_2O$	0.025	0.00125		
Organic components				
Nicotinic acid	0.5	0.025	500	10
Pyridoxine–HCl	0.5	0.025		
Thiamine–HCl	0.1	0.005		
Glycine	2.0	0.1		
Myo-Inositol	100	5.0		
Sucrose	20,000–30,000	20–30	Added directly	
Agar[c]	7000–8000	7–8	Added directly	
pH	5.8–5.9			

[a]Dissolve separately and added to the rest of the components
[b]Should be dissolved in hot sterile water separately and combined together to get a yellow solution
[c]Agar is added after adjusting the pH. All solutions are prepared in sterile distilled water

2. Culture multiplication: MS-B medium supplemented with 30 g/L sucrose, 0.46 μM kinetin and 4.90 μM indole 3-butyric acid (IBA) and solidified with 7 g/L agar.
3. For rooting: MS-B medium containing 30 g/L sucrose (*see* **Note 4**).

2.2. Micropropagation of *Ipomoea mauritiana*

2.2.1. Selection of Mother Plants

Use healthy, vigorous mother plants growing in the nursery or field (*see* **Note 1**).

2.2.2. Sources of Explants

1. Vigorous and healthy 20- to 30-cm tall shoots (*see* **Note 2**).

2.2.3. Culture Media and Culture Conditions

1. The culture initiation medium, rooting medium and the conditions are same as in **Subheading 2.1.4**.
2. For shoot multiplication use MS basal medium supplemented with 30 g/L sucrose, 2.32 μM kinetin and 2.22 μM N_6-Benzyladenine (BA) solidified with 7 g/L agar (Table 1) (*see* **Note 4**).

3. Methods

The described method allows an efficient cloning of thousands of uniform plantlets from a single explant isolated from a selected mother plant.

3.1. Micropropagation of Holostemma ada-kodien

For micropropagation, the steps taken are preparation of explant, culture initiation, multiplication, rooting (either in vitro or *ex vitro*), hardening, and field establishment.

3.1.1. Preparation of Explant

1. Thoroughly clean and pre-surface sterilize the explants before moving the plant to the laminar airflow.
2. Detach the pre-conditioned shoots (*see* **Note 3**; **Subheading 2.1.3**) from the mother plant and wash under running tap water for 15 min.
3. Remove leaves and leave a small portion of the petiole intact to protect the meristematic tissue in the leaf axil. Wash the twig with dilute Tween-20 (2 drops in 100 mL) and use a brush to remove all the dirt sticking to it. Excise shoot and separate 3 to 4 cm long shoot tips and nodal segments.
4. Treat with 4% sodium hypochlorite (NaOCl) and Tween-20 (2 drops/100 mL) for 2 min and rinse 4–5 times with sterile distilled water.

5. Under aseptic conditions, trim explants to about 1.5-cm long segments (*see* **Note 5**) and treat with 4% NaOCl for 2 min under constant agitation. Wash 5–6 times in sterile double distilled water and inoculate.

3.1.2. Culture Initiation

1. Culture the surface sterilized explants in MS basal medium containing 20 g/L sucrose solidified with 7 g/L agar in order to screen out the aseptic cultures for further work (*see* **Note 6, Subheading 2.1.4**).

3.1.3. Shoot Multiplication

1. Transfer aseptic cultures showing axillary bud break to MS solid medium supplemented with 0.46 μM kinetin, and 4.90 μM IBA and 30 g/L sucrose in 350 mL culture bottles containing 50 mL medium.
2. Solidify the medium using 8 g/L agar (*see* **Note 7, Subheading 2.1.4**).

3.1.4. In Vitro *and Ex Vitro Rooting*

Both in vitro and *ex vitro* rooting is possible in *H. ada-kodien.*

1. Excise 1.5 to 2 cm long shoots from multiplying cultures and culture on MS basal medium without any growth regulators for root induction (refer *see* **Subheading 2.1.4**).
2. For *ex vitro* rooting, maintain cultures at room temperature (30 ± 2°C) for 2 d with slightly loosened bottle caps.
3. Excise 2 to 3 cm long single shoots from the shoot clumps or grown up shoots. Plant shoots in coarse, moist sand and keep in a humid chamber with high relative humidity in diffused light for 7–15 d with occasional misting for facilitating *ex vitro* rooting (*see* **Note 8** and **9**).

3.1.5. Hardening, Transplanting and Acclimatization

1. Remove in vitro rooted plants carefully without damaging the roots and clean them very well in tap water to remove all traces of media and agar.
2. Plant individual plantlets in polythene bags containing moist coarse sand and keep in a humid chamber to maintain high humidity for 1 wk with occasional misting.
3. Remove plants from the humidity chamber after 15 d and place in the shade for 14 d before transfer to the field (*see* **Note 8** and **9**).

These methods allow for clonal multiplication of nearly 250,000 plantlets from a single explant after 10 subculture cycles in 4–5 wk intervals with 90% field establishment (**Fig. 1**) under extreme care (*see* **Note 10**).

3.2. Micropropagation of Ipomoea mauritiana

The method consists of preparation of explant, culture initiation, multiplication, rooting, hardening, and field establishment.

Fig. 1. Micropropagation in *Holostemma ada-kodien* (**a**) Axillary bud initiation from nodal explant. (**b**) Multiple shoot induction. (**c**) In vitro rooted shoots. (**d**) *Ex vitro* rooted shoots. (**e**) Micropropagated plants ready for field transfer. (**f**) Tuber production in *ex vitro* rooted plants 3 mo after planting in field.

3.2.1. Preparation of Explant

1. Collect shoots and cut into single node segments and shoot tips.
2. Clean plants thoroughly, sterilize the surface before use.
3. Remove leaves from the twigs and leave a small portion of the petiole intact to protect the meristematic tissue in the leaf a xil.
4. Treat the explant with a fungicide, copper oxy-chloride (0.3%) for 30 min followed by single wash in distilled water.
5. Cut the shoot into approximately 3-cm long shoot tips and nodal segments and keep in dilute Tween-20 (2 drops in 100 mL) for 10 min.
6. Rinse once with 70% alcohol for 30 s. Then treat with 4% sodium hypochlorite (NaOCl) and Tween-20 (2 drops/100 mL) for 2 min and rinse 4–5 times with sterile distilled water.
7. Under aseptic conditions, trim the explants to 1.5 to 2 cm long segments (*see* **Note 5**) and treat with 4% NaOCl for 2 min under constant agitation. Wash in sterile double distilled water 5–6 times.

3.2.2. Culture Initiation

1. Culture the surface sterilized explants in MS basal medium containing 20 g/L sucrose solidified with 7 g/L agar (*see* **Note 6**).

3.2.3. Shoot Multiplication

1. Transfer the aseptic cultures with axillary bud break to MS solid medium supplemented with 3.32 μM kinetin, 2.22 μM N_6-benzyl adenine (BA) and 30g/L sucrose (*see* **Note 7**).

3.2.4. In Vitro *and Ex Vitro Rooting*

Both in vitro and *ex vitro* rooting are successful in *I. mauritiana* also.

1. Excise 1.5 to 2 cm long shoots from the multiplied shoot cultures and culture on to MS basal medium for root induction.
2. For *ex vitro* rooting, take out the culture bottles with in vitro cultures from the growth rooms and keep under room temperature (30 ± 2°C) for 1 d with the caps of bottles slightly released.
3. Excise approximately 3 cm long single shoots from the multiplied cultures and plant in coarse moist sand. Maintain in a humid chamber with high humidity in diffused light for 10 d with occasional misting (*see* **Note 8, 9**).

3.2.5. Hardening Transplanting and Acclimatization

1. After 20 d of growth in rooting medium, whole rooted plants are suitable for successful planting out.
2. Take care while removing the plantlets from the vessel and clean roots to remove the medium adhered to them.

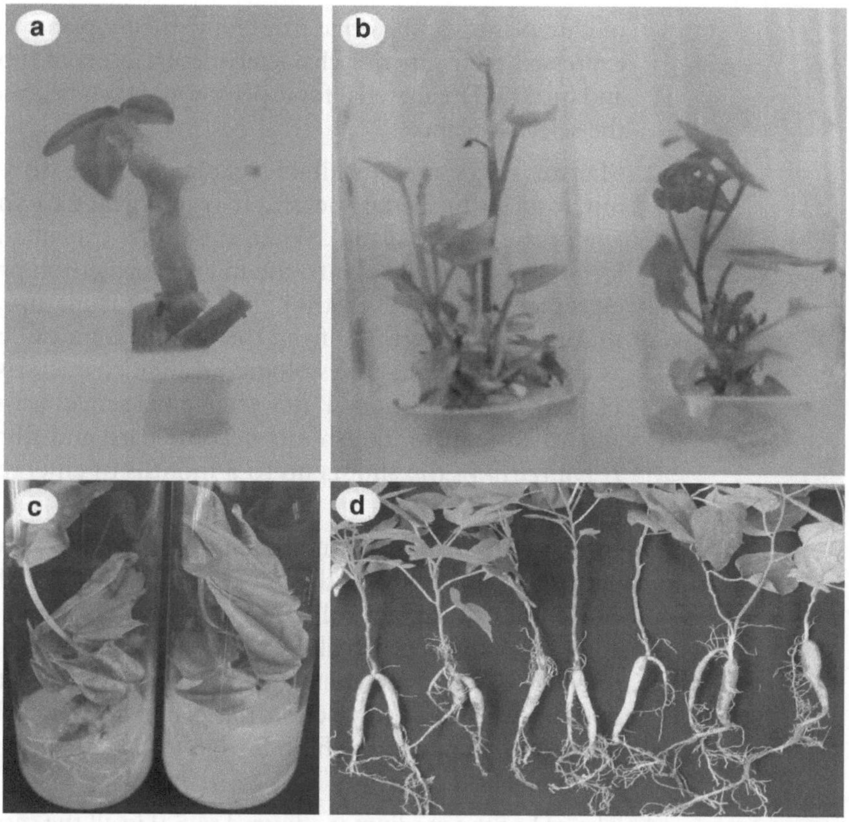

Fig. 2. Micropropagation in *Iopmoea mauritiana* (**a**) Axillary bud initiation from nodal explant. (**b**) Multiple shoot induction. (**c**) In vitro rooted shoots. (**d**) Tuber production in micropropagated plants 3-months after planting in field.

3. Plant in polythene bags containing coarse moist sand and keep in a humid chamber in diffused light for 15 d with occasional misting.
4. Remove plants from the humidity chamber after 15 d and transfer to a shade house for an additional 2 wk before transfer to the field (*see* **Note 8** and **9**).

The methods described facilitate the efficient clonal multiplication of nearly 300,000 plantlets from a single explant after 10 subculture cycles of 4 to 5 wk intervals with 90% field establishment (*see* **Note 10; Fig. 2**).

4. Notes

1. Plants with high yield of good quality tuber should be selected as mother plant in both the species and they should be labeled and maintained for future use. However, selection

of elite plants can be done from fully mature plants that have expressed their superior characteristics in terms of tuber yield and quality. The material for explant source can be raised from the selected plants.

2. Mature twigs are unsuitable explant source for culture initiation in both the species. It is important to know the age of the source plants. When the initial explants are collected from juvenile plants, the micropropagation process is extremely simple. Even though the plant is mature, young shoots freshly emerging from the mature stem can be used, as this tissue retains physiologic juvenile characteristics. In *H. ada-kodien*, shoots 1 mo after emergence are suitable and when mature, the shoots become hard and fibrous. In *I. mauritiana* the shoots up to 3 months of emergence are suitable.
3. In *H. ada-kodien*, pre-conditioning of the primary explant is performed by removing the upper most part of the shoot tip 2–3 d prior to explant collection and this will help in accelerating the activation of axillary buds and help in faster in vitro response. This is recommended as a step to break the bud dormancy. During this period the plants are sprayed with a fungicide, Copper oxy-chloride (0.3%), in order to control contamination.
4. The pH of the medium is adjusted to 5.9 in all the cases, prior to the addition of agar. After melting the agar the media is poured in suitable culture vessels and autoclaved at 121°C and 1.5 kg/cm^2 pressure for 20 min.
5. The shoot should be cut into nodal segments ~ 3 to 4 cm long in such a way so as to leave between 1 and 1.5 cm above and 2.5 to 3 cm below the axillary bud. During final trimming prior to inoculation, the explant should be cut so as to retain at least 0.5 cm above and 1 cm below the bud. The portion of the petiole that is kept intact for the protection of the bud, from the chemicals used as surface sterilants, should be cut and removed.
6. Use of basal medium with reduced levels of sucrose (20 g/L) for culture initiation is adopted in this procedure to eliminate the contaminated cultures and to reduce the loss on growth regulators. Use 25 mm × 150 mm culture tubes with 10 mL medium for initiation and 350 mL borosilicate bottles with 30 mL medium for multiplication.
7. The new shoots formed in vitro after 4–5 wk of culture of primary explants are excised and cut into 1 to 1.5 cm long nodal segments and shoot tips and sub-cultured into 350 mL glass bottles (5 explants/bottle) containing 30 mL of the shoot multiplication medium. In *H. ada-kodien* the leaves are opposite and each explant consists of two axillary buds.

In culture initiation medium, both buds emerge out in 30% of the cultures and in the rest only one axillary bud grow. Shoot multiplication is achieved through both shoot elongation and multiplication. Single node explants excised from the shoots originated from primary cultures often grow into single unbranched shoots or at times develop multiple shoots. These shoots could be further dissected and used to initiate new cultures and thus multiplication is possible via node culture and is the simplest method of in vitro propagation. In *I. mauritiana* shoot multiplication is mostly through multiple shoot formation. The medium used promoted shoot growth without callus formation in both the species and hence may be free from the risk of induced genetic irregularity. Stabilization of in vitro response is attained within 4 to 5 subculture cycles and the cultures can be maintained for many years. In case of insufficient elongation of shoots (stunted growth) resulting from continuous cycles of culture in the multiplication medium, it is recommended to transfer the cultures to growth regulator free basal medium, so that the accumulated growth regulator effect can be reduced and the cultures behave normally.

8. The rooting and acclimatization of in vitro formed shoots are very important steps in mass production of viable plantlets by micropropagation. *Ex vitro* rooting of in vitro formed shoots is more cost-effective, whereas in vitro rooting is labor-intensive, time and energy consuming, and more expensive. *Ex vitro* rooting in a nursery allows rooting and acclimatization to be achieved in one step.
9. The humid chamber is made of an iron structure covered by polythene sheet, which is maintained in a 75% nursery shade house. The relative humidity ranged from 70 to 80%. Misting is performed two times per wk. The established plants are transferred to the nursery within 7–15 d. The polythene covering of the humid chamber is removed partially so that the plants are slowly exposed to the normal nursery conditions. Humidity is gradually decreased during this period. New growth will be seen in the established plants. Plants can be transferred to the field after 1 mo.
10. The successful use of the described micropropagation protocol requires certain general practices, which are:
    (a) Continuous observation of cultures is utmost desirable for the detection of contaminated cultures and other culture malformations like vitrification, yellowing etc.
    (b) A certain percentage of microbial contamination is unavoidable at the culture initiation stage (the procedure described is capable of giving culture establishment to the tune of 80% during bright sunny days), but utmost

care should be taken at the time of culture transfer in the multiplication stage to avoid cross contamination, that would otherwise lead to serious loses of in vitro cultures and overall reduction in total number of micropropagated plants.

(c) Timely subculture (4 to 5 wk interval) is recommended to avoid loss of cultures resulting from the presence of dead tissue in the stock cultures and fungal growth in bottle caps.

Acknowledgments

The authors would like to thank Sir Dorabji TATA Trust, Mumbai, India for the financial support.

References

1. Anonymous (1959). The Wealth of India – A Dictionary of Indian Raw Materials and Industrial Products. Raw Materials Vol. V (H–K), Council of Scientific and Industrial Research (CSIR), New Delhi, India, p. 111
2. Anonymous (2002). The Wealth of India – A Dictionary of Indian Raw Materials and Industrial Products. Raw Materials Vol. 3 (D–I), National Institute of Science Communication (NISCOM), CSIR, New Delhi, India, p. 294
3. Sivarajan, V. V. and Balachandran, I. (1994). Ayurvedic Drugs and their Plant Sources, Oxford and IBH Publishing Co. Pvt. Ltd., New Delhi, India, pp. 195–196, 510–511
4. Rastogi, P.R., Mehrotra, B.N., and Srivastava, M. (1993). Compendium of Indian medicinal plants. Vol: 3. Central Drug Research Institute, Lucknow, India, pp. 346–347
5. Rastogi, P.R., Mehrotra, B.N., and Srivastava, M. (1990). Compendium of Indian medicinal plants. Vol: 1. Central Drug Research Institute, Lucknow, India, pp. 226–227
6. Rastogi, P.R., Mehrotra, B.N., and Srivastava, M. (1991). Compendium of Indian medicinal plants. Vol: 3. Central Drug Research Institute, Lucknow, India, pp. 392
7. Martin, K. P. (2002). Rapid propagation of *Holostemma ada-kodien* Schult., a rare medicinal plant, through axillary bud multiplication and indirect organogenesis. *Plant Cell Rep.* **21(2)**: 112–117
8. Martin, K. P. (2003). Plant regeneration through somatic embryogenesis on *Holostemma ada-kodien*, a rare medicinal plant. *Plant Cell Tissue Organ Cult.* **72(1)**: 79–82
9. Sudha, C.G., Krishnan, P.N. and Pushpangadan, P. (1998). *In vitro* propagation of *Holostemma annulare*. K. Schum, a rare medicinal plant. *In Vitro Cell. Dev. Biol.* **20(4)**: 34–57
10. Sudha, C.G., Krishnan, P.N. and Pushpangadan, P. (2002). Regeneration of plants from *in vitro* root segments of *Holostemma annulare*. Schum – a rare medicinal plant. *Curr. Sci.* **78**: 503
11. Geetha S.P., Indira Balachandran and P. N. Ravindran (2007). *In vitro* shoot culture and micro-tuber induction in Jivanti (*Holostemma ada-kodien* Schult.). In: Proceedings National Symposium on 'Plant Biotechnology: New Frontiers', Kukreja AK et al. (eds) 18–20 November, 2005, Central Institute of Medicinal and Aromatic Plants (CIMAP), Lucknow, India, pp. 202–206
12. Priya, L., Anita, S., Rajmohan, K., and Alex, S. (2007). *In vitro* shoot regeneration in *Ipomoea mauritiana* Linn. In: Kesavachandran, R., Nazeem, P. A., Girija, D., John, P. S., and Peter, K. V. (eds.). Recent Trends in Horticultural Biotechnology, New India Publishing Agency, New Delhi, India, pp. 291–293
13. Murashige, T. and Skoog, F. (1962). A revised medium for rapid growth and bioassays with tobacco tissue culture. *Physiol. Plant.* **15**: 473–497

Chapter 8

In Vitro Production of Gymnemic Acid from *Gymnema sylvestre* (Retz) R. Br. Ex Roemer and Schultes Through Callus Culture Under Abiotic Stress Conditions

Abdul Bakrudeen Ali Ahmed, Adhikarla Suryanarayana Rao, and Mandali Venkateswara Rao

Summary

Plant secondary metabolites have enormous potential for research and new drug development. Many secondary metabolites have a complex and unique structure and their production is often enhanced by biotic and abiotic stress conditions. Gymnemic acid ($C_{43}H_{68}O_{14}$), a pentacyclic triterpenoid isolated from the leaves of *Gymnema sylvestre,* exhibits potent inhibitory effect on diabetes. The gymnemic acid content is determined by chromatographic methods: Camag HPTLC system equipped with a sample applicator Linomat IV and TLC scanner and integration software CAT 4.0. In HPLC C_{18} (ODS) reverse phase column; water 486 UV detector; mobile phase, water/methanol (35:65, HPLC grade) + 0.1% acetic acid are used. Sample (20 μL) is applied with a flow rate of 1 mL/min and read at 230 nm with UV detector. The production of gymnemic acid is significantly higher in callus treated with 2,4-dichloro phenoxy acetic acid (2,4-D) and kinetin (KN). The blue light increases gymnemic acid accumulation upto 4.4-fold as compared with fluorescent light treatment and out of which 2.8 is found in leaves. Gymnemic acid is isolated from callus, grown under stress conditions followed by preparative TLC, simple and reproducible character based on HPTLC and high performance liquid chromatography.

Key words: *Gymnema sylvestre*, Asclepiadaceae, Callus culture, Auxins, Cytokinins, Blue light, Gymnemic acid, TLC, HPTLC, HPLC, Diabetes

1. Introduction

Gymnema sylvestre is used to cure madhumeha (glycosuria) and other urinary disorders. Traditionally, it is recommended to patients suffering from diabetes mellitus, to treat stomach ailments, and as a diuretic *(1)*. In Sri Lanka, the plant is utilized

S. Mohan Jain and Praveen K. Saxena (eds.), *Methods in Molecular Biology, Protocols for* In Vitro *Cultures and Secondary Metabolite Analysis of Aromatic and Medicinal Plants, vol. 547*
© Humana Press, a part of Springer Science+Business Media, LLC 2009
DOI: 10.1007/978-1-60327-287-2_8

to cure bone fractures. In Japan, it is widely used as a health food in tea bags, tablets, beverages, and confectioneries *(2)*. Glycoside of gymnemic acid may block the absorption of sugar from the intestine and sweet taste of sugars. Plant extract also increases the number of insulin producing cells in pancreas and balance insulin level. Gymnemic acid mixture *(3)* of triterpene glucuronides inhibits absorption of glucose in small intestine *(4)*, suppresses hyperglycaemia *(5)*, and decreases body weight in rats *(6)*. Gymnemic acid (GA) inhibits glucose-stimulated gastric inhibitory peptide secretion in human *(7)*.

The conventional propagation of *Gymnema sylvestre* has been hampered because of poor seed viability, low germination rate, and poor rooting ability of vegetative cuttings. As a result, its commercial exploitation could not be achieved. Therefore, the use of in vitro propagation methods would accelerate large-scale plant multiplication and conservation. In vitro studies have demonstrated the formation of highest number of multiple shoots (57.2 shoots/explant) on Murashige and Skoog (MS) medium supplemented with 1.0 mg/L BA, 0.5 mg/L KN, 0.1 mg/L NAA, 100 mg/L malt extract, 100 mg/L citric acid, and high root induction rate on MS medium containing 3.0 mg/L BA *(8)*. Similarly, somatic embryogenesis induction is successful by culturing hypocotyl, cotyledon, and leaf explants excised from *G. sylvestre* seedlings on MS medium amended with 0.5–5.0 μM 2, 4-D, 0.5–2.0 μM BA and 2% sucrose. The embryo germination and plantlet formation were achieved in fresh embryogenic maturation medium *(9)*. Recent reports indicate that in vitro callus has active compounds, gymnemic acid, and gymnemagenin in sufficiently large amounts *(10, 11)*. Factors such as exogenous plant growth regulators, shaking speed of the culture, and pH of the medium, play important roles in growth and gymnemic acid production in cell suspension culture *(12)*. Sucrose, inoculum density, auxins, and aeration are important to cell growth in a bioreactor *(13)*. This chapter describes a protocol for in vitro culture and the production of gymnemic acid from *G. sylvestre*.

2. Materials

2.1. Source of Plant

Gymnema sylvestre (6-yr old) collected from Pachamalai hills, Tamil Nadu, India are available from the Medicinal Plants Garden of Department of Plant Science at Bharathidasan University Tiruchirappalli in Tamil Nadu, India.

2.2. Explants Selection and Sterilization

1. Wash young leaf explants in running tap water for 20 min (*see* **Note 1**).
2. Wash them first with 2% Teepol solution *(v/v)* (Reckitt Benckiser; India) followed by sterile distilled water (*see* **Note 2**).
3. Under aseptic condition, immerse washed explants in freshly prepared 70% alcohol *(v/v)* for 15 s and followed by rinsing with sterile distilled water (*see* **Note 3**).
4. Disinfect explants with 3% sodium hypochlorite *(v/v)* for 2 min (*see* **Note 4**).
5. Final wash explants with 0.1% mercuric chloride *(w/v)* for 2 min and rinse them in sterile distilled water (*see* **Note 5**).

2.3. Preparation of Culture Media

1. MS medium (Murashige and Skoog medium) (*14*) is very suitable for the callus induction.
2. Prepare stock solutions of macronutrients, micronutrients, vitamins, and amino acids.

Alternatively, ready-made culture media (Sigma; St. Louis, MO) may be used.

3. Add appropriate amount of stock solutions with 30 g/L sucrose (Hi-media; India) while preparing the culture medium (*see* **Note 6**).
4. Callus induction: Induce callus from leaf explants by adding different concentrations of plant growth regulators (auxins 0.5–5.0 mg/L): indole-3-acetic acid (IAA), 1-naphthalene acetic acid (NAA), and 2, 4-D, (cytokinins 0.2–2.0 mg/L), BA, and KN.
5. Prepare stock solutions of plant growth regulators auxins: 2, 4-D, NAA, IAA, indole-3-butyric acid (IBA); cytokinins: 6-benzyl amino purine (BA) and 6-furfuryl amino purine (KN) (*see* **Note 7**).

2.4. Culture Conditions

1. Adjust pH to 5.7–5.8 before autoclaving and solidify with 0.8% agar *(w/v)* (Bacteriologic grade) (Hi media; India).
2. Sterilize the culture medium by autoclaving at 1.06 kg/cm and 121°C for 2 min.
3. Maintain all cultures at 25 ± 2°C under photoperiod of 16-h light/8-h drak with light intensity 45–50 μmol/m^2/s from cool white fluorescent tubes (Philips; India) with 55–60% relative humidity.

2.5. Physical Parameters

1. Give different light treatments of blue light (40 W, 455–495 nm), green light (40 W, 495–575 nm), red light (40 W, 625–780 nm,), and white fluorescent light (40 W, 380–780 nm) with photoperiod of 16-h light/8-h dark.

2. Expose cultures to various temperatures 20°C, 30°C, and 35°C in an orbital shaker (Remi Instruments Ltd.; India), 25 ± 2°C (control culture).
3. Give different treatments of photoperiod regimens: 4-h light/20-h dark, 8-h light /16-h dark, 12-h light/12-h dark, 20-h light/4-h dark with controls of 24-h light/24 h dark and 16-h light/8-h dark (*see* **Note 8**).

2.6. Chemical Parameters

1. Add different sucrose concentrations (2%, 4%, 5%, 6%), control (3% sucrose) to the culture medium.
2. Prepare stock solutions of NH_4 NO_3 (1–4 mM) and maintain at 4 ± 2°C.

2.7. Phytochemical Screening for Gymnemic Acid

1. In thin layer chromatography (TLC), use coating material silica gel at 1:2 *(w/v)* silica gel/water for gymnemic acid separation.
2. Coat suspension on the carrier glass plate (20 × 10 cm) using a stahl type spreader 0.5-mm thick (*15*) (*see* **Note 9**).
3. Separate extracts by using mobile phase solvent system consists of isopropyl alcohol:chloroform: methanol: Acetic acid (5:3:1:0.5).
4. Vanillin sulphuric acid reagent – dissolve 1 g vanillin in 90 mL ethanol. Add 5.0 mL acetic anhydride and 5.0 mL sulphuric acid to the solution.
5. High-performance thin layer chromatography (HPTLC) using Camag HPTLC system (CAMAG; Switzerland) equipped with a sample applicator Linomat IV, thin rough plate, development chamber, TLC scanner III, and Integration software 4.0.
6. Use an aluminium plate (10 × 10 cm) pre-coated with silica gel 60F 254 as an adsorbent.
7. Determination of gymnemic acid: Use Mobile phase—HPLC-grade water:methanol (35:65) and 0.1% acetic acid.
8. Add sample: 20 μL at flow rate of 1.0 mL/min. Read at 230 nm with a UV spectrophotometer to detect the gymnemic acid.

3. Methods

3.1. Establishment of Callus Culture

1. Prepare stock solutions of MS, B5, SH, and WPM macronutrients, micronutrients, vitamins, and amino acids and use the required amount during medium preparation.

2. Callus initiation and nature of callus were better in MS medium than in other media B5, SH, and WPM, **Table 1** (*see* **Note 10**).

Table 1
Callus Biomass Production from Leaf Explant of *Gymnema sylvestre* on Various Culture Media

	B5 medium	SH medium	MS medium	WPM medium
2, 4-D (mg/L)	DW (mg/L)	DW (mg/L)	DW (mg/L)	DW (mg/L)
0.5	65de – WFC	55^{d} – BFC	95^{bc} – GFC	–
1.0	72^{bc} – BFC	62^{c} – BFC	102^{b} – GCC	–
1.5	84^{a} – BFC	74^{a} – BFC	119^{a} – GCC	–
2.0	75^{b} – BBC	66^{b} – BCC	90^{d} – GLC	–
2.5	67^{d} – WWC	52de – BCC	85de – WFC	–
3.0	54^{f} – WFC	43^{f} – BFC	73^{f} – WFC	–
3.5	45^{g} – WFC	39^{fg} – BFC	64^{fg} – WWC	–
4.0	41^{gh} – WWC	32^{h} – BFC	58^{h} – WWC	–
4.5	38^{i} – WWC	20^{i} – BFC	51^{hi} – WWC	–
5.0	26^{j} – WWC	18^{ij} – BFC	48^{j} – WWC	–
0.5	62de – BFC	58^{bc} – BCC	95^{d} – GFC	–
1.0	79^{ab} – BFC	62^{b} – BFC	116^{a} – GCC	–
1.5	84^{a} – BFC	69^{a} – BFC	107^{b} – GCC	–
2.0	72^{c} – WWC	55^{d} – BWC	98^{bc} – LGFC	–
2.5	64^{d} – WWC	51de – BWC	85^{e} – LGFC	–
3.0	58^{f} – WWC	45^{f} – BFC	82^{ef} – WFC	–
3.5	49^{g} – WWC	38^{g} – BFC	74^{g} – WFC	–
4.0	42^{h} – WWC	32^{h} – WWC	61^{h} – WWC	–
4.5	39^{hi} – WC	28^{hi} – WFC	41^{i} – WWC	–
5.0	31^{j} – WWC	25^{j} – WFC	32^{ij} – WWC	–

abbr: BCC, black compact callus; BFC, black friable callus; GCC, green compact callus; GFC, green friable callus; WWC, white watery callus; WFC, white friable callus; DW, dry weight

Note: Values are mean of 30 replicates per treatment and repeated thrice. Values with the same superscript are not significantly different at 5% probability level according to DMRT

3. To obtain disease free cultures, surface sterilization for 15 s, with 70% ethyl alcohol, sodium hypochlorite 3% *(v/v)* for 2 min and 0.1% mercuric chloride *(w/v)* for 2 min is appropriate for inducing callus cultures.
4. Callus formation from different explants initiates in 12–15 d of culture with leaf explants with 94.5% (*see* **Note** 11).
5. Callus does not induce from explants on the culture medium without plant growth regulators even after 20 d of culture.
6. Leaf explants placed in medium supplemented with various growth regulators show different morphological changes such as, green compact, green friable, white friable, white loose, and brown compact callus, when the cultures are maintained for a very long time (after 60 d).
7. Harvest proliferated callus from 45-d-old cultures and take fresh weight.
8. Dry harvested fresh callus in an Oven (Universal Biochemical; India) at 40°C for 24 h and take dry weight in a balance.
9. The combination of auxins and cytokinins is highly productive in callus production, when compared with auxins alone (*see* **Note** 12).
10. Callus growth curve: Measure callus fresh weight at various time intervals (15, 25, 35, 45, and 55 d) (*see* **Note** 13).
11. Weigh callus sample to determine the callus growth rate at every 10-d time interval.
12. During the lag phase, there is little callus growth of *G. sylvestre* until the day 15 of culture. The exponential growth phase begins on the day 15 and continues up to day 35day of culture when the stationary growth phase begins. During 10 d of stationary phase, the cell density increases 4.4-fold. The amount of gymnemic acid synthesis increases thrice, when compared to lag phase.
13. With different treatments; light (blue, green, red, white, and dark), carbon source (2%, 3%, 4%, 5%, and 6%), photoperiod (4 h, 8 h, 12 h, 20 h, 24 h light period), nitrogen source (1 mM, 2 mM, 3 mM, and 4 mM), and temperature (20°C, 30°C, 35°C), the fresh and dry weight of calli are calculated at 15, 25, 35, 45, and 55 d.
14. The optimum concentrations of 1.5 mg/L auxin 2, 4-D and 0.5 mg/L cytokinin KN shows synergistic effect on callus growth under chemical and physical stress conditions. Under blue light, green compact callus form and produce highest biomass followed by red and green light (*see* **Note** 14).
15. Among the various temperature treatments (20°C, 25°C ± 2°C, 30°C, and 35°C) at 30°C highest proliferation of green friable callus is seen.

16. 5% sucrose enhances the green compact callus formation and the dry weight is 164 mg/L after 45 d of culture.
17. The highest dry weight biomass of 159 mg/L was observed in the 12h photoperiod and showed more the green compact callus than other photoperiods.
18. MS medium supplemented with 1.5 mg/L 2, 4-D, 0.5 mg/L KN, and 3 mM NH_4NO_3 is suitable for callus biomass production.

3.2. Phytochemical Screening and Identification of Gymnemic Acid

1. Collect in vivo healthy leaves and dry them under shadow conditions.
2. Harvest in vitro callus and dry in oven at 40°C for 24 h.
3. Homogenize dry callus and dry in vivo leaves with pure methanol five times.
4. Centrifuge the collected methanol samples at 5000 rpm for 10 min.
5. Use supernatant for phytochemical analysis.

3.2.1. Standard Sample Preparation

1. Dissolve 1 mg gymnemic acid (Yoshikawa group; Kyoto Pharmaceutical University, Kyoto, Japan) in 1 mL methanol.
2. Use for further phytochemical analysis.

3.2.2. Separation of Gymnemic Acid by Thin Layer Chromatography

1. Corresponding bands of same R_f value as authentic sample of gymnemic acid are observed in all the samples analysed.
2. Separate extracts by using mobile phase solvent system consisting of isopropyl alcohol:chloroform methanol:acetic acid (5:3:1:0.5).
3. After the chromatographic run is over, dry the chromatogram at the room temperature and spray with a specific reagent (vanillin sulphuric acid reagent) for the detection of compounds.
4. Prepare freshly vanillin sulphuric acid reagent before starting to analyse the samples to determine gymnemic acid (*see* **Note 15**).
5. Standard gymnemic acid show only single band. However, in callus cultured cells an additional band is observed by spraying with vanillin sulphuric acid reagent having an R_f value (0.44) greater than gymnemic acid (0.43).

3.2.3. Separation of Gymnemic Acid by High Performance Thin Layer Chromatography

1. For these experiments, the mobile phase is isopropyl alcohol: chloroform:methanol:acetic acid (5:3:1:0.5) (*see* **Note 16**).
2. Allow solvent to run up to 80 mm and scan chromatograms at 200 nm. Use 1.0 mg/mL methanol solution of gymnemic acid as a reference standard compound (**Fig. 1a**).

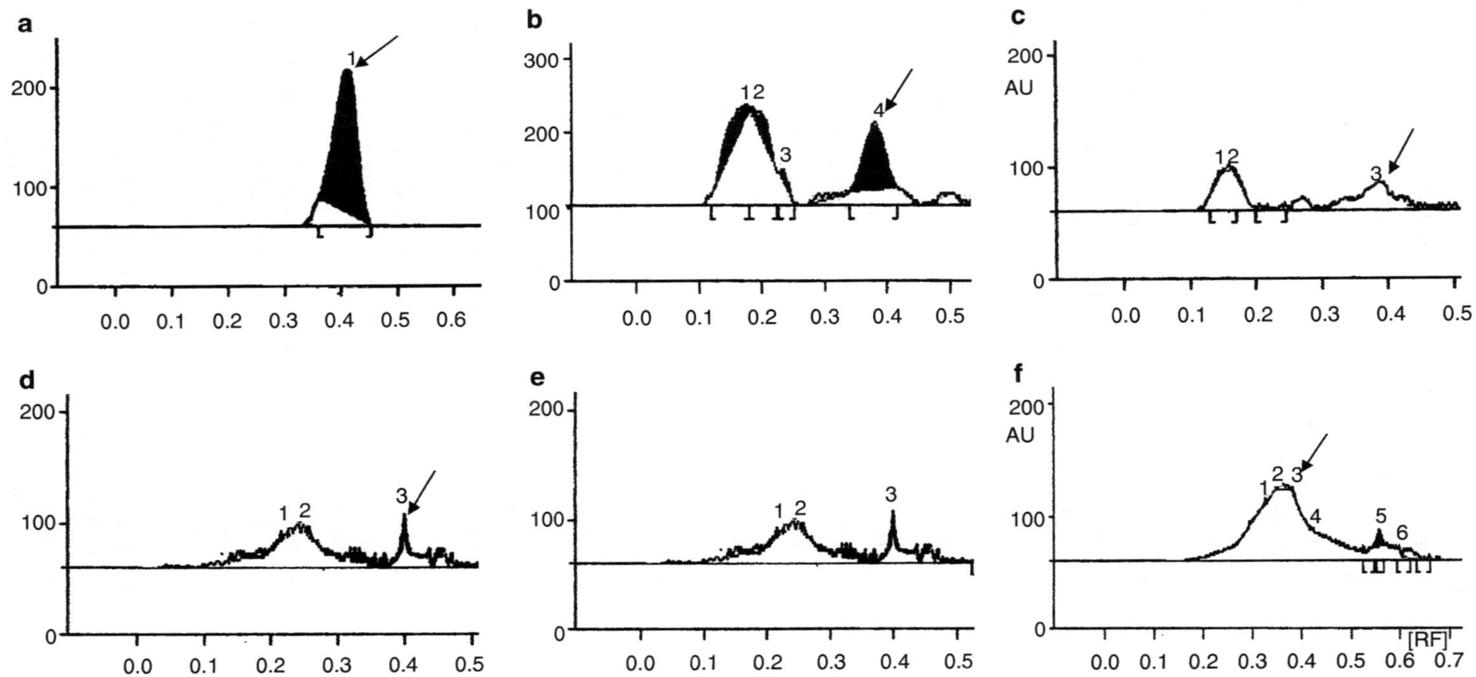

Fig. 1. **(a)** Standard gymnemic acid; **(b)** Gymnemic acid in leaf; **(c)** Gymnemic acid in MS + 1.5 mg/L 2,4-D + 0.5 mg/L KN after 25 D; **(d)** Gymnemic acid in MS + 1.5 mg/L 2,4-D + 0.5 mg/L KN after 35 d; **(e)** Gymnemic acid in MS + 1.5 mg/L 2,4-D + 0.5 mg/L KN after 45 d; and **(f)** Gymnemic acid in MS + 1.5 mg/L 2,4-D + 0.5 mg/L KN after 55 d.

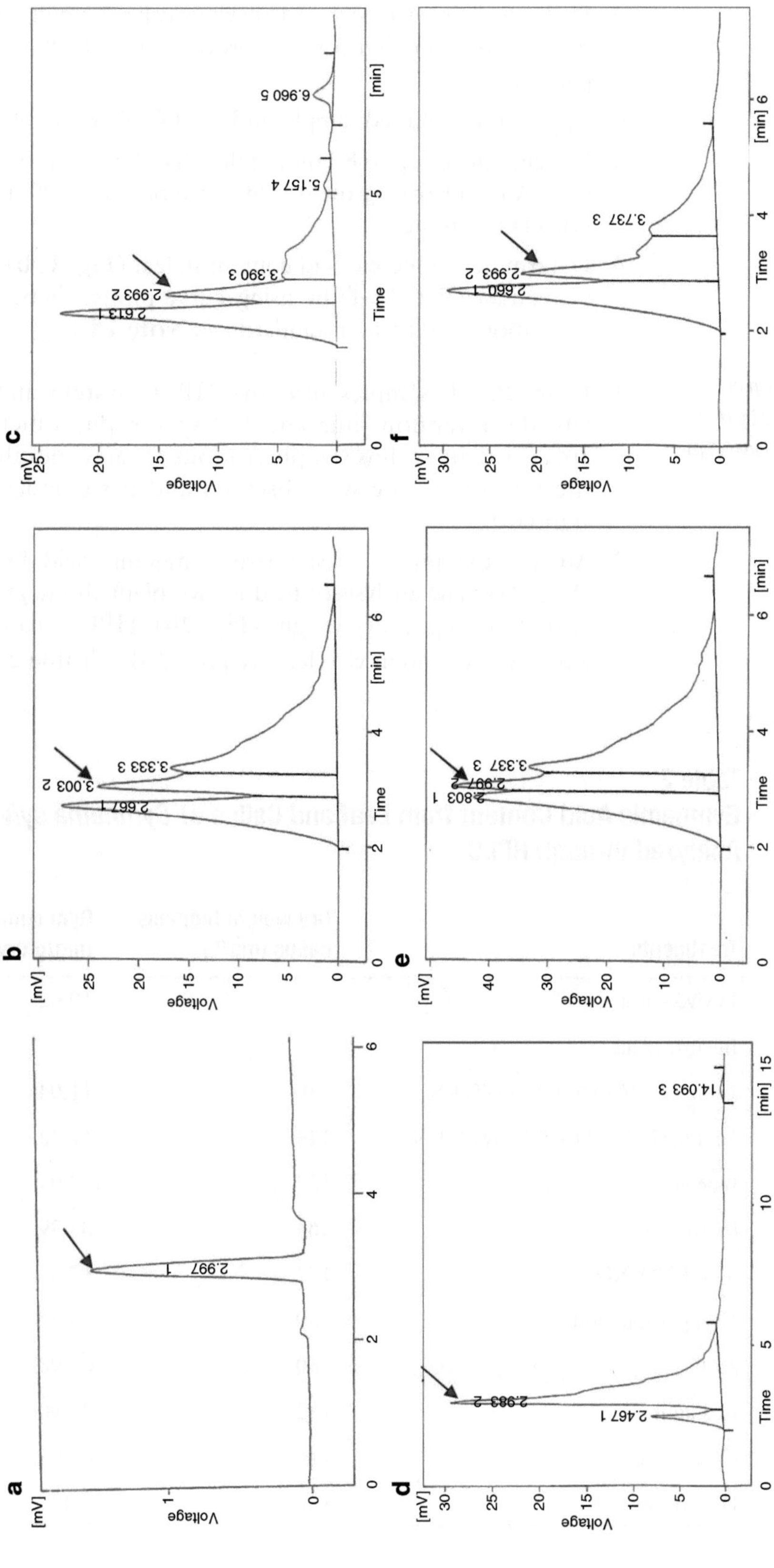

Fig. 2. **(a)** Standard gymnemic acid; **(b)** In vivo leaf; **(c)** MS + 1.5 mg/L 2,4-D + 0.5 mg/L KN; **(d)** Blue light treatment; **(e)** 5% sucrose treatment; and **(f)** 12-h photo period treatment.

3. Shake well in vivo leaf and in vitro callus sample and filter through 0.4 μm filter paper to prevent the test solution of any particles.
4. Apply 20 μL filtered sample on HPTLC plate (*see* **Note 17**).
5. Develop plates up to 80 mm under chamber saturation condition. After air-drying the solvent, scan plates at 200 nm at UV reflectance mode.
6. Determine gymnemic acid content in leaf (**Fig. 1, b**) and callus extract (**Fig. 1c–f**) by using curve plotted between concentrations and area of standard (*see* **Note 18**).

3.2.4. Separation of Gymnemic Acid by High Performance Liquid Chromatography

1. Load 20 μL samples into the HPLC system and calculate the retention time and UV spectrophotometer peak area. Leaf and callus sample readout at same mobile phase, the retention time was observed and it's compared with standard.
2. Analyse the purity of standard gymnemic acid by HPLC (**Fig. 2a**). The analysis of field grown plant shows gymnemic acid 19.52 mg/g dry weight (**Fig. 2b**). HPLC analysis of in vitro calluses and in vivo leaf are provided in **Table 2**.

Table 2
Gymnemic Acid Content from Leaf and Callus of *Gymnema sylvestre*-Analyzed through HPLC

Treatments	Dry weight biomass callus (mg/L)	Gymnemic acid (mg/g dry weight)
In vivo – leaf	–	19.52
In vitro callus		
1.0 mg/L NAA + 1.5 mg/L KN	139	11.04
1.5 mg/L 2, 4-D + 0.5 mg/L KN	144	12.22
Blue light	172	53.94
5% sucrose	164	33.39
3 mM NH_4NO_3	152	17.34
12-h photoperiod	159	26.27
30°C	150	02.90
Red light	122	08.90
Green light	116	03.07
Dark room	132	0.00

3. The highest gymnemic acid yield from leaf derived callus (12.22 mg/g dry weight, **Fig. 2c**) wass obtained on the MS medium containing 1.5 mg/L 2, 4-D, 0.5 mg/L KN with blue light treatment after 45 ds (53.94 mg/g dry weight callus, **Fig. 2d**) (*see* **Note 19**).
4. Gymnemic acid yields in other treatments: 5% sucrose (33.39 mg/g dry weight, **Fig. 2e**), 12-h photoperiod (26.27 mg/g dry weight, **Fig. 2f**), red light (8.90 mg/g dry weight), green light (3.07 mg/g dry weight), 30°C (2.9 mg/g dry weight), and 3 mM NH_4NO_3 treatment (17.34 mg/g dry weight).

4. Notes

1. Mature leaf, stem, and petiole explants and very young explants are unsuitable, possibly because of their negligible content of meristematic cells.
2. Two drops of Teepol solution is added to rinse explants for 3 min and the long rinsing time is deleterious to explant for callus induction.
3. To obtain disease free culture surface sterilization for 15 s with 70% ethyl alcohol is suitable and ethanol disinfection exceeding 30 s is lethal to leaf explant.
4. 3% sodium hypochlorite treatment for 2 min is appropriate and exceeding to this treatment time the leaf chlorophyll begins to release from the leaf explants.
5. The explant sterilization with 0.1% mercuric chloride *(w/v)* for 2 min is appropriate for culture proliferation and maintenance.
6. The storage of all stock solutions in amber coloured glass bottles is most ideal at 4 ± 2°C.
7. The plant growth regulators auxins (2,4-D, NAA, IAA, and IBA) and cytokinins (BA and KN) are prepared by dissolving in a minimal quantity of 1N NaOH/1N Hcl. Mix to final concentration of 1 mg/10mL by adding distilled water.
8. The photoperiod effects to induce CO_2 concentration inside the vessel and the daily exchange in turn affects the growth of callus under CO_2 non enriched condition.
9. The freshly coated plates are left on the tray until the transparency of the layer disappear and stack in drying rack and heat in a vertical position for 30 min at 110°C in order to activate the plates.

10. Among the different media tested, MS media is the best for callus induction followed by B5, SH, and WPM medium. The callus sprouting on B5, SH, and WPM medium shows only limited development even if they are maintaining for long period in culture.
11. The leaf induces green compact callus, whereas petiole and stem produce white friable and white loose callus.
12. The combination of 2,4-D with KN induce green compact callus. The MS medium supplemented with 1.5 mg/L 2,4-D and 0.5 mg/L KN induced the highest biomass of callus (F.W – 1.56 g/L, D.W – 144 mg/L) after 45 d of culture.
13. Callus formation initiates within 8–12 d of culture; lag growth phase (15 d), exponential growth phase (25–35 d), stationary growth phase (45 d), and decline growth phase at 55 d onward.
14. Green and red light induce green friable callus but dry weight of callus is drastically reduced as compared with blue light.
15. Sample preparation and development of suitable mobile phase of solvent system are two important steps in developing the analytical procedure which becomes more significant for herbal drugs because of their affinity towards various solvents.
16. After run, dry the chromatogram at the room temperature and spray with vanillin-sulphuric acid reagent for the detection of triterpenoids.
17. The advantage of this HPTLC system is that ten samples can be applied at a single time with standard gymnemic acid.
18. The calibration curve is linear, further correlation coefficient indicates good linearity between concentration and area. This method allows reliable quantification of gymnemic acid and provides good resolution from other constituents of *Gymnema sylvestre*. To ascertain the purity of peak in test sample, compare its in vivo and in vitro reflectance spectrum with standard, which provides test sample purity.
19. Cryptochrome is highly sensitive to blue light and closer to UV-B in wave length. In vitro callus under blue light, more P_r is transformed to P_{fr} than under white light and the phenylalanine ammonia lyase activity in the cultures is higher than that under white light.

Acknowledgments

We are thankful to Prof. Shigenobu Arihara and Kazuko Yoshikawa, Kyoto Pharmaceutical University, Kyoto, Japan for providing gymnemic acid standard; Mr. S. Govindu, Technician,

Central Electrochemical Research Institute, Karaikudi, Tamil Nadu, for carrying out the HPLC analysis; and Mr. Manimaran, Lecturer, JSS college of Pharmacy, Ooty, for the help rendered during HPTLC analysis.

References

1. Gharpurey, K.G. (1926). *Gymnema sylvestre* in the treatment of diabetes. *Indian Med Gaz.* **61**, 155.
2. Yoshikawa, M., Murakami, T., Kadoya, M., Yuhao, Y., Murakami, N., Yamahara, J., and Matsuda, H. (1997). Medicinal food stuffs IX the inhibitor of glucose absorption from the leaves of *Gymnema sylvestre* R.Br., (Asclepiadaceae): structures of Gymnemosides a and b. *Chem Pharm. Bull.* **45**, 1671–1676.
3. Imoto, T., Yamamoto, F., Miyasaka, A., and Hatano, H. (1991). High-performance liquid chromatography-atmospheric pressure ionization mass spectrometry of gymnemic acids. *J. Chromatogr.* **557**, 383–389.
4. Shimizu, K., Iino, A., Nakajima, J., Tanaka, K., Nakajyo, S., Urakawa, N., Atsuchi, M., Wada, T., and Yamashita, C. (1997). Suppression of glucose absorbtion by some fraction from *Gymnema sylvestre* leaves. *J. Vet. Med. Sci.* **59**, 245–251.
5. Sugihara, Y., Nojima, H., Matsuda, H., Murakami, T., Yoshikawa, M., and Kimura, I. (2000). Antihyperglycemic effects of gymnemic acid IV, a compound derived from *Gymnema sylvestre* leaves in streptozotocin-diabetic mice. *J. Asian Nat. Prod. Res.* **2**, 321–327.
6. Terasawa, H., Miyoshi, M., and Imoto, T. (1994). Effects of long-term administration of *Gymnema sylvestre* watery-extract on variations of body weight, plasma glucose, serum triglyceride, total cholesterol and insulin in Wistar fatty rats. *Yonago Acta. Med.* **37**, 117–127.
7. Kurihara, Y. (1969). Antisweet activity of Gymnemic acid A1 and its derivatives. *Life Sci.* **8**, 537–543.
8. Komalavalli, N., and Rao, M.V. (2000). *In vitro* micropropagation of *Gymnema sylvestre* – a multipurpose medicinal plant. *Plant Cell Tiss. Org. Cult.* **61**, 97–105.
9. Kumar, H.G., Murthy, H.N., and Paek, K.Y. (2002). Somatic embryogenesis in *Gymnema sylvestre. Plant Cell Tiss. Org. Cult.* **71**, 85–88.
10. Gopi, C., and Vatsala, T.M. (2006). *In vitro* studies on effects of plant growth regulators on callus and suspension culture biomass yield from *Gymnema sylvestre* R.Br. *African J. Biotech.* **5**, 1215–1219.
11. Kanetkar, P.V., Singhal, R.S., Laddha, K.S., and Kamat, M.Y. (2006). Extraction and quantification of gymnemic acids through gymnemagenin from callus cultures of *Gymnema sylvestre. Phytochem. Anal.* **17**, 409–413.
12. Devi, C.S., Murugesh, S., and Srinivasan, V.M. (2006). Gymnemic acid production in suspension cell cultures of *Gymnema sylvestre. J. Appl. Sci.* **6**, 2263–2268.
13. Lee, E.J., Mobin, M., Hahn, E.J., and Peak, K.Y. (2006). Effects of sucrose, inoculum density, auxins, and aeration volume on cell growth of *Gymnema sylvestre. J. Plant Biol.* **49**, 427–431.
14. Murashige, T., and Skoog, F. (1962). A revised medium for rapid growth and bioassays with tobacco tissue culture. *Physiol. Plant.* **15**, 473–497.
15. Stahl, E. (1969). *Thin layer chromatography.* Springer, New York.

Chapter 9

Establishment of Plant Regeneration and Cryopreservation System from Zygotic Embryo-Derived Embryogenic Cell Suspension Cultures of *Ranunculus kazusensis*

Suk Weon Kim and Myung Jin Oh

Summary

This chapter describes culture conditions for high-frequency plant regeneration via somatic embryogenesis and cryopreservation from cell suspension cultures of *Ranunculus kazusensis*. Zygotic embryos form white nodular structures and pale-yellow calli at a frequency of 84.9% on half-strength Schenk and Hildebrandt (SH) medium supplemented with 0.1 mg/L 2,4-dichlorophenoxyacetic acid (2,4- D). However, the frequency of white nodular structure and off-white callus formation decreases to 25% with an increasing concentration of 2,4- D up to 10 mg/L cell suspension cultures are established from zygotic embryo-derived pale-yellow calli using half-strength SH medium supplemented with 0.1 mg/L 2,4- D. Upon plating onto half-strength SH basal medium, over 90% cell aggregates give rise to numerous somatic embryos and develop into plantlets. Regenerated plantlets are transplanted to pots filled with soil and grown to maturity at 90% survival rate in a growth chamber. Furthermore, we have developed the cryopreservation system using embryogenic cell suspension cultures of *Ranunculus kazusensis*. The re-growth rate of cryopreserved cells in 20% glycerol and 10% dimethylsulfoxide (DMSO) is 10% and 28.3%, respectively. These results show that DMSO is more effective cryoprotectant than glycerol in long-term preservation of embryogenic cell suspension cultures. The plant regeneration and cryopreservation system established in this study could be applied for mass propagation and *ex situ* conservation of this plant species.

Key words: Cryopreservation, Plant regeneration, *Ranunculus kazusensis*, Somatic embryogenesis, Suspension culture

1. Introduction

Ranunculus kazusensis Makino, herbaceous water plant, belongs to family *Ranunculaceae*. This species is native to Korea, and is threatened to extinction as a result of habitat loss. The family Ranunculaceae consists of approximately 2500 described species distributed

S. Mohan Jain and Praveen K. Saxena (eds.), *Methods in Molecular Biology, Protocols for* In Vitro *Cultures and Secondary Metabolite Analysis of Aromatic and Medicinal Plants, vol. 547*
© Humana Press, a part of Springer Science+Business Media, LLC 2009
DOI: 10.1007/978-1-60327-287-2_9

among 59 genera *(1)*. Crude extracts of *Ranunculus* spp. have fungicidal *(2)*, insecticidal *(3)*, analgesic, and anti-inflammatory activities *(4)*. A few studies on plant regeneration in the genus *Ranunculus* have been conducted, including plant regeneration via organogenesis *(5)* and somatic embryogenesis of *R. asiaticus (6, 7)*, and plant regeneration via anther cultures of *R. japonicus (8)*.

Somatic cell mutations induced from a long-term subcultures could lead to the loss of genetic uniformity of the regenerated plants or somaclonal variation *(9)*. Cryopreservation is considered as the most ideal method for long-term preservation of plant germplasm including in vitro cultures *(10)*. A number of cryopreservation techniques have been developed to allow plant tissues to acquire cold hardiness prior to cooling in liquid nitrogen. The most commonly used procedure is the slow-cooling method (also known as two-step freezing), based on Withers *(11)*, that involves conditioning of tissues at low levels of cryoprotectant, followed by slow cooling. The major drawback of the slow-cooling method is the requirement of expensive equipment (i.e., a programmable freezer) that may not be available in many laboratories. An alternative of this method, use Nalgene freezing containers and a deep-freezer (–70°C) for gradual cooling of samples *(12)*.

In vitro plant regeneration systems may provide an alternative means for mass proliferation and *ex situ* conservation of endangered plant species. This chapter describes culture conditions and cryopreservation for high-frequency plant regeneration via somatic embryogenesis from zygotic embryo-derived cell suspension cultures of *R. kazusensis*.

2. Materials

2.1. Plant Materials and Culture Condition

1. Collect mature seeds of *Ranunculus kazusensis* Makino from the fields.
2. Half-strength SH *(13)* medium contains SH inorganic salts, 5 mg/L thiamine·HCl, 5 mg/L nicotinic acid, 0.5 mg/L pyridoxine·HCl, 1000 mg/L *myo*-inositol, 30 g/L sucrose, and 4 g/L gelrite.
3. The composition of B5 medium *(14)* is Gamborg inorganic salts supplemented with 10 mg/L thiamine·HCl, 1 mg/L nicotinic acid, 1 mg/L pyridoxine·HCl, 100 mg/L *myo*-inositol, 20 g/L sucrose, and 4 g/L gelrite.
4. 2, 4-D. Stock solution Dissolve 100 mg/L 2,4-D (Sigma; St. Louis, MO) in 1N NaOH solution. Add distilled water to raise volume to 100 mL, and store at –4°C in the refrigerator. Add stock solution to the tissue culture medium and raise the

final volume of the medium according to the requirement (*see* **Note 1**).

5. Prepare stock solutions of other plant growth regulators (NAA, IAA, kinetin) in the same way 2,4- D stock solution is prepared.
6. For surface-sterilization of seeds, dilute commercial bleach solution (~4%) with sterile distilled water at 0.4% *(v/v)* sodium hypochlorite solution (*see* **Note 2**).
7. All cultures are maintained at 25°C in the dark or in the light (30 μmol/m^2/s from cool-white fluorescent lamps with a 16-h photoperiod).

2.2. Cryopreservation of Embryogenic Cell Suspension Cultures

1. Dilute dimethyl sulfoxide (DMSO) (Sigma; St. Louis, MO) to 50% *(v/v)* by adding distilled water. Filter-sterilize (pore size, 0.22 μm) it and store at room temperature (*see* **Note 3**).
2. Add distilled water to dilute glycerol (Sigma; St. Louis, MO) to 50% *(v/v)* and sterilize by using filter (pore size, 0.22 μm), and store at room temperature.
3. Dissolve 2,3,5-triphenly tetrazolium chloride (Sigma; St. Louis, MO) in DMSO solution. Dilute with distilled water at 0.18 M and store in single use aliquots at −20°C (*see* **Note 4**).
4. Nalgene freezing container (Nalgene; Cryo 1C).
5. 1.8-mL Cryotube vial (Nunc).

3. Methods

3.1. Preparation of Culture Media and Callus Induction

1. Prepare SH or B5 culture media using media powder (Sigma). Add sucrose, plant growth regulators, and gelrite in the culture medium and sterilize by autoclaving at 121°C, 1.05 kg/cm^2 (15–20 psi, for 15 min). Pour 25 mL cool medium to 40–50°C in Petri dishes.
2. For surface sterilization of seeds, put mature seeds of *R. kazusensis* into a 15-mL Falcon tube. After brief washing with 70% EtOH for 1 min, rinse seeds 3–4 times with sterile distilled water followed by 0.4% sodium hypochlorite solution treatment containing a few drops of Tween-20 for 20 min with occasional agitation. Rinse seeds 3–4 times additional with sterile distilled water to get rid of hypochlorite solution. Dry sterilized seeds with sterile filter paper (Toyo/70 mm).
3. Dissect longitudinally surface-sterilized seeds. Approximately 2- to 3-mm long, and culture in half strength SH basal medium at 25°C in the dark. After 2 wk of incubation, collect the elongated zygotic embryos from dissected seeds (*see* **Note 5**).

4. To examine the effect of 2,4- D on embryogenic callus and somatic embryo formation, place the zygotic embryos onto half-strength SH medium supplemented with 0, 0.1, 0.3, 1, 3, or 10 mgL/L 2,4- D (*see* **Note 6**). Each treatment consists of ten explants per dish with three replicates. After 8 wk of culture, determine the frequency of explants producing white nodular structures and pale-yellow calli for each treatment (*see* **Note 7**). The results are shown in **Figs. 1** and **2**.

3.2. Establishment of Embryogenic Cell Suspension Cultures and Plant Regeneration

1. Use initial pale-yellow calli, approximately 1 g, maintained on half-strength SH medium amended with 0.1 mg/L 2,4- D to establish embryogenic cell suspension culture. Carefully disintegrate the pale-yellow calli, with sterile forceps and transfer in 250-mL Erlenmeyer flask containing 20 mL half-strength SH liquid medium containing 0.1 mg/L 2,4- D. Maintain cell suspension cultures on a gyratory shaker, 100 rpm, at 25°C in the dark.

Fig. 1. Plant regeneration of *Ranunculus kazusensis* Makino via somatic embryogenesis. (**a**) Zygotic embryos; (**b**) Globular structure and callus formation from zygotic embryo; (**c**) Plant regeneration via somatic embryogenesis; (**d**) Establishment of embryogenic cell suspension cultures; (**e**) Plantlets development from cell aggregates; (**f**) Rooting of plantlets; (**g**) Multiple plant regeneration from embryogenic cell suspension culture; (**h**) Successful soil transfer of regenerated plants, (**i**) Flowering of regenerated plants. Scale bars represent 2 mm (**a**, **b**, **c**, **e**, **f**, **i**), 200 μm (**d**), and 2 cm (**g**, **h**).

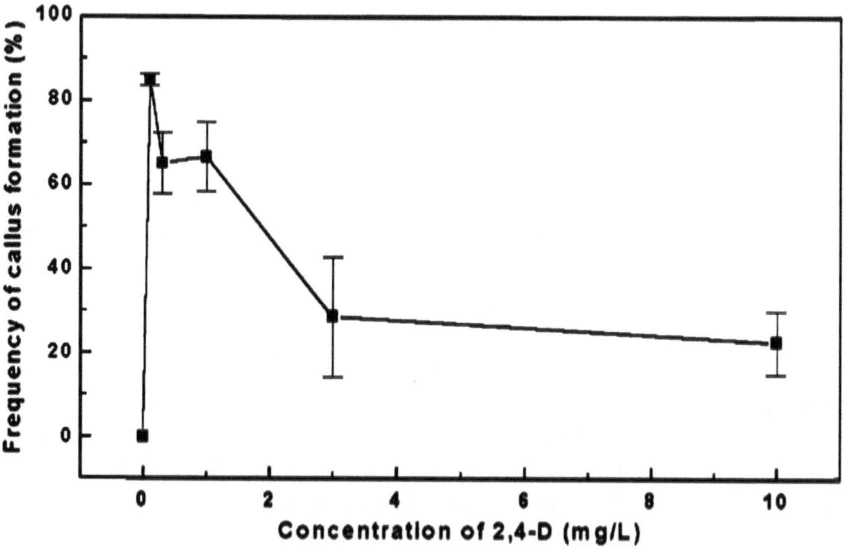

Fig. 2. Effect of 2,4- D on embryogenic callus formation on zygotic embryos of *Ranunculus kazusensis* Makino. Each treatment consisted of ten explants with three replicates. *Vertical bars* represent SD.

2. For stabilization of cell suspension cultures, add 10 mL fresh liquid medium in cell suspension cultures at 1-wk intervals. Observe cells repeatedly under the inverted microscope during the culture. After 1 wk of growth in cell suspension culture, add 10 mL fresh liquid medium without removing the old medium (*see* **Note 8**).
3. After an additional 2 wk of culture, transfer 5 mL cell suspension cultures in old medium to 250-mL Erlenmeyer flasks containing 50 mL half-strength SH liquid medium supplemented with 0.1 mg/L 2,4- D (*see* **Note 9**). After 2 rounds of subculture at 3-week intervals, use cell suspension cultures for plant regeneration and cryopreservation experiments. When the culture period is extended from 3 to 4 wk after subculture, the cultured cells begin to degenerate into brownish cells and cells lose embryogenic potential.
4. To regenerate whole plants, collect cell aggregates, 1- to 2-mm size, from 2-wk old cell suspension cultures using a stainless steel mesh, pore size, 1 mm, and rinse with half-strength SH basal liquid medium. Transfer cell aggregates onto half-strength SH solidified basal medium. Each treatment consists of 25 explants per dish with 5 replicates. After 4 wk of culture in the light, 30 μmol/m^2/s from cool-white fluorescent lamps with a 16-h photoperiod, determine the frequency of plantlet formation via somatic embryogenesis by counting the number of green regenerated plantlets. Over 90% cell aggregates form white somatic embryos and most of them germinate and develop into green plantlets.

5. Transfer rooted plantlets to pots filled with soil, and maintain in a growth chamber for 4 wk, 25°C day/22°C night, 80 μmol/m^2/s from cool-white fluorescent lamps with a 16-h photoperiod (*see* **Note 10**). The survival rate of regenerated plantlets transferred to soil is 90%.
6. After soil acclimatization, add more soil but do not cover the growing plantlets. And then fill up the pot with tap water near the top of the pot. After about 1 mo, plants will begin flowering. The results are shown in **Fig. 1**.

3.3. Cryopreservation of Embryogenic Cell Suspension Cultures and Plant Regeneration

1. Use embryogenic cell suspension cultures maintained on half-strength SH medium supplemented with 0.1 mg/L 2,4-D or B5 medium supplemented with 0.1 mg/L IAA, 0.1 mg/L NAA, 1.5 mg/L 2,4-D, 0.25 mg/L kinetin, and 20 g/L sucrose for freezing. Transfer 1-wk-old cells in 50-mL Falcon tubes. After centrifugation at 1000 rpm (180 g) for 5 min, remove liquid medium and resuspend cells in 20 mL fresh liquid medium.
2. Transfer one mL cell suspension to cryovial, 1.8 mL, Nunc, and add filter-sterilized 50% DMSO and 50% glycerol to each cryovials. Adjust final concentrations of DMSO and glycerol up 0, 5, 10, 15, and 20% (*see* **Note 11**). Seal cryovials with parafilms and keep them on ice bath for 1 h to improve the penetration of cryoprotectants.
3. Transfer pre-chilled cells in a freezing container supplemented with isopropanol, and left frozen at –70°C freezer for 4 h (*see* **Note 12**). Put cryovial of frozen cells in storage cane and plunge into the liquid nitrogen (–196°C) immediately.
4. After 1 wk of storage in the liquid nitrogen, remove cryovials and thaw at 40°C in a water bath for 90 s until cells are defrozen. Rapid thawing (60–90 s at 40°C) reduces or prevents the formation of damaging ice crystals within cells during rehydration. To minimize the toxic effect of cryoprotectants during prolonged exposure, remove cryoprotectants carefully and quickly. Transfer cells quickly to the solid culture medium overlaying a sterile filter paper and dry out remaining cryoprotectants *(15)*. After 1 h, transfer cell aggregates, approximately 1 mm diameter, to fresh B5 medium supplemented with 0.1 mg/L IAA, 0.1 mg/L NAA, 1.5 mg/L 2,4-D , 0.25 mg/L kinetin 20 g/L sucrose, and 4 g/L Gelrite. Each treatment consists of 25 explants per dish with three replicates.
5. After 4 wk of culture in the dark at 25°C, determine the viability of cryopreserved cells with 2, 3, 5-triphenyl tetrazolium chloride (TTC) reduction assay *(16)*. Transfer cells to Petri dishes, add 0.18 M TTC solution, incubate at 25°C in the dark for 16 h, and count the number of red cell aggregates. Examine re-growth rate of cryopreserved cells after 4 wk of culture in the dark at 25°C. The result is shown in **Table 1**.

Table 1
Effect of Cryoprotectants on Regrowth of Cryopreserved *R. kazuensis* Cells after 4 wk of Culture on B5 Medium

Treatment	Frequency of viability (%)	Treatment	Frequency of viability (%)
No freeze control	100	Slow freeze + no cryoprotectants	0
No freeze + 20% glycerol	100	No freeze + 20% DMSO	100
Slow freeze + 5% glycerol	1.1 ± 1.9	Slow freeze + 5% DMSO	20 ± 3.3
Slow freeze + 10% glycerol	2.2 ± 1.9	Slow freeze + 10% DMSO	10 ± 3.3
Slow freeze + 15% glycerol	8.9 ± 1.9	Slow freeze + 15% DMSO	23.9 ± 3.5
Slow freeze + 20% glycerol	11.1 ± 1.9	Slow freeze + 20% DMSO	27.8 ± 5.1

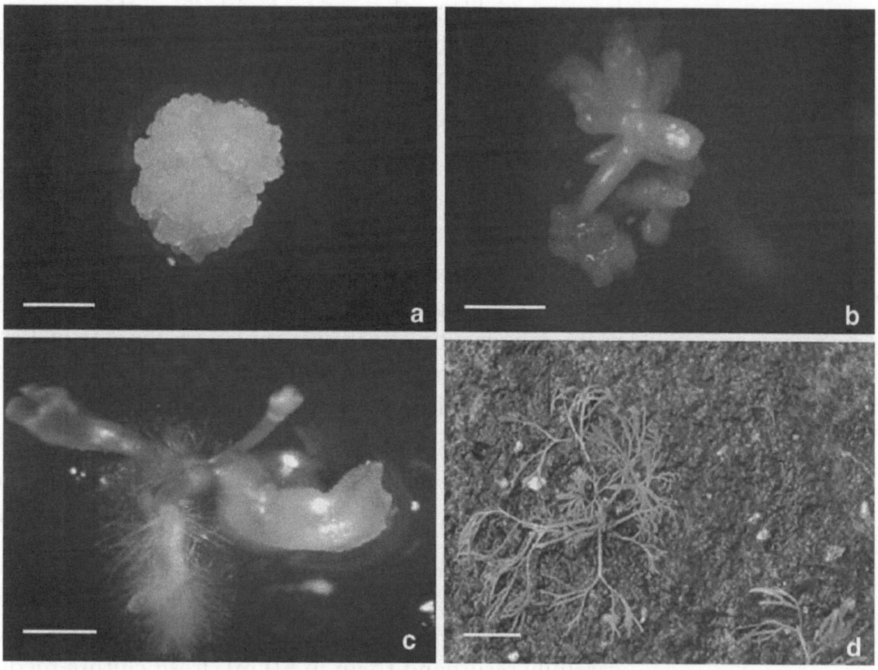

Fig. 3. Plant regeneration of cryopreserved *R. kazuensis* cells after 4 weeks of culture on V medium. (**a**) Cryopreserved cells with 15% DMSO. (**b**) Plant regeneration from cryopreserved callus. (**c**) Rooting of plantlets. (**d**) Successful soil transfer of regenerated plants. Scale bars represent 1 mm (**a–c**), 2 cm (**d**).

6. Subculture re-grown cells at 4-wk interval and transfer onto the half-strength SH basal medium. After 4 wk of culture in the light 30 $\mu mol/m^2/s^1$ from cool-white fluorescent lamps with a 16-h photoperiod, numerous plantlets are regenerated. Results are shown in **Fig. 3**.

4. Notes

1. Unless mentioned otherwise, all solutions should be prepared in water that has a resistivity of 18.2 MΩ-cm. This standard is referred to as "distilled water" in this text.
2. To avoid bacterial and fungal contamination, which is detrimental to plant tissue culture, explants are surface-sterilized before they are used to establish in vitro axenic cultures. The bleach solution, most common disinfectant, is strong oxidants to destroy any microbial contamination. Thus the use with high concentration and over-exposure may injure plant tissues.
3. DMSO and glycerol are best stored at room temperature in a desiccator. Buy small bottles as it may decline in quality after opening. Although not directly toxic, DMSO is a very powerful solvent and is able to rapidly penetrate intact skin (leaving a fishy or garlicky taste in your mouth). As a result, there is a potential hazard associated with using this compound. It is very important to avoid contact with DMSO and dispose of any wastes containing DMSO properly.
4. 0.18 M TTC solution is best stored at –20°C. Do not re-use after opening.
5. Avoid any damage to the zygotic embryos while dissecting longitudinally (no. 11).
6. The culture media for zygotic embryo should be dispensed following enough cooling down in order to prevent incomplete dehydration. Excessive moisture content in Petri dish inhibits embryo growth and callus formation and may also contaminate cultures.
7. The frequency of embryogenic callus formation is determined by counting the explants producing white nodular structures and pale-yellow calluses by observation with microscopy or naked eye.
8. Observe cell divisions under the inverted microscope.
9. For subculture, pipet 1 mL small cell aggregates, and then carefully resuspend them in 50 mL fresh culture medium.
10. Regenerated plantlets are successfully rooted without auxin treatment.
11. All processes are performed under the aseptic condition. Record the appropriate information about the cells in your cell repository records. Fully detail in these records the culture's storage conditions, including all of the following information: culture identity, date frozen, freezing medium and freezing method used. Additional culture information—especially its origin, history, growth parameters, special

characteristics and applications—is also helpful and should be included whenever possible.

12. Store at room temperature when not in use. Replace isopropanol after every fifth use.

Acknowledgments

This work was supported by a grant from KRIBB Research Initiative Program and a grant (BDW0800723) from the Korea Science and Engineering Foundation. The authors are thankful to the Plant Diversity Research Center of Korea for providing plant materials.

References

1. Tamura, M. (1993). Ranunculaceae, in the families and genera of vascular plants, flowering plants dicotyledons (Kubitski, K., ed.), vol II. Springer, Berlin, pp. 563–583.
2. Qasem, J. R. (1996). Fungicidal activity of *Ranunculus asiaticus* and other weeds against *Fusarium oxysporum* f. sp. *lycopersici*. *Ann Appl Biol* **128**, 533–540.
3. Bhattacharyya, P. R., Nath, S. C., and Bordoloi, D. N. (1993). Insecticidal activity of *Ranunculus sceleratus* (L.) against *Drosophila melanogaster* and *Tribolium castaneum*. *Indian J Exp Biol.* **31**, 85–86.
4. Cao, B. J., Meng, Q. Y., and Ji, N. (1992). Analgesic and anti-inflammatory effects of *Ranunculus japonicus* extract. *Planta Med* **58**, 496–498.
5. Pugliesi, C., Rabaglio, M., Cecconi, F., and Baroncelli, S. (1992). Plant regeneration from tissue cultures of Persian buttercup (*Ranunculus asiaticus* L.). *Plant Cell Tissue Organ Cult* **28**, 125–128.
6. Beruto, M., and Debergh, P. (1992). Somatic embryogenesis in *Ranunculus asiaticus* L. hybr. thalamus cultivated in vitro. *Plant Cell Tissue Organ Cult* 29, 161–165.
7. Beruto, M., and Debergh, P. (2004). Micropropagation of *Ranunculus asiaticus*: a review and perspectives. *Plant Cell Tissue Organ Cult* **77**, 221–230.
8. Ko, J. A., Kim, Y. S., Kim, M. J., and Eun, J. S. (1994). Immature pollenderived plant regeneration in anther cultures of *Ranunculus japonicus* Thunb. *Korean J Plant Biotechnol* **21**, 293–298.
9. Larkin, P. J., and Scowcroft, W. R. (1981). Somaclonal variation: A novel source of variability from cell cultures for plant improvement. *Theor. Appl. Genet.* **60**, 197–214.
10. Benson, E. E., Lynch, P. T., and Jones, J. (1992). The detection of lipid peroxidation products in cryoprotected and frozen rice cells: consequences for post thaw survival. *Plant Sci* **85**, 107–114.
11. Withers, L.A., and King, P.J. (1979). Proline: a novel cryoprotectant for the freeze preservation of cultured cells of *Zea mays* L. *Plant Physiol.* **64**, 675–678
12. Vendrame, W. A., Holliday, C. P., Montello, P. M., Smith, D. R., and Merkle, S. A. (2001). Cryopreservation of yellow-poplar (*Liriodendron tulipifera*) and sweetgum (*Liquidambar* spp.) embryogenic cultures. *New Forests* **21**, 283–292
13. Schenk, R. U., and Hildebrandt, A. (1972). Medium and techniques for induction and growth of monocotyledonous and dicotyledonous plant cell cultures. *Can J Bot* **50**, 199–204.
14. Gamborg, O. L., Miller, R. A., and Ojima, K. (1968). Nutrient requirements of suspension cultures of soybean root cells. *Exp Cell Res* **50**, 148–151.
15. Kartha, K. K., Leung, N. L., Gaudet-LaPrairie, P., and Constabel, F. (1982). Cryopreservation of periwinkle, *Catharanthus roseus* cells cultured in vitro. *Plant Cell Reports* **1**, 135–138.
16. Steponkus, P. L., and Lanphear, F. O. (1967). Refinement of the triphenyl tetrzolium chloride method of determining cold injury. *Plant Physiol.* **42**, 1423–1426.

Chapter 10

In Vitro Culture and Secondary Metabolite Isolation in Bryophytes

Aneta Sabovljevic, Marko Sabovljevic, and Nebojsa Jockovic

Summary

Bryophytes, the second largest group of land plants, are extremely rich in terpenoids, phenols, glycosides, and fatty acids. Although bryophytes could be used in medicine, their chemistry is not very well known and the problem remains to obtain enough quantity of same species for analysis. In vitro cultivation of bryophytes is the most appropriate way for large biomass production and isolate of numerous useful compounds showing some interesting biologic activities. This technique is also useful in developmental, cellular, molecular, biochemical, and eco-physiologic studies.

Key words: Bryophytes, In vitro culture, Mosses, Hepatics, Secondary products

1. Introduction

Bryophytes are the second largest group of land plants, after the flowering plants, with 15,000–25,000 species (*(1, 2)* and many intraspecific taxa *(3)*). They settled all known ecosystems from desert till arctic with exception of salt water. However, the biomass productivity of bryophytes varies from one to another ecosystem. In general, bryophytes consist of three very diverse groups: hornworts, liverworts, and mosses. Their size varies from 0.5 mm (e.g., tiny leafy liverworts) to 70 cm (e.g., moss Dawsonia superba).

The use of bryophytes as medicinal and aromatic plants is sporadic. Also, ethno-bryology or chemical constituents of bryophytes are not elaborated very well *(4, 5)*. Bryophytes are extremely rich in terpenoids, phenols (flavonoids and bibenzyl

S. Mohan Jain and Praveen K. Saxena (eds.), *Methods in Molecular Biology, Protocols for* In Vitro *Cultures and Secondary Metabolite Analysis of Aromatic and Medicinal Plants, vol. 547*
© Humana Press, a part of Springer Science+Business Media, LLC 2009
DOI: 10.1007/978-1-60327-287-2_10

derivatives), glycosides, and fatty acids, but also in some rare aromatic compounds. Bryophytes are considered as a "remarkable reservoir" of new, natural products or secondary compounds, many of which have shown interesting biologic activity. These activities can be presented as: antimicrobial, antifungal, cytotoxic, antitumor, vasopressin (VP) antagonist, cardiotonic, allergy causing, irritancy and tumor effecting, insect anti-feedant, insecticidal, molluscicidal, piscicidal, plant growth regulatory, superoxide anion radical release inhibition and 5-lipoxygenase, calmodulin, hyaluronidase, and cyclooxygenase inhibition features of bryophytes. Some latest results also predict beneficial influence of bryophytes in AIDS therapy (some bibenzyles of liverworts).

Considering small size of most bryophytes and mixed growth with other organisms, axenic cultures are a need for developing production of sufficiently biomass of taxonomic clean entities to use in further biological and pharmaceutical studies *(6)*.

Although culturing plant tissues and organs under axenic conditions was first established in bryophytes, these small plants have not received favorable attention as research experimental material *(7)*. Therefore, most studies of plant morphogenesis are now being done on vascular plants. Apart from economic consideration of experimental work with bryophytes, many physiologic, biochemical, molecular biologic and ecologic, as well as other problems could be studied more easily in bryophytes rather than in vascular plants *(8–14)*.

2. Materials

2.1. Sterilization of Gametophytes and Sporophytes

1. Gametophytes or sporophytes of selected bryophyte species (Atrichum undulatum (Hedw.) P. Beauv., Brachytecium velutinum (Hedw.) Schimp., Bryum argenteum Hedw., Campylopus oesterdianus (C. Müll.) Mitt., Dicranum scoparium Hedw., Hypnum cupressiforme Hedw., Lunularia cruciata (L.) Dumort., Marchantia polymorpha L., Oreas martiana (Hoppe et Hornsch.) Brid., Sphagnum palustre L.) field- or greenhouse-grown plants (18 ± 2°C, 10/14 h light to darkness).
2. 15 mm × 90 mm Petri dishes.
3. 50- to 250-mL Erlenmeyer flasks.
4. Autoclaved distilled water.
5. Sterile forceps.
6. Sodium hypochlorite solution (8% active chlorine).

2.2. Spore Germination

1. 15 mm × 90 mm Petri dishes).
2. Murashige and Skoog (MS) *(15)* medium stocks (MS stock I, II, and III) prepare as described in **Table 1** and store in the refrigerator at 4°C (*see* **Note 1**).
3. Long (16 h day/8 h light) or short (8 h day/16 h night) day conditions (light intensity 47 µmol/m^2/s).

Table 1
Composition of the Murashige and Skoog (MS)[a] and BCD[b] medium

Chemical constituents	Concentration MS (mg/l)	BCD (mg/l)
Minor inorganic nutrients		
NH_4NO_3	33,000	–
KNO_3	38,000	1,010
$CaCl_2{\cdot}2H_20$	8800	147
$MgSO_4{\cdot}7H_2O$	7400	250
KH_2PO_4	3400	250
$CaCl_2{\cdot}6H_2O$[b]	–	219
KI	166	28
H_3BO_3	1240	614
$MnSO_4{\cdot}4H_2O$	4460	–
$MnCl_2{\cdot}4H_2O$	–	389
$ZnSO_4{\cdot}7H_2O$	1720	55
$Na_2MoO_4{\cdot}2H_2O$	50	25
$CuSO_4{\cdot}5H_2O$	5	55
$CoCl_2{\cdot}6H_2O$	5	55
$AlK(SO_4)_2{\cdot}12H_2O$	–	110
KBr	–	28
LiCl	–	28
$SnCl_2{\cdot}2H_2O$	–	28
$NiCl_2{\cdot}6H_2O$	–	59
Iron source		
$FeSO_4{\cdot}7H_2O$	5560	12.5
$Na_2EDTA_2H_2O$	7460	–

(continued)

Table 1
(continued)

Chemical constituents	Concentration MS (mg/l)	BCD (mg/l)
Organic supplement		
myo-Inositol	20,000	–
Nicotinic acid	100	1
Pyridoxine-HCl	100	–
Thiamine HCl	100	0.5
Glycine	400	–
p-aminobenzoic acid	–	247
di-ammonium (+) tartarate	–	920
Carbon source		
Glucose	–	5,000
Sucrose	As per the experiment	5,000

[a]After dissolving all the stock solutions in enough deionized water make up to 1 L and adjust the pH to 6.5 with 4 M KOH (before addition of agar). Add 8 g/L agar for solid medium and autoclave at 114°C for 25 min

[b]Either way you will need to add $CaCl_2$ to 1 mM immediately before pouring plates. $CaCl_2$ may be kept as a 1 M stock solution (sterilized by autoclaving)

2.3. In Vitro *Establishing of Primary and Secondary Protonema; Gametophore Induction; Bryophyte Multiplication* In Vitro*; Growth Regulation of Bryophytes* In Vitro

1. Murashige and Skoog (MS) *(15)* medium stocks (MS stock I, II, and III) prepare as described in **Table 1** and store at 4°C in the refrigerator (*see* **Note 1**).
2. BCD medium stocks (B, C, D stock, $CaCl_2$ solution) prepare according to the **Table 1** and store at 4°C *(16–19)* (*see* **Note 1**).
3. Indole butyric acid (IBA) (Duchefa Biochemie BV; Haarlem, The Netherlands). Prepare 100 μM stock solution and store in the freezer at –20°C (*see* **Note 2**).
4. 6-Benzylaminopurine (BAP) (Duchefa Biochemie BV; Haarlem, The Netherlands).
5. Prepare stock solution (100 μM) and store in the freezer at –20°C (*see* **Note 2**).

2.4. Secondary Metabolite Isolation from Bryophytes (HPLC Analysis of Flavonoids)

1. 5 g (dry weight) of selected bryophyte species powder.
2. High-performance liquid chromatorgraphy (HPLC) grade methanol (*see* **Note 3**).
3. HPLC grade water (*see* **Note 3**).
4. HPLC analytical unit (Aparatus) (Gilson), with binary pump (Agilent; cat. no. G1312A) and diode-array detector (Agilent; cat. no. 1315B).

5. 25 × 0.46 cm, 5 μm particle size column (Spherisorb ODS2 column with a C18 ODS guard column (Waters; Milford, MA).
6. Mobile phases: Solvent A: water-formic acid (19:1). Solvent B: methanol.

3. Methods

3.1. In Vitro *Bryophyte Culture Establishment*

1. Collect bryophyte plants with or without sporophytes grown in nature or under controlled, greenhouse conditions, 18 ± 2°C, 10/14 h light to darkness.
2. Collect plants with healthy, immature sporophytes that contain spores, wash them thoroughly in running tap water for 5 min.
3. Surface sterilize unopened sporophytes in 10–19% sodium hypochlorite solution and 8% active chlorine for 5–10 min. Time can vary depending on the species (*see* **Note 4**).
4. Rinse them 3 times in cold, sterile distilled water for 5 min.
5. Open the sporophyte capsules aseptically (under the laminar flow bench) and transfer spores with a sterile needle to Petri dishes containing 20 mL solid nutrient MS medium with half strength MS micro and macro salts, without sugar, and phytohormones, pH 6.0.
6. Incubate cultures at 25 ± 2°C for 1–3 wk in the long day conditions (16/8 h light to darkness), and 47 μmol/m^2/s photoperiod. Within 1–3 wk spores start germination. Protonema and caulonema should develop on the same medium within 1–4 wk.
7. When plants are collected without sporophytes, take apical shoots of plants and remove them from soil and/or detritus wash them thoroughly in running tap water for 30 min.
8. Surface sterilize small gametophyte apical shoots in 7–10% sodium hypochlorite solution containing a few drops of liquid soap for 5 min. Concentration of sodium hypochlorite solution can vary, depending of the species and its filoid, bryophyte leaf morphology and size. Hence, for apical shoot sterilization better use few different commercial bleach solutions (*see* **Note 4**).
9. Rinse sterilized apical shoots three times in cold, sterile distilled water.
10. Transfer 10–20 apical shoots using a sterile forceps to Petri dishes containing 20 mL MS medium containing half strength MS micro- and macro-salts, without sugar and phytohormones, pH 6.0.

11. Incubate cultures at 25 ± 2°C for 4–8 wk in the long day condition, 16/8 h day length 47 μmol/m²/s photoperiod. Protonema and caulonema develop on the same medium within 4 wk.

3.2. In Vitro Protonema Culture and Bud Induction

1. Primary and secondary protonema and caulonema develop on the same medium between 1 and 4 wk after spore germination on half-strength MS medium.
2. Transfer protonema/caulonema on MS medium containing 7.5–15 g/l sucrose.
3. In many bryophyte species, bud induction occurs spontaneously on protonema/caulonema culture.
4. In case when bud induction fails during 4–8 wk period, add IBA: 0.1–1 μM and BAP: 0.03–0.1 μM in the MS media (*see* **Note 5**).
5. Gametophyte shoots will grow spontaneously after bud induction. When shoots reach 0.5- to –1-cm size, transfer to 50% MS medium, or 50% MS medium supplemented with 7.5–15 g/l sucrose (**Figs. 1** and **2**).

3.3. In Vitro Multiplication of Bryophytes

1. Collect actively growing bryophyte shoots from the solid cultures, cut them into 1- to 2-cm long segments and inoculate 10–20 of them into Petri dishes containing 20 mL 50% MS solid medium supplemented with 7.5–15 g/l sucrose.
2. Cultures should grow on solid medium in Petri dishes at 25 ± 2°C in the long day condition, 16/8 h day length and 47 μmol/m²/s photoperiod (*see* **Note 6**).
3. Maintain cultures by subculturing at 4-wk intervals (*see* **Note 7**) (**Figs. 1** and **2**).

3.4. Liquid Culture of Gametophytes

1. Collect healthy, gametophyte shoots from 4- to 8-wk old actively growing cultures and initiate suspension cultures for the production of flavonoids (*see* **Note 8**).
2. Inoculate 20 shoots, 1- to 2-cm long, in 100- to 250-mL Erlenmeyer flasks containing 50–100 mL 50% MS medium supplemented with 7.5 g/l sucrose.
3. Incubate cultures on a shaker at 50–100 rpm in the long day conditions, at 25°C.
4. After 6–8 wk, assess the growth of gametophyte shoots in terms of fresh weight, dry weight, growth index and content of flavonoids. Fresh weight of shoots should be measured directly after the end of experiment. Measure dry weight of shoots after drying shoots at room temperature for 3–5 d, until the moisture content in shoots reaches approximately 10%. Record shoot growth index of shoot multiplication that represents the number of new grown shoots starting from one shoot explant.

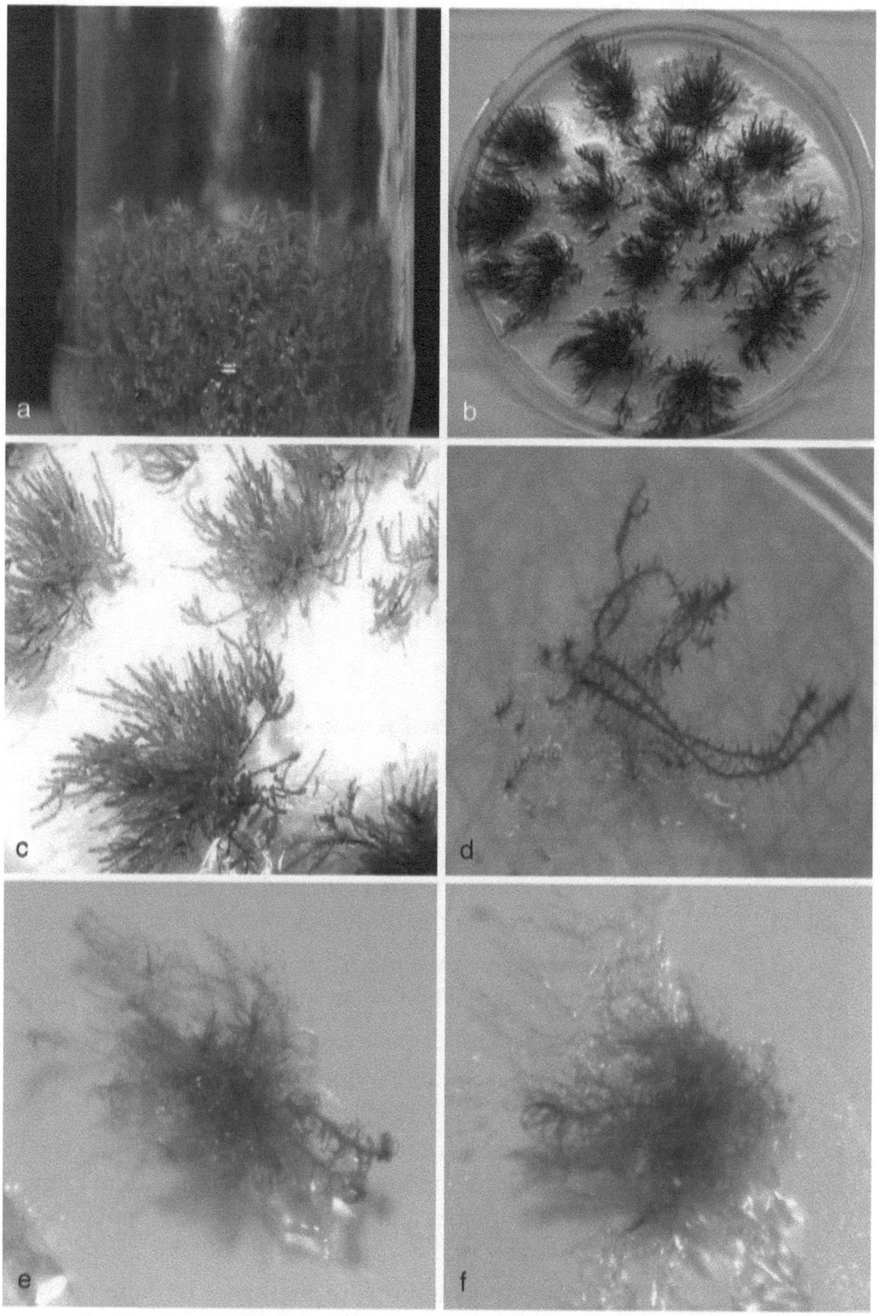

Fig. 1. In vitro culture of various bryophytes: (**a**) Atrichum undulatum; (**b, c**) Bryum argenteum; (**d**) Campylopus oesterdianus; (**e, f**) Dicranum scoparium.

3.5. Preparation of Bryophyte Extract for Estimation of Flavonoids

1. Use 5.0 g dry weight of selected bryophyte species powder for flavonoid extraction. Dry bryophyte plant material (gametophyte shoots) at room temperature and grind to make powder (*see* **Note 9**).
2. Extract the powdered material with methanol (175 and 125 ml) in two steps stirring with a magnetic stirrer. After extraction,

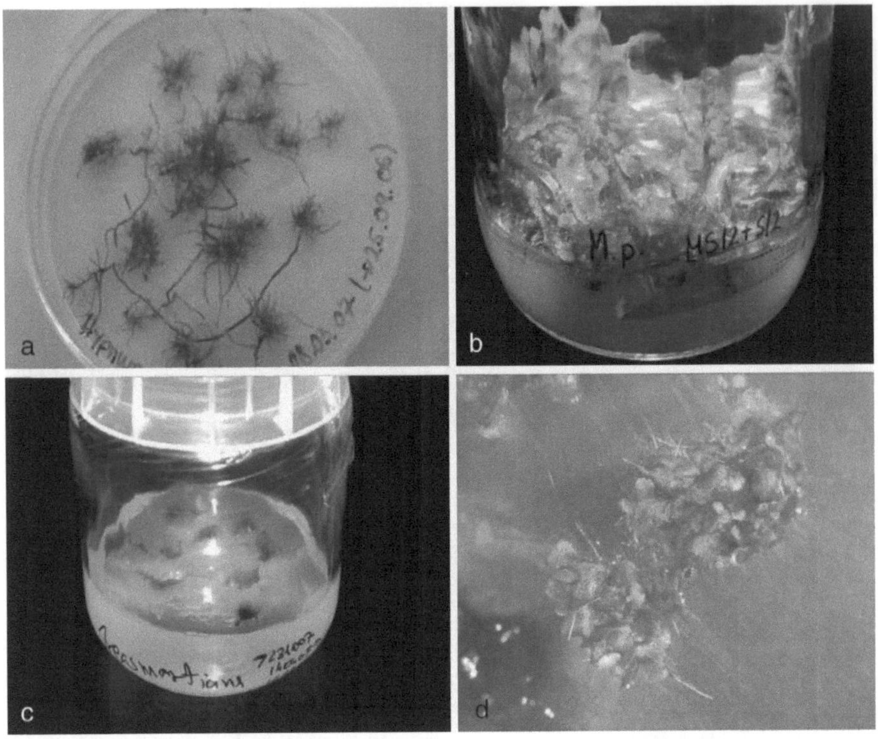

Fig. 2. In vitro culture of various bryophytes: (**a**) Hypnum cupressiforme; (**b**) Marchantia polymorpha; (**c**) Oreas martiana; (**d**) talous protonema of Sphagnum palustre

reduce volume of two methanol extracts under low pressure until it becomes thick with viscous consistency.

3. Dissolve the viscous methanol extract residue with 20 mL 2N chloridric acid. The flavonoids transfer to its ion phase.
4. Purify methanol extracts over a solid phase extraction column. Conditions: C18 Bond Elut cartridge with methanol and 2 N chloridric acid before applying the methanol extract.
5. Load methanol solution onto the column under the vacuum. The phenolic compounds and chlorophylls bind to the stationary column phase, and discard the filtrate containing residual compounds.
6. Wash out the column with methanol to elute phenolic compounds while chlorophylls remain on column. Elute contains the purified extract.
7. Separate purified methanol extracts by HPLC on reverse phase (RP) column. The peaks are detected using ultraviolet (UV) spectra. Confirm the presence of various compounds, such as phenolcarbonic acids or flavoinoids by comparison with retention time (Rt) and UV spectra of standard substances.

3.6. Conditions for HPLC Analysis

1. Perform chromatography on the Gilson analytical unit, consisting of vacuum degasser, binary pump, auto sampler,

temperature-regulated column compartment and diode array detector (DAD).

2. Perform analyses with Spherisorb ODS2 column, 25.0 × 0.46 cm; 5 μm particle size. The mobile phase is (a) aqueous formic acid solution: (b) methanol (19:1). Separate the flavonoid compound fractions by gradient elution as follows: 5% B for 3 min; 15% B for 10 min; 25% B for 12 min; 30% B for 10 min; 35% B for 4 min; 45% B for 5 min; 50% B for 3 min; 55% B for 3 min; 70% B for 6 min; 75% B for 4 min; 100% B for 2 min; at a flow rate of 0.9 ml/min. The detector monitored the flow at 280 nm, 320 nm and 350 nm. Set the sample injection at 20 μl. Perform three injections for each sample.
3. Identify peaks on the basis of their retention time values and UV spectra by comparison with those of standard solution (**Figs. 3** and **4**). Confirm peak identity by spiking the extracts with pure standards.

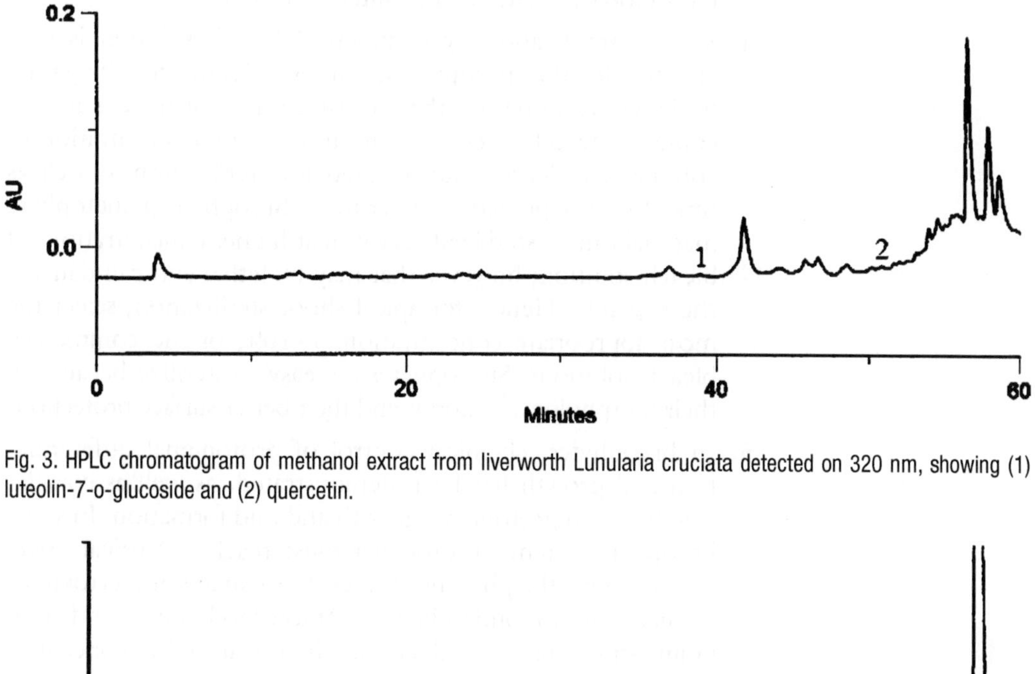

Fig. 3. HPLC chromatogram of methanol extract from liverworth Lunularia cruciata detected on 320 nm, showing (1) luteolin-7-o-glucoside and (2) quercetin.

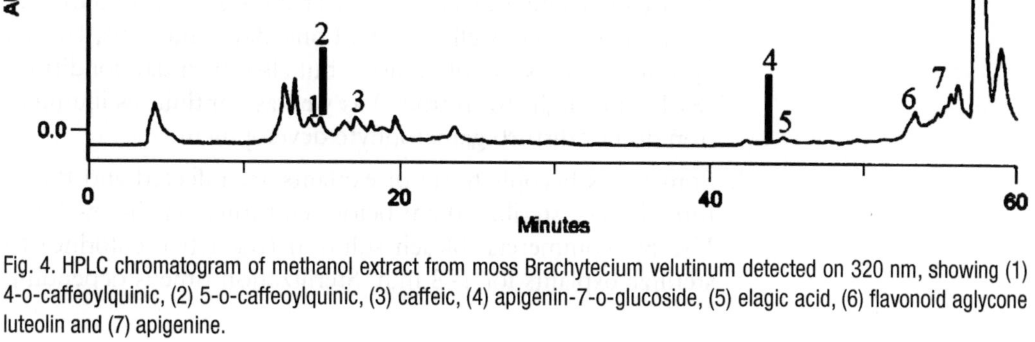

Fig. 4. HPLC chromatogram of methanol extract from moss Brachytecium velutinum detected on 320 nm, showing (1) 4-o-caffeoylquinic, (2) 5-o-caffeoylquinic, (3) caffeic, (4) apigenin-7-o-glucoside, (5) elagic acid, (6) flavonoid aglycone luteolin and (7) apigenine.

4. Notes

1. MS and BCD media preparation: Make stock solutions of major inorganic nutrients, minor inorganic elements, iron source, vitamins, and individual plant growth regulators (**Table 1**). For BCD medium, prepare 1 M $CaCl_2$ stock solution, sterilized by autoclaving. Stock solutions should be kept in the refrigerator at 4°C. Check them frequently and discard if precipitation occurs. For growth regulators IBA and BAP, make small stocks, concentration 100 μM and keep them in the refrigerator for short time, or at –20°C.
2. Plant growth regulator stock solutions (100 μM) should be prepared and kept at –20°C.
3. All solvents should be of HPLC grade; solvents and solutions to be analyzed should be filtered through 0.45 μM polytetrafluoroethylene (PTFE) filters before use.
4. Surface sterilization by commercial bleach solution is very effective for the sporophytes, but less effective for the gametophytes. Considering the fact that leaves of mosses mostly contain one cell layer without cuticles, the concentration of commercial bleach solution used for sterilization as well as time for this procedure is critical. Bryophyte gametophyte material can be sterilized very well at higher concentrations of bleach solution; however that may be lethal and can damage the explants. Hence, for apical shoot sterilization, select the most appropriate concentration, 7–10%, of the commercial bleach solutions. Sporophytes are easy to sterilize because of their morphology, anatomy, and their better surface protection.
5. In bryophytes, chemical control of protonemal differentiation and growth has been demonstrated, as well as correlation between protonemal growth and bud formation. In some bryophyte-systems, protonema must reach a "critical" size, and in others the phytohormones, both auxins and cytokinins are necessary for bud induction. At low levels auxins and cytokinins stimulate bud induction, whereas at higher concentrations they cause complete inhibition of bud formation.
6. In vitro bryophytes usually grow at 25 ± 2°C, although they can also grow very well at 20°C. Long-day conditions, 16/8 h light to darkness, are often used, but also short day conditions (8/16 h of light to darkness) as well as continuous illumination do not disturb gametophyte development.
7. Sometimes bryophyte shoot explants are infected and therefore, surface sterilize them before culturing on the medium. Use 5% commercial bleach solution (8% active chlorine) to sterilize explants for 1–3 min. Sterilization time is depending

on the species, plant material size, morphology and anatomy. Subsequently wash explants thoroughly with sterilized distilled water several times and culture on the nutrient medium.

8. Growth rate in bryophyte liquid cultures mainly depends on medium composition and physical factors like aeration, CO_2 content, light/dark conditions. For maintaining gametophyte liquid culture, use healthy and young 1- to 2-cm long gametophyte shoot explants. Do not subculture liquid cultures before 6-wk intervals in order to get optimal proliferation rate and biomass production.
9. Dry bryophyte shoots at room temperature, maximum at 40°C to a moisture content of about 10%. Higher temperatures than 40°C are not used that may result in the loss of flavonoids.

Acknowledgments

This work was supported by the Serbian Ministry of Science. Grants No: 143031 and 143015.

References

1. Gradstein, S.R., Churchill, S.P., and Salaar, A.N. (2001). Guide to the Bryophytes of Tropical America. Mem. N. Y. Bot. Gard. **86**, 1–577.
2. Crum, H. (2001). Structural Diversity of Bryophytes. University of Michigan Herbarium, Ann Arbor, MI, pp. 379.
3. Glime, J.M. (2007). Bryophyte Ecology. Vol. 1. Physiological Ecology. Ebook sponsored by Michigan Technological University and the International Association of Bryologists. Accessed on 2008.01.17 at http://www.bryoecol.mtu.edu/.
4. Sabovljevic, A. and Sabovljevic, M. (2008). Bryophytes, a source of bioactive and new compounds, in Recent Progress in Medicinal Plants Vol. 22: Phytopharmacology and Therapeutic values IV, (Govil, J.N., and Singh, V.K., eds.), Studium Press LLC, TX, pp. 9–25.
5. Sabovljevic, A., Sokovic, M., Sabovljevic, M. and Grubisic, D. (2006). Antimicrobial activity of Bryum argenteum. Fitoterapia **77**, 144–145.
6. Duckett, J.G., Burch, J., Fletcher, P.W., Matcham, H.W., Read, D.J., Russel, A.J. and Pressel, S. (2004). In vitro cultivation of bryophytes: a review of practicalities, problems, progress and promise. J. Bryol. **26**, 3–20.
7. Sabovljevic, M., Bijelovic, A., and Dragicevic, I. (2003). In vitro culture of mosses: Aloina aloides (K.F. Schultz) Kindb., Brachytecium velutinum (Hedw.) B.S. & G., Ceratodon purpureus (Hedw.) Brid., Eurinchium praelongum (Hedw.) B.S. & G. and Grimmia pulvinata (Hedw.) Sm. Turk. J. Bot. **27**, 441–446.
8. Bijelovic, A., Sabovljevic, M., Grubisic, D., and Konjevic, R. (2004). Phytohormone influence on the morphogenesis of two mosses (Bryum argenteum Hedw. and Atrichum undulatum (Hedw.) P.Beauv.). Isr. J. Plant Sci. **52**, 31–36.
9. Sabovljevic, A., Sabovljevic, M., Grubisic, D., and Konjevic, R. (2005). The effect of sugars on development of two moss species (Bryum argenteum and Atrichum undulatum) during in vitro culture. Belg. J. Bot. **138**, 79–84.
10. Bijelovic, A. and Sabovljevic, M. (2003). Callus induction and plant regeneration in the moss Aloina aloides (Shultz) Kindb. (Pottiaceae, Bryopsida). Arch. Biol. Sci. Belgrade **55**, 77–80.
11. Cvetic, T., Sabovljevic, M., Sabovljevic, A., and Grubisic, D. (2005). In vitro culture and apogamy-alternative pathway in the life cycle of the moss Amblystegium serpens (Amblystegiaceae). Arch. Biol. Sci. Belgrade **57**, 267–272.
12. Sabovljevic, A., Cvetic, T., and Sabovljevic, M. (2006). Establishment and development

of the Catherine`s moss Atrichum undulatum (Hedw.) P.Beauv. (Polytrichaceae) in in vitro conditions. Arch. Biol. Sci. Belgrade **58**, 87–93.

13. Cvetic, T., Sabovljevic, A., Sabovljevic, M., and Grubisic, D. (2007). Development of the moss Pogonatum urnigerum (Hedw.) P. Beauv. Under in vitro conditions. Arch. Biol. Sci. Belgrade **59**, 57–62.
14. Sabovljevic, M. and Sabovljevic, A. (2007). Contribution to the coastal bryophytes of the Northern Mediterranean: Are there halophytes among bryophytes? Phytologia Balcanica **13**, 131–135.
15. Murashige, T. and Skoog, F. (1962). A revised medium for rapid growth and bioassays with tobacco tissue cultures. Physiol. Plant. **15**, 473–497.
16. Ashton, N.W. and Cove, D.J. (1977). The isolation and preliminary characterisation of auxotrophic and analogue resistant mutants of the moss, Physcomitrella patens. Mol. Gen. Genet. **154**, 87–95.
17. Ashton, N.W., Grimsley, N.H., and Cove, D.J. (1979). Analysis of gametophytic development in the moss, Physcomitrella patens, using auxin and cytokinin resistant mutants. Planta **144**, 427–435.
18. Grimsley, N.H., Ashton, N.W., and Cove, D.J. (1977). The production of somatic hybrids by protoplast fusion in the moss, Physcomitrella patens. Mol. Gen. Genet. **154**, 97–100.
19. Knight, C.D., Cove, D.J., Cuming, A.C., and Quatrano, R.S. (2002). Moss gene technology, in Molecular Plant Biology Vol. 2 (Gilmartin, P.M., Bowler, C., eds.), Oxford University Press, Oxford, pp. 285–301.

Chapter 11

Micropropagation and In Vitro Conservation of Vanilla (*Vanilla planifolia* Andrews)

Minoo Divakaran and K. Nirmal Babu

Summary

Vanilla (*Vanilla planifolia* Andrews (*syn. V. fragrans* Salisb.), a source of natural vanillin, plays a major positive role in the economy of several countries. A native to the Central America, its primary gene pool is threatened by deforestation and over collection that has resulted in disappearance of natural habitats and wild species. Therefore, multiplication and conservation of vanilla diversity is of paramount importance because of its narrow genetic base. It plays an important role in the production of disease free planting material for commercial cultivation. Simple protocols for micropropagation, in vitro conservation and synthetic seed production are described in this chapter which could further be applied to other related vanilla species as well.

Key words: Embryo rescue, In vitro conservation, Micropropagation, Plant regeneration, Seed culture, Somaclones, Synthetic seeds, Vanilla

1. Introduction

Vanilla planifolia Andrews (*syn. V. fragrans* Salisb.) is an important crop in the flavor industry and a commercially important orchid spice (**Fig. 1**). Vanilla is the fermented and cured fruit of the orchid *V. planifolia* and used extensively to flavor ice creams, chocolates, beverages, cakes, custard, puddings, and other confectionery. It is also used in perfumery and to some extent in medicine as a nerve stimulant. The fragrance and flavor of vanilla beans result primarily from vanillin. The highly purified vanillin is

S. Mohan Jain and Praveen K. Saxena (eds.), *Methods in Molecular Biology, Protocols for* In Vitro *Cultures and Secondary Metabolite Analysis of Aromatic and Medicinal Plants, vol. 547*
© Humana Press, a part of Springer Science+Business Media, LLC 2009
DOI: 10.1007/978-1-60327-287-2_11

Fig. 1. Vanilla vine with pods.

widely used as a chemical intermediate in the synthesis of numerous pharmaceutical products *(1)*.

Although cultivated throughout the tropics, its natural populations in South Mexico – the most critical sources of novel genetic diversity are on the verge of disappearance *(2)*. The rapid destruction of vanilla natural habitats has compelled to search for alternate methods of conservation, multiplication, and genetic variability. In addition, a great threat of losing vanilla plantations is looming due to common diseases are of fungal origin *viz.*, foot rot and wilting caused by *Phytophthora meadii*, *Fusarium oxysporum*, *Calospora vanillae*, *Sclerotium* rot, and lack of genetic diversity in the available gene pool *(3)*.

Micropropagation will certainly be a great help in large-scale production of disease-free planting materials of elite *V. planifolia* genotypes *(4)*. The need for conserving different species before they extinct has also been emphasized *(2)* and thus in vitro conservation would be an ideal approach to augment the *ex situ* conventional methods. Cryopreservation of pollen will help to overcome the availability of viable pollen due to asynchronous flowering for interspecific hybridization, while conserving the haploid gene pool.

This protocol describes an efficient plant regeneration system, a prerequisite to exploiting somaclonal variation, in vitro selection, development of novel genotypes through transgenic pathway etc., which was extended to micropropagate diverse vanilla species overcoming the species/clonal specific nature of various protocols.

2. Materials

2.1. Micropropgation

2.1.1. Plant Materials

Excise explants viz., nodal segments and shoot tips from field-grown vines of *Vanilla planifolia*, *V. andamanica*, *V. tahitensis*, *V. aphylla*, and *V. pilifera*.

2.2. Glasswares

1. 500-mL Borosilicate conical flasks.
2. 2.250-mL Glass bottles.
3. 22 cm × 3.5 cm Culture tubes.
4. 250-mL conical flasks

2.3. Plant Growth Regulators

1. Auxins-∝-naphthalene acetic acid (NAA).
2. 0–3.0 mg/L Indole-3-butyric acid (IBA).
3. Cytokinins: – 6-Benzylaminopurine (BA) and 0–3.0 mg/L 6-fufurylamino purine (kinetin) (Sigma; St. louis, MO)

2.4. Gelling Agents

1. 7.0 g/L Bacteriological grade agar (Qualigens; India.

2.5. Culture Medium

1. Murashige and Skoog (MS) medium *(5)* basal medium (**Table 1**).
2. Stocks solutions for macronutrients, micronutrients, vitamins, amino acids, and plant growth regulators (**Table 1**).
3. Sucrose 20–30 g/L.
4. Adjust pH to 5.8 before adding agar, melt agar in microwave oven before adding to the medium, and autoclave the media at 121° C at 16 psi for 20 min.

2.6. Incubation Conditions

1. Incubate cultures at 22 ± 2°C for a 14-h photoperiod with 35 μmol/m^2/s of light provided by cool white fluorescent tubes.

2.7. In Vitro and Cryopreservation

2.7.1. In Vitro Conservation

Use in vitro regenerated shoot buds as propagules for conservation by slow growth induction.

2.7.2. Production of Synthetic Seeds

Use in vitro regenerated shoot buds, protocorms, and callus for encapsulation. Sodium alginate and Calcium chloride (Sigma) for preparation of synthetic seeds

2.7.3. Long-Term Storage by Cryopreservation

Synseeds containing miniaturized shoot tips, protocorms, pollen, and liquid nitrogen.

Table 1
Composition of Murashige and Skoog[a] Basal Medium

Composition		Concentration (mg/l)	Strength	Quantity[b]
A. Macronutrients				
Ammonium nitrate	NH_4NO_3	1650		
Potassium nitrate	KNO_3	1900		
[a]Calcium chloride	$CaCl_2{\cdot}2H_2O$	440		
Potassium orthophosphate	KH_2PO_4	170	20X	50 mL
Magnesium sulphate	$MgSO_4{\cdot}7H_2O$	370		
B. Micronutrients				
[a]Sodium EDTA	Na_2EDTA	37.30		
[a]Ferrous sulphate	$FeSO_4{\cdot}7H_2O$	27.80		
Boric acid	H_3BO_3	6.20		
Manganese sulphate	$MnSO_4{\cdot}4H_2O$	22.30		
Potassium iodide	KI	0.83		
Zinc sulphate	$ZnSO_4{\cdot}7H_2O$	8.60		
Sodium molybdate	$Na_2MoO_4{\cdot}2H_2O$	0.25		
[a]Copper sulphate	$CuSO_4{\cdot}5H_2O$	0.025	100X	10 mL
[a]Cobalt chloride	$CoCl_2{\cdot}6H_2O$	0.025		
C. Vitamins				
Myo-inositol	$C_6H_{12}0_6$	100.00		
Thiamine HCl	$C_{12}H_{17}ClN_4OS{\cdot}HCl$	0.10	100X	10 mL
Nicotinic acid	$C_6H_5NO_2$	0.50		
Pyridoxine HCl	$C_8H_{11}NO_3{\cdot}HCl$	0.50	100X	10 mL
D. Amino acid				
Glycine	$C_2H_5NO_2$	2.00		
E. Myo-inositol		100 mg	100X	10 mL
F. Growth regulators				
1. 2,4-D		50 mg/200 mL		As required
2. NAA		50 mg/200 mL		
3. BA		50 mg/200 mL		
4. Kinetin		50 mg/200 mL		

[a] Dissolve separately before mixing to the final stock
[b] For preparation of 1 L medium

3. Methods

3.1. In Vitro *Seed Germination and Embryo Rescue (3,6,7)*

1. Collect 4- to 7-mo-old mature pods. Sterilize the surface under the laminar air flow by dipping in 95% ethanol and flaming it.
2. Split open capsules to remove seeds and transfer them to sterilized culture medium.
3. Both solid as well as liquid media can be used for seed germination and culture; MS solidified medium supplemented with 0.5 mg/L kinetin, 20 g/L sucrose; and. continuously rotate liquid cultures on a rotary shaker at 200 rpm
4. Transfer germinating seeds to multiplication medium, MS medium amended with 1 mg/L BA and 0.5 mg/L IBA **(Fig. 2)**.

3.2. Micropropagation (3,8)

1. Collect shoot tips/nodal segments from healthy *V. planifolia*, *V. aphylla*, *V. pilifera*, and *V. andamanica* plants.
2. Wash and clean explants with 100 mL/L teepol () and surface sterilize with 1 g/L mercuric chloride for 5–7 min in the laminar flow chamber.
3. Follow with a thorough washing with sterilized, double distilled water (to remove any traces of mercuric chloride) and make a fresh basal cut before inoculation in MS medium containing 1 mg/L BA and 0.5 mg/L IBA.
4. Transfer cultures to fresh culture medium of same composition at 20-d intervals. Incubate cultures at 22 ± 2° C with a photoperiod of 14 h with a irradiance of 35 μmol/m^2/s provided by cool white fluorescent tubes.
5. After 5–7 mo, each explant culture may be multiplied successfully in 1:12 ratio **(Fig .3)**.

3.3. In Vitro *Rooting*

1. In vitro plants produce roots at every node of the cultured shoots.
2. Prolific root induction: Transfer 2.0 cm long shoots to the basal MS medium devoid of plant growth regulators. The rooting initiates at day 3 of shoot culture **(Fig. 4)**.

Fig. 2. Initiation of germination from minute black seeds, indirect and direct seed germination.

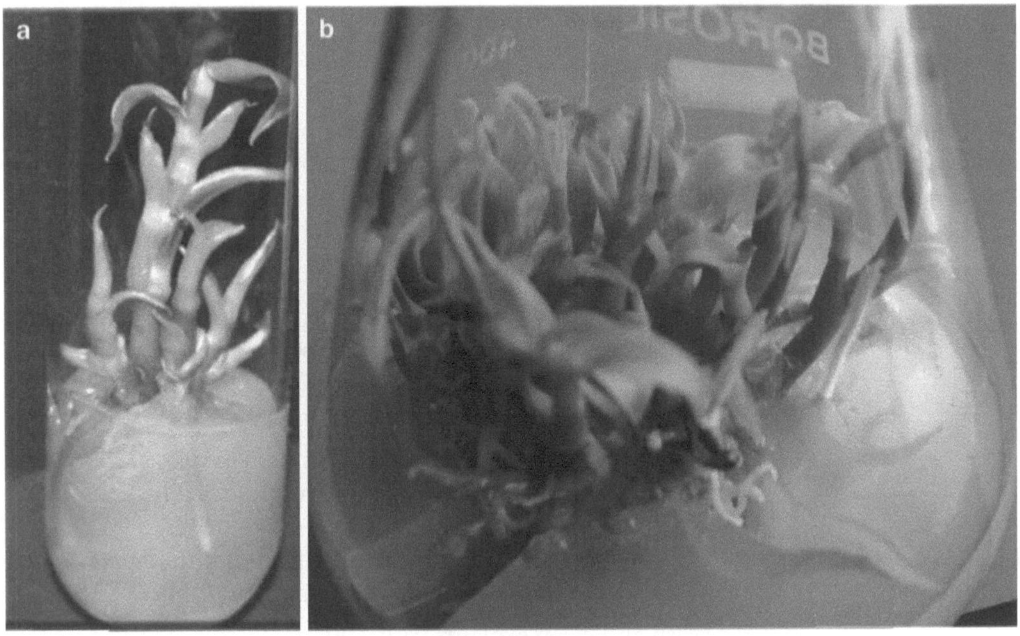

Fig. 3. Induction of multiple shoots and subsequent proliferation.

Fig. 4. Root induction from the base as well as from the nodes.

3. Harden each well rooted shoot, as mentioned below, and plant out to the field or for next cycle of multiplication.
4. For next cycle of shoot multiplication, re-culture 2-cm long shoots or single nodes by cutting shoots having 2–3 nodes.

3.4. Callus Induction

1. Culture explants, excised from in vitro shoots, roots, or protocorms for callus induction.
2. Transfer explants to MS medium supplemented with 1 mg/L BA and 0.5 mg/L NAA.
3. Initially keep cultures in the dark for 2–5 d for callus induction.
4. Transfer callus cultures to the fresh medium of same composition at 20-d intervals.

3.5. Plant Regeneration (3)

1 Transfer fresh callus to the MS medium containing 1 mg/L BA and 0.5 mg/L IBA. Subculture on fresh culture medium at 20-d intervals

2. Callus differentiates into protocorms and shoots with roots in approximatel 3–6 mo (**Fig. 5**).
3. Separate and transfer the well developed shoots with roots for hardening and planting out or for next cycle of multiplication.

3.6. Production of Synthetic Seeds (3)

1. Suspend propagules viz., in vitro developed protocorms or shoot buds (0.5–1.0 cm) in a matrix of MS medium containing 4% *(w/v)* sodium alginate.
2. Drop explants together with matrix, drop by drop in 1.036 g/150 mL $CaCl_2 \cdot 2H_2O$ solution with a Pasteur pipet with a widened hole for easy passage of propagules. Make sure that each bead contains one explant.
3. Leave explants in $CaCl_2 \cdot 2H_2O$ solution for 30–40 min. Shake at 100 rpm for proper bead formation.
4. The beads are recovered by decanting $CaCl_2$ solution. Wash twice in sterile water (**Fig. 6**).
5. The synthetic seeds or "synseeds" thus produced can be stored up to 10 mo in sterile water with 80% germination rate at 22 ± 2° C.
6. Transfer synthetic seeds for germination in the multiplication medium, MS medium amended with 1 mg/L BA and 0.5 mg/L IBA.

Fig. 5. Different stages in plant regeneration from callus.

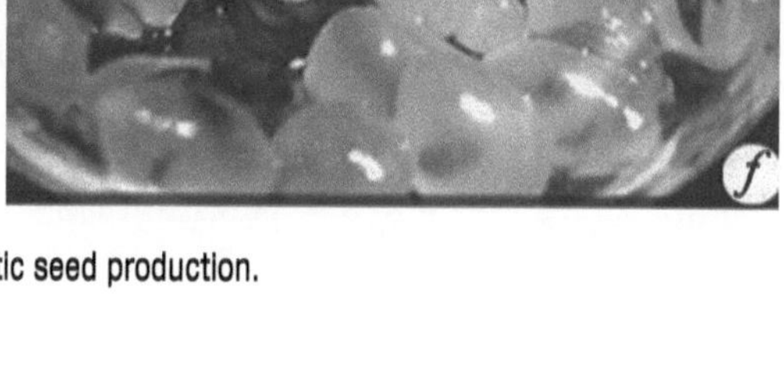

Fig. 6. Synthetic seed production.

3.7. Germplasm Conservation (7,9)

3.7.1. Medium Term In Vitro Conservation by Slow Growth

1 Slow growth can be induced in vanilla cultures by reducing the carbon source, reduction of basal media concentration, addition of mannitol, and minimizing evaporation loss to increase subculture intervals substantially.
2. Transfer in vitro-grown vanilla shoots, 2-cm long, to MS medium containing 15 g/L sucrose and mannitol, respectively without plant growth regulators.
3. Seal the culture vessels with poly propylene caps or parafilm to minimize evaporation and maintain them at 22 ± 2° C for a 12-h photoperiod with a reduced light intensity of 30 μmol/m²/s.
4. These shoot cultures with minimal growth can be maintained in vitro for 7–10 yr with yearly subculture on the fresh medium (**Fig. 7**).
5. For normal growth, transfer plantlets in MS medium containing 1.0 mg/L BA and 0.5 mg/L IBA at 22 ± 2°C under 35 μmol/m²/s light intensity.

3.7.2. Long-Term Storage of Encapsulated Shoot Tips and Pollen by Cryopreservation

Pretreatment and Dehydration

1. Carefully choose small, 4–5 mm diameter, uniform circular beads for cryopreservation.
2. Stepwise preculture beads on MS liquid medium amended with increasing sucrose concentrations of 0.1, 0.3, 0.5, 0.7, and 1.0 M for 5 d, with 1 d for each concentration.
3. Dehydrate the pretreated beads on a sterilized filter paper in a 90-mm Petri dish and air dry for 1–10 h in laminar hood at room temperature.
4. Determine optimum air dry time by bead germination tests.

Cryopreservation

1. Transfer ten dehydrated beads per 2.0-mL cryo vials.
2. Rapid freezing is done by plunging the cryo vials directly in the liquid nitrogen.
3. Samples can be stored in liquid nitrogen for 10 yrs.

Thawing, Post Culture and Regeneration of the Whole Plant

1. Thaw samples by immersing vials for 5–10 min directly in a water bath at 40°C.
2. Culture thawed beads in a 90-mm petri dish containing 25-mL solidified MS medium containing 30 g/L sucrose, 1 mg/L BAP, and 0.5 mg/L IBA in the darkness initially for 7 d and then move them under a light intensity, of 35 μmol/m²/s for stock shoots to grow. The recovery rate is 50–70%.
3. After 2 wk, shoot buds that emerge out from the synthetic seed coat are subcultured on the fresh MS basal medium for proper growth of the shoot and root system and finally plant them for hardening and field transfer.

Fig. 7. In vitro conservation by slow growth. The shorter shoots (L) indicate minimal growth with root system while the longer shoots (R) show growth in normal medium requiring frequent subcultures.

3.8. Hardening and Planting Out

1. In vitro plantlets with well developed roots are washed with care with running water in order to remove the traces of medium sticking to the roots. Care should be taken to prevent any damage to roots while cleaning roots.
2. Dip them in 0.3% Dithane-M45 for 5–10 min and transplant in poly bags containing a mixture of garden soil, sand and vermiculite in equal proportions (1:1:1).
3. Keep the transplanted plantlets in a humid chamber, relative humidity 70–80% and at a light of intensity 25–30 μmol/ m^2/s for 3–4 wk for hardening and establishment. The hardened plants can be kept in the nursery for one year and finally plant them in the field.

4. Notes

1. While flaming pods, the process can be repeated to avoid contamination and care should be taken to avoid any damage to seeds that may have an adverse affect on seed viability.
2. Contaminated liquid culture should be completely discarded and autoclave it before throwing away. Hence, for embryo rescue and in case of inter specific hybridization, solid medium helps in saving unaffected seeds.
3. Most vanilla stocks are infected with one or more viruses. Therefore, use virus diagnostic kits, enzyme-linked immunosorbent assay and reverse-transcriptase polymerase chain reaction to detect virus infected stock plants.
4. Make sure that the stock plant material is completely virus free
5. When full strength MS medium containing 30 g/L sucrose is used for in vitro conservation experiments, shoot cultures grow rapidly and within 180 d culture vessels are full of shoots; medium is exhausted; and shoot cultures dry up. For preventing this condition, modify culture medium by reducing sucrose concentration to 10–15 mg/L, half strength nutrients, and add 10–15 mg/L mannitol. This change would result in reduced shoot growth rate 2–3 cm in 360 d and highest survival rate at least 80%.

References

1. Purseglove, J.W., Brown, E.G., Green, C.L., Robbins, S.R.J. (1981). Spices Vol 2 pp 644–735, Tropical Agricultural Series. Longman Inc., New York.
2. Lubinsky, P. (2003). Conservation of wild vanilla. In Proc. Vanilla 2003. 1st International Congress, Nov. 11–12, Princeton, NJ, USA.
3. Minoo, D. (2002). Seedling and somaclonal variation and their characterization in Vanilla. Ph.D Thesis. Calicut University, Kerala, India.
4. Cervera, E. and Madrigal, R., (1981). *In vitro* propagation (*Vanilla planifolia* A.). *Environ. Exp. Biol.* **21**, 441.
5. Murashige, T. and Skoog, F. (1962). A revised medium for rapid growth and bioassays with tobacco tissue cultures. *Physiologia Plantarum* **15**, 473–497.
6. Minoo, D., Nirmal Babu, K., Ravindran, P.N., and Peter, K.V. (2006). Inter specific hybridization in vanilla and molecular characterization of hybrids and selfed progenies using RAPD and AFLP markers. *Scientia Horticulturae.* **108**, 414–422.
7. Minoo, D., Sajina, A., Nirmal Babu, K., Ravindran, P.N. (1997). Ovule culture of vanilla and its potential in crop improvement. In Edison S, Ramana KV, Sasikumar B, Nirmal Babu K and Santhosh Eapen J. (eds.). Biotechnology of Spices, Medicinal and Aromatic Plants, Indian Society for Spices, Calicut, India, pp **112**–118.
8. Minoo, D., Nirmal Babu, K., and Peter, K.V. (2006). Conservation of Vanilla species – *in vitro. Scientia Horticulturae* **110**, 175–180.
9. Ravindran, P.N., Nirmal Babu, K., Saji, K.V., Geetha, S.P., Praveen, K., and Yamuna, G. (2004). Conservation of Spices genetic resources in *in-vitro* gene banks. ICAR Project report. Indian Institute of Spices Research, Calicut, Kerala, India, pp. 81.

Chapter 12

Protocol for In Vitro Regeneration and Marker Glycoside Assessment in *Swertia chirata* Buch-Ham

Sushma Koul, K. A. Suri, P. Dutt, M. Sambyal, A. Ahuja, and M.K. Kaul

Summary

We have developed a rapid in vitro propagation system, via axillary shoot formation from nodal explants of *Swertia chirata* Buch Ham. Culture medium supplemented with 2 mg/L. BAP is best for direct shoot regeneration initially formed adventitious buds from axils of the nodal explants after 30 days. The reduced BAP concentration 0.5 mg/L proliferate shoots effectively. Kept the number of hyperhydrated shoots to minimal and induced on an average 22–38 shoots per flask (4.3 cm average length). The regenerated shoots (5- to 6-m long) formed roots very well in Murashige and Skoog (MS) medium devoid of any growth regulator and followed by acclimatization of plants in pre-sterilized sand containing 1% *Trichoderma viride* and *Azatobactor chrococcum* as bioinoculants. The regenerated plants don't show any genomic alterations. This protocol also outlines procedure of assessment of marker iridoid glycosides (amarogentin and amaroswerin) from callus, roots, multiple shoots, regenerated plants, and mother plant. High propagation frequency, reproducibility of procedure, molecular, and phenotypic and chemical stability ensures the efficiency of the developed protocol.

Key words: *Swertia chirata* Buch-Ham, Micropropagation, *Trichoderma viride*, *Azatobactor chrococcum*, Iridoid glycoside, Amarogentin, Amaroswerin

1. Introduction

Genus *Swertia* (Family Gentianaceae) is a rich source of Xanthonoids, flavanoids, glycosides, and terpenoids *(1)*. Plant herb is extensively used in Ayurvedic and folk medicine for treatment of hepatitis, cholecystitis, pneumonia, dysentery, scabies, neurasthenia, and as bitter tonic and febrifuge *(1)*. This species differs from other Indian *Swertia* species because of the presence of two main

S. Mohan Jain and Praveen K. Saxena (eds.), *Methods in Molecular Biology, Protocols for* In Vitro *Cultures and Secondary Metabolite Analysis of Aromatic and Medicinal Plants, vol. 547*
© Humana Press, a part of Springer Science+Business Media, LLC 2009
DOI: 10.1007/978-1-60327-287-2_12

pharmacologic active bitter iridoid glycosides – amarogentin and amaroswerin *(2)*. Plants grow wild at an altitude of 1200–1600 m in small pockets of Darjeeling, Kumaon, and Chotanagpur hills. *Swertia chirata* once abundantly growing herb exported from India is now being imported from the neighboring countries. It has a huge demand on both domestic and international markets as well as in the herbal industry, which is estimated to grow at 10% annually. *Swertia chirata* plant population is rapidly declining due to overexploitation, habitat destruction, destructive harvesting, and inadequate attempts for the replenishment of wild stock of the plant species. The International Union for Conservation of Nature and Natural resource (IUCN) has categorized *S. chirata* as a critically endangered species *(3)*. In the market *Andrographis paniculata* Nees., *Swertia alata* Royle, and *Swertia angustifolia* Buch-Ham are used as major adulterants to *S. chirata (4)*. Domestication of plants using conventional techniques has not yet been successful *(5)*. A reduced span of seed viability and low germination rate restrict its propagation through seeds *(6)*. Plant materials are procured over vast geographic distances with low-cost production, which have problems in quality assurance including incorrectly labeled material, improper storage conditions, biotic contamination, and pesticide residues. Therefore, mass propagation of the critically endangered plant species has been developed. This chapter describes fast high-througput regeneration strategy and in vitro shoot culture as an alternative source of iridoid glycosides using biotechnologic approaches.

2. Materials

2.1. Materials Required for Tissue Culture

1. *Swertia chirata* was collected from Darjeeling area of North-East Himalayas in June.
2. Murashig and Skoog (MS) *(7)* basal medium supplemented with different plant growth regulators (Merck Co.; Schurcharcdt, Germany and Sigma Chemicals Co.; St. Louis, MO).
3. Borosil glasswares and sterile disposable Petri dishes (Tarson India Ltd.).
4. Laminar Air Flow (Thermadyne).
5. Orbital Shaker (Orbitek-L) (Scigenics Biotech Ltd.; India).
6. 1-L Growtech (Tarson; India).

2.2. Materials for PCR Analysis

1. Refrigerated Centrifuge and Master Cycler Gradient (Eppendrof; Germany).
2. Random Decamer Oligonucleotides from Operon Inc. USA.
3. Taq DNA polymerase and 100-bp ladder (Banglore Genei Pvt. Ltd.; India).

4. Gel Doc system (Amesham; USA).
5. $MgCl_2$ and assay buffers (Biotools).

2.3. Materials Required for Chemical Analysis

1. Laboratory grade ethanol (E. Merck; Mumbai, India).
2. Biochemicals (E. Merck Co.; Schurcharcdt, Germany; Sigma; St. Louis, MO; Ranbaxy Laboratories Ltd.; India; Himedia; India; and Loba-Chemie; India).
3. High-performance liquid chromatography (HPLC) grade methanol (Rankem, Ranbaxy Chemicals Ltd.; India) and analytical grade acetic acid (E. Merck; Mumbai, India).
4. Prepare HPLC grade water with a Milli-Q water purification system.
5. JASCO DIP-360 polarimeter for optical rotations.
6. IR spectra on a JASCO IR-700 spectrophotometer.
7. Mass spectra on a JEOL AX-500 mass spectrometer.
8. Glycerol is used as the matrix for FABMS measurement.
9. JEOL FX-100 spectrometer for ^1H and ^{13}C NMR spectra.
10. Shimadzu HPLC system for quantification of iridoid glycosides.

3. Methods

3.1. Establishment of Cultures

3.1.1. Initiation of Aseptic Cultures

1. To establish aseptic cultures shoot portions are treated with Tween-20 for 2 min followed by thorough washing under running water.
2. Separate shoot apices and nodal segments, are surface sterilize with 0.1% mercuric chloride for 3 min. Washed three times with sterilized distilled water (*see* **Note 1**).
3. Culture explants on 0.8% agar solidified MS *(7)* nutrient medium containing 3% sucrose and different combinations of auxin and cytokinins (**Table 1**).
4. Add growth regulators in the medium and adjust pH to 5.8 with 1N NaOH or 1N HCL prior to autoclaving. Media are gelled with 0.8% agar and sterilize for 20 min at 121°C (15 psi pressure).
5. Incubate cultures at 22 ± 2°C for a 16 h-photoperiod with a light intensity of 40 μmol/m²/s.
6. For initiation of cultures, culture 25 explants per experiment and repeat four times.
7. After 8 wk of initial culture, calculate the frequency rate of explant response as percentage explants producing shoot buds per explants (**Table 1**).

Table 1
Influence of Plant Growth Regulators on Shoot Induction Capability in Nodal Explants of Swertia chirata Buch Ham

MS + PGR (mg/l)	Percentage response	Mean shoots per explant	Mean shoot length (cm)	Morphogenetic response
BA 1.0	72	3.3 ± 0.65	2.3 ± 0.27	Moderate shoot differentiation
BA 2.0	83	5.8 ± 0.54	3.5 ± 0.29	Prolific shoot differentiation
KN 1.0	2	0.57 ± 0.06	0.5 ± 0.03	Poorly developing shoot
KN 2.0	5	1.12 ± 0.12	0.4 ± 0.03	Moderate shoot differentiation
NAA 1.0	50	2.73 ± 0.18	1.7 ± 0.09	Moderate shoot differentiation
IAA 1.0	22	0.33 ± 0.03	2.1 ± 0.10	Poorly developing shoot
IBA 1.0	7	0.23 ± 0.03	0.3 ± 0.02	Moderate shoot differentiation
2,4-D 1.0	33	–	–	Pale nodular callus at cut end
BA1.0 + NAA 1.0	76	3.77 ± 0.46	4.2 ± 0.47	Moderate shoot differentiation
BA1.0 + IAA 1.0	31	0.67 ± 0.09	3.1 ± 0.39	Poorly developing shoot
BA1.0 + IBA 1.0	15	0.45 ± 0.03	1.7 ± 0.15	Poorly developing shoot
BA1.0 + 2,4-D 1.0	54	0.12 ± 0.009	0.7 ± 0.02	Green callus at cut end
KN 1.0NAA 1.0	12	0.43 ± 0.06	1.5 ± 0.08	Poorly developing shoot
KN 1.0IAA 1.0	9	0.31 ± 0.02	0.8 ± 0.04	Poorly developing shoot
KN 1.0IBA 1.0	–	–	–	–
KN 1.02,4-D 1.0	17	–	–	Brownish callus at cut end

Results are mean ± S.E of four experiments (each treatment in every experiment had 25 replicates)

8. Results are mean ± SE of four experiments. Each treatment in every experiment has 25 replicates.

3.1.2. Initiation of Different Morphotypes

1. Multiple axillary shoots initiated from nodal explants on solid MS medium with BAP 2.0 mg/L were subsequently maintained in same basal medium with BAP 0.5 mg/L served as basic stock cultures for further experimentation.
2. Axenic leaves from shoots growing on MS basal medium with BAP 0.5 mg/L were transferred to various combinations of medium for initiation of morphologically different cultures.
3. MS medium with 0.5 mg/L NAA induced de novo root formation whereas BAP 0.5 mg/L inducted shoots (**Fig. 1**) but 1.0 mg/L BAP in combination with 2,4-D 1.0 mg/L triggered profuse callus induction (**Fig. 2**). Same hormonal combinations were used to maintain the morphotypes.

Fig. 1. (**a–g**) Micropropagation procedure.

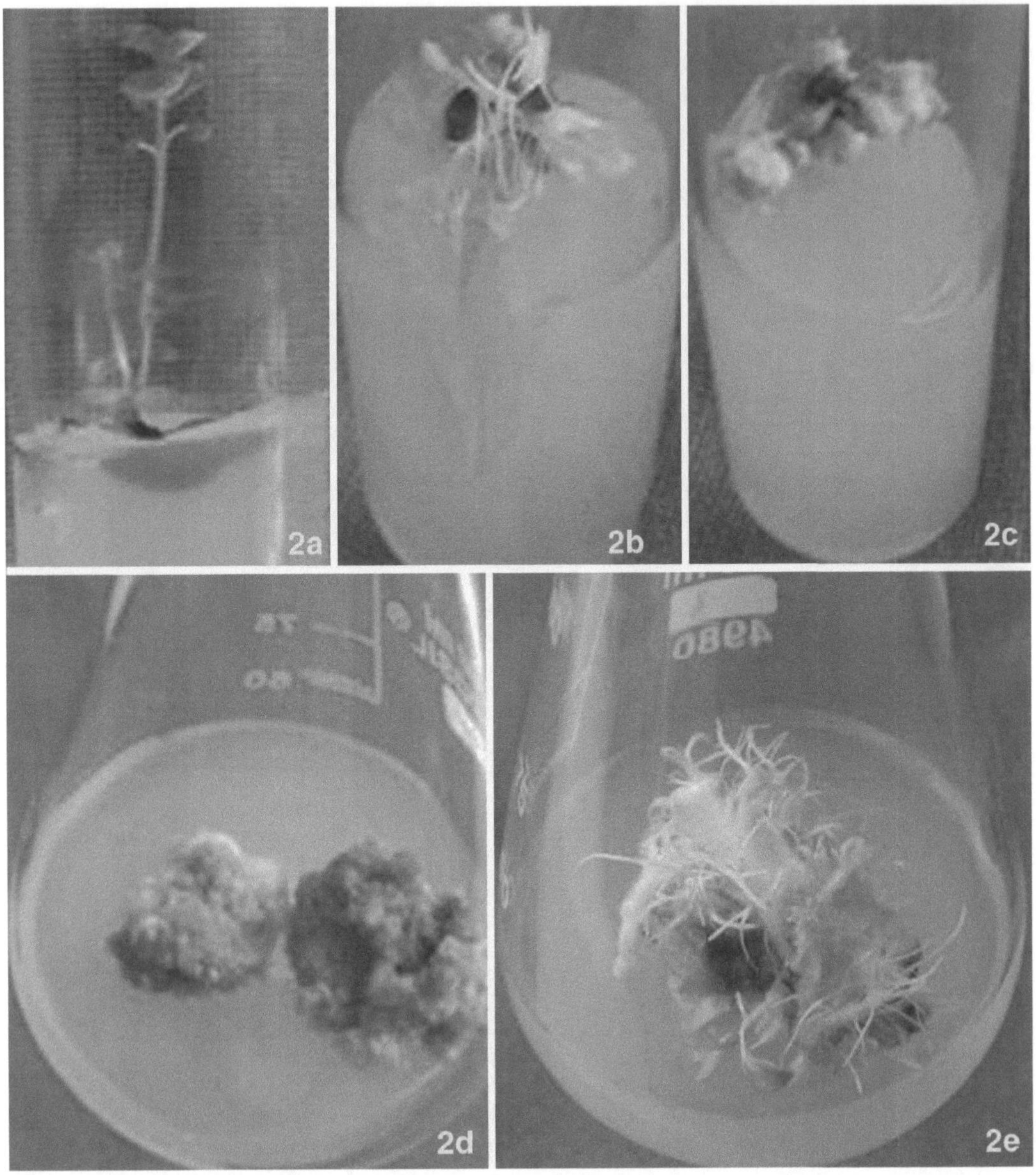

Fig. 2. (**a–e**) Development of different morphotypes.

Fig. 3. (**a–c**) Upscaling studies.

4. To find out suitable morphological form for upscaling, callus, roots, and shoot cultures (**Fig. 2** and **3**) were screened for growth and iridoid glycoside production. Cultures were harvested at 10 d intervals up to 60 d and shoots were found to produce higher amarogentin and amaroswerin.

3.2. Micropropagation

3.2.1. Establishment, Maintenance, and Root Formation

1. Transfer shoot cultures initiated on either 2.0 mg/L BAP or different concentrations of BAP and NAA fortified MS medium (**Table 2**) for further maintenance and multiple shoot induction (*see* **Note 2**).
2. Shoot buds differentiated on 2.0 mg/L BAP on transfer to BAP and NAA alone or in different combinations, exhibit differential response (**Table 2**) Multiplication rate is scored as number of shoots per culture vessel
3. Significance is calculated as per Duncan's multiple range test ($P \leq 0.05$) and same letter is marked on values which are not significantly different.

Table 2
Effect of BAP and NAA on Shoot Proliferation Capabilities in *Swertia chirata* Buch Ham

MS + PGR(mg/l) combination	Number of shoots per culture	Shoot length (cm)
BAP 0.5	38.58^{a}	4.36 ± 1.76
BAP 1.0	27.45^{a}	4.94 ± 1.54
BAP 2.0	21.36^{a}	3.36 ± 1.34
NAA 0.5	3.62^{c}	4.67 ± 1.29
NAA1.0	9.44^{bc}	5.22 ± 2.13
NAA2.0	6.38^{bc}	4.82 ± 1.45
BAP0.5 + NAA0.5	23.28^{a}	4.91 ± 2.10
BAP0.5 + NAA1.0	18.74^{ab}	4.23 ± 1.32
BAP0.5NAA2	13.14^{b}	4.31 ± 1.30
BAP1.0 NAA0.5	11.36^{b}	4.19 ± 1.11
BAP1.0 NAA1	9.83^{b}	4.27 ± 1.34
BAP1.0 NAA2	5.37^{c}	4.15 ± 1.38
BAP2.0 NAA0.5	7.21^{bc}	3.21 ± 0.87
BAP2.0 NAA1	5.23^{c}	3.27 ± 0.01
BAP2.0 NAA2	3.24^{c}	3.93 ± 1.06

Note: Values bearing same letter are not significantly different as per Duncan's multiple range test ($P \leq 0.05$)

Table 3
Effect of growth regulator concentration on rooting

MS + (mg/L)	Rooting (%)	Rooting Period (weeks)	Morphology
0	80	8	White thick well developed root system
IBA			
0.5	30	0	Whitish thin short roots
1.0	60	8	
2.0	62	8	
5.0	60	8	
NAA			
0.5	50	10	Brownish sparsely developed roots
1.0	50	10	
2.0	60	10	
5.0			

4. In 0.5 mg/L BAP prolific shoots without any intervening callus and minimal hyperhydrated shoots are formed.
5. Axillary shoot proliferation remains consistent for several passages with an average 7- to 9-fold increase in MS medium with 0.5 mg/L BAP. The same medium is further used for maintenance and proliferation of *Swertia chirata* shoot cultures.
6. For root initiation cut 4- to –6-cm long shoots and transfer to full strength MS medium amended with 3% sucrose and various levels of NAA, IBA, and hormone free medium (**Table 3**).
7. Root growth is prolific and thick in auxin free medium, whitish and short in IBA, and brownish sparsely developed in NAA (**Table 3**). However, best rooting 80–85% occurs in the medium devoid of plant growth regulators. Generally, —between 5 and 8 roots emerge from the basal end of the propagule. Medium devoid of any growth regulator is regularly used for rooting.

3.2.2. Hardening and Field Transplantation

1. Eight-week old rooted plantlets are removed from culture flasks, wash thoroughly with water to remove agar and transfer onto different substrates, soilrite, garden soil, vermiculite, peat moss and sand, in various proportions and combinations. Sand is the best substrate with 41% survival rate (*see* **Note 3**).

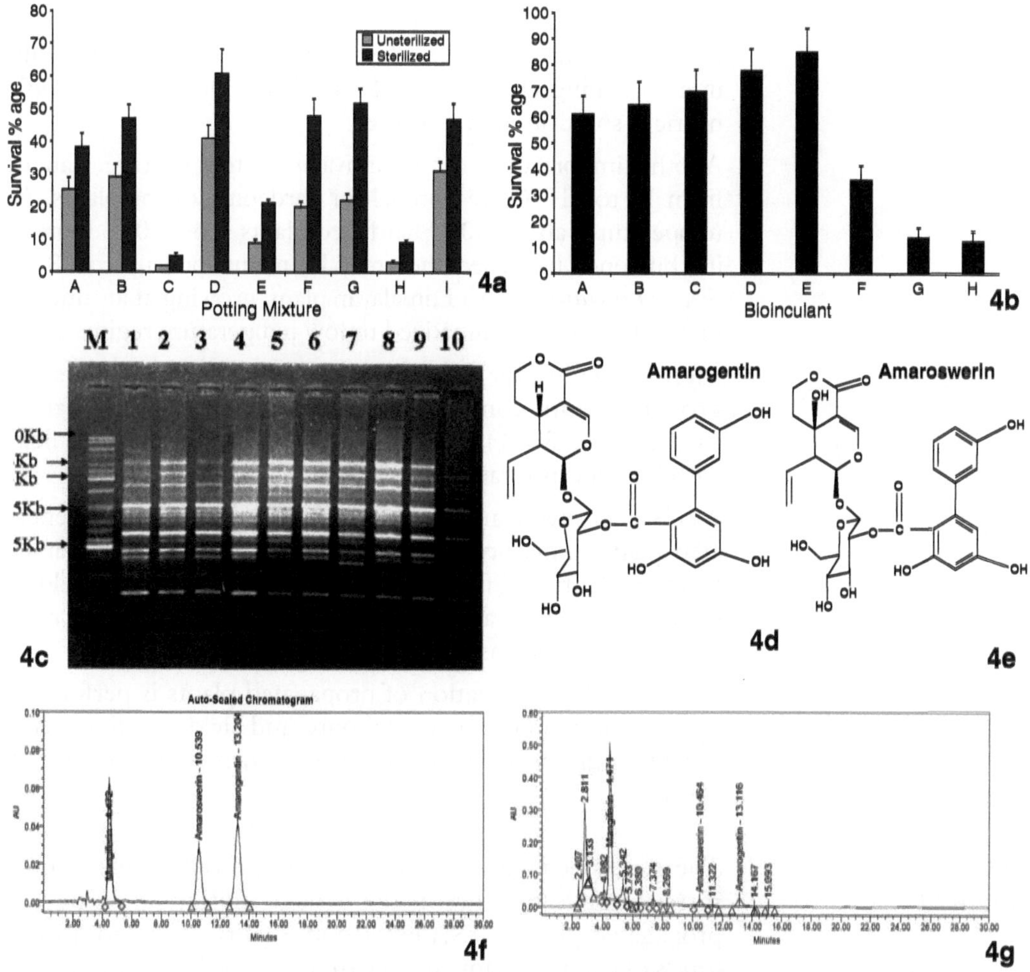

Fig. 4. (**a**) Effect of potting mixture: (A) soilrite (B) vermiculite (C) peatmoss (D) sand (E) garden soil (F) soilrite + garden soil (G) vermiculite + garden soil (H) peatmoss + garden soil (I) Sand + garden soil (**b**) Effect of various bioinoculants: (A) control (B) 1% trichoderma (**C**) 1% azatobactor (D) 0.5% Tri + 0.5% aza (E)1% Tri + 1% aza (F) 1% *Bacillus* (G) 1% *Pseudomonas* (H) 1% *P. indica* (**c**) Comparative RAPD profiling (primer OPK 13) of in vitro raised plantlets of . *Lanes 1*-2, Mmother plant; *Lanes 3—10*: tissue culture raised plants (**d, e**) structural formula (**f, g**) Chemoprofiling.

2. To remove harmful microbes, sterilize sand by autoclaving at 15 psi, 121°C. This step enhances the survival rate up to 61% (**Fig. 4a**). Use sand for further experimentation as transplantation substrate.
3. Inoculation with selected bacteria at the time of transplantation provided the first line of defense to tissue culture raised plants. To investigate effect of bioinoculats sterilized sand is mixed with 1% *(w/w)* mycelium microbes *Bacillus subtilis, Pseudomonas spp., Azatobacter chrococcum*, and *Trichoderma viride* individually or in combinations (**Fig. 4b**) *Azatobacter chrococcum* and *Trichoderma viride,* when used in combination at 1% each, increase the survival percentage from 61 to 85% (80–85%)).

4. Plants are maintained at 18 ± 2°C temperature under 16–h dark/8-h light photoperiod with 80–90% relative humidity in a hardening chamber for 4 wk. Water plants with Knop's nutrient solution once per week.
5. Another important factor in survival is the temperature. Ranges from 15 to –17°C are optimal for hardening of the plants. At temperatures above 20°C hardened plants show 100% mortality, but once hardened they may be maintained up to 28°C. *Swertia chirata* being a Himalayan plant, growing at an altitude up to 3000 m, is acclimatized to low temperature regime *(8)*.
6. After hardening for 4 to –6 wk in sand (**Fig. 1d**), plants are grown in pots containing a mixture of soil:sand: farm-yard-manure(3:1:1) for 8 wk and then transferred to the field with 90% survival rate (**Figs 1a–g**).
7. Assess phenotypic and chemical integrity of in vitro regenerated plants. Phenotypically regenerated plants are similar to parental plants and flower profusely in the greenhouse. Flowers are tetramerous and gamopetalous with two glands in each depression at the base of the petal (**Fig. 1h**).
8. Chemical characterization of propagated plants is performed by analyzing shoots in greenhouse and field conditions for two bitter glycosides amarogentin and amaroswerin of *Swertia chirata*.

3.2.3. Molecular Characterization

1. The success of any micropropagation procedure with commercial implication lies in the genetic uniformity of micropropagated plants. DNA extraction is carried out from fresh leaves of parental plant and in vitro developed hardened plants using procedure of Doyle and Doyle *(9)*.
2. RAPD profiles are developed using Sets A and E, with each of 20 random decamer oligonucleotides.
3. Polymerase chain reaction (PCR) reaction mixture 25 μL contains 2 μL DNA extract; 2.5 μL 10X Tris-HCl, pH 9.0; 0.5 μl dNTP mix; 2 μL primer,;1.5 μL $MgCl_2$; and 1.0 μL Taq polymerase.
4. DNA amplification is done in Master Cycler Gradient. After an initial heating at 95°C for 1.5 min final extension cycle is conducted at 72°C for 5 min. The amplification products are resolved by electrophoresis in 1.5% agrose gel run on 1X TAE buffer at 5 v/cm for 3 h and document on Gel Doc System.
5. Repeat all PCR reactions three times to check reproducibility. The assessment of genetic fidelity of the in vitro developed plants is performed by RAPD analysis by using OPK primers. Randomly selected in vitro regenerated plants and the donor

mother plant display monometric banding pattern confirming genetic integrity of the regenerated plants (**Fig. 4c**).

3.3. Development of Shake Cultures

3.3.1. Establishment of Cultures

1. Shoots cultures of *Swerita chirata* isolated from in vitro plantlets grown on MS minerals and vitamins medium. The growth and proliferation medium is added with 100 mg/l, *myo*-inositol, 0.5 mg/l BAP, and 3% sucrose without agar.
2. Each flask is inoculated with ten shoots in 250-mL flasks containing 30 mL medium and 20 segments in 500-mL flasks added with 50-mL culture medium.
3. Incubate cultures at 25 ± 2°C on a rotary shaker, 70 rpm, with 16 h light 40 μmol/m^2/s followed by 8 h dark regime.
4. Shoots in the liquid medium grow faster because of disappearance of apical dominance. The optimal time for shoot culture growth is 20 d.
5. Shoot cultures are routinely sub-cultured on this growth medium to sustain stock plants at the end of exponential growth phase. The proliferation stage in liquid cultures is carried out in 250- to –500-mL Erlenmeyer flasks (**Fig. 3a**).
6. For upscaling of shoot biomass production remove shoot cultures from 250- to 500-ml Erlenmeyer flasks and transfer into 1-L growtech (*see* **Note 4**) (**Fig. 3b**).

3.4. Chemical Analysis

3.4.1. Identification of Marker Compound

3.4.1.1. Extraction and Fractionation (10)

1. Extract defatted air-dried and milled aerial parts of 20 kg *S. chirata* with methanol at room temperature. The extract is concentrated to 1/6 under reduced pressure at 50 ± 5°C. Allow to stand overnight at room temperature.
2. Filter out separated precipitate Fraction A and dilute the filtrate with water to separate out gummy mass Fraction B. The clear aqueous solution is extracted with $CHCl_3$ Fraction C and *n*-butanol Fraction D (*see* **Note 5**).

3.4.2. Isolation of Magniferin

1. 1-g solid from Fraction A on repeated crystallizations from 20% aqueous ethanol yields magniferin.

3.4.3. Characterization of Magniferin

1. MP: 265–267°C (lit. m.p. 270–271°C), $[\alpha]_D$ + 30.40 (~ 0.5 20% aqueous ethanol).The structure of magniferin has been further confirmed on the basis of spectral data IR; ^{1}H NMR; ^{13}C NMR & MS
2. IR (KBr): 3358, 1681, 1609, 1454, and 1404 cm^{-1}
3. ^{1}H NMR (DMSO-d_6): δ 2.50–5.50 m (Ssugar protons), 6.39s (4-H), 6.88s (5-H), 7.40 s (8-H), 13.78s (1-OH)
4. ^{13}C NMR (DMSO-d_6): δ 161.6 (C-1), 107.4 (C-2), 163.6 (C-3), 93.4 (C-4), 156.2 (C-4a), 150.8 (C-4b), 102.6 (C-5), 153.8 (C-6), 143.5 (C-7), 108.2 (C-8), 111.8 (C-8a), 101.3

(C-8b), 179.0 (C = O), 81.2 (C-1′), 73.2 (C-2′), 70.6 (C-3′ and C-4′), 78.9 (C-5′), 61.4 (C-6′)

5. FABMS: *(m/z)* 423 $(M + H)^+$, 259

3.4.4. Isolation of Amarogentin and Amaroswerin

1. Column chromatography over silica gel, 60–120 mesh of fraction D (40 g) on elution with $CHCl_3$–MeOH mixtures (19:1 and 9:1) yielded semi purified fractions coded as D-1, containing 70–75% pure amarogentin and D-2 containing 60–65% pure Amaroswerin.
2. Pure molecules (i.e., amarogentin and amaroswerin) have been obtained by column chromatography over silica gel, 100–200 mesh followed by recrystallisation from methanol.
3. The structures of the pure molecules have been further confirmed on the basis of spectral data (i.e., ^{1}H NMR; ^{13}C NMR (*see* **Note 8**) and MS).

3.4.5. Amarogentin

1. MP: 230-32°, $[\alpha]_D$-109.2° (c 0.25, MeOH) [lit. mp 229-30°, $[\alpha]_D$-116.6° (MeOH)
2. ^{1}H NMR (Acetone-d_6: δ 3.80m (6′-H_2O, 4.30d (J = 7 Hz, 1′-H), 4.77 dd (J = 9.8 Hz, H_2O-4.30 d) (J = 7 Hz, 1′-H), 4.77 dd (J = 9.8 Hz, 2′-H), 5.10–5.70m (1-H, 8-H, 10-H2), 6.26d (J = 2 Hz, 6″-H), 6.40d (J = 2 Hz, 4″-H), 6.80m (2″-H, 4″ -H, 6″ -H), 7.24t (J = 8 Hz, 5″ -H), 7.46d (J = 2Hz, 3-H), 8.40br (OH), 9.36br (OH), 11.76s (OH)
3. ^{13}C NMR (CD_3OD): δ 97.0 (C-1), 153.5 (C-3), 105.3 (C-4), 28.2 (C-5), 25.5 (C-6), 69.4 (C-7), 132.5 (C-8), 43.1 (C-9), 121.0 (C-10), 167.4 (C-11), 96.6 (C-1′), 74.7 (C-2), 74.4 (C-3′), 71.4 (C-4′), 78.0(C-5′), 62.3 (C-6′), 148.3 (C-1″), 103.9 (C-2″), 165.6, (C-3″), 103.1 (C-4″), 163.5 (C-5″), 112.7 (C-6″), 146.2 (C-1″), 114.5 (C-2″), 157.1 (C-3″), 116.3 (C-4″), 129.2 (C-5″), 121.0 (C-6″), 171.2 (O–CO–)
4. FABMS: *m/z* 587 (M+H)+, 391, 229

3.4.6. Amaroswerin

1. MP $[\alpha]_D$-24.4° (c 0.76, MeOH) [lit. $[\alpha]_D$-13° (MeOH);
2. ^{1}H NMR (Acetone-d_6: δ 3.80m (6′-H_2O, 4.28d (J = 7 Hz, 1′-H), 4.74dd (J = 9.8 Hz, 2′-H), 5.12–5.52m (8-H and 10-H_2), 5.60d (J = 1.5 Hz, 1-H), 6.26d (J = 2Hz, 6″-H), 6.38d (J = 2 Hz, 4″-H), 6.80m (2″-H, 4″-H,6″ -H), 7.24t (J = 8 Hz, 5″ -H), 7.46s (3-H)
3. ^{13}C NMR (CD_3OD): δ98.3 (C-1), 153.8 (C-3), 109.4 (C-4), 65.8 (C-5), 33.4), (C-6, 64.36,(C-11), 97.6 (C-1′), 75.0 (C-2′), 74.9 (C-3′), 71.3 (C-4′), 78.4 (C-5′), 62.3 (C-6′), 148.6 (C-1″), 103.3 (C-2″), 166.3 (C-3″), 103.3 (C-4″), 163.0 (C-5″), 112.7 (C-6″), 146.5 (C-1″), 114.6 (C-2″),

157.5 (C-3″), 116.5 (C-4″), 129.4 (C-5″), 121.0 (C-6″), 171.9 (O–CO–)

4. FABMS: *m/z* 625 (M + Na)+, 391,229

3.5. Extraction of Tissue Culture Samples for Quantification

1. For chemical analysis of callus, shoots cultures, and shoots from green house at different stages are air-dried to remove the moisture.
2. Use 1-g dry powdered tissue to carry out cold extraction for 2 h on a magnetic stirrer with 50% aqueous ethanol. Filter the extract and repeat extraction procedure three times. Combine all the three extracts and concentrate on a Rota evaporator under reduced pressure. Three samples of the each treatment are processed in the similar way.
3. Weigh extracts and carry out quantitative estimation of the extract by HPLC. 10 μL extract in HPLC grade MeOH is injected in Rp column of Shimadzu class-VP instrument. 2% acetic acid in HPLC water: MEOH in the ratio of 60:40 at flow rate of 0.7 mL/min is used as carrier solvent. Maintain column temperature at 30°C and a wave length of 271 nm.

3.6. Quantification by HPLC

1. Prepare samples by dissolving magniferin 2.14 mg, amarogentin 2.32 mg and amaroswerin 2.23 mg in 25 mL methanol. 5-, 10-, 15-, 20-, 25-, and 30-μL reference standard solutions are used for making standard curve of each standard separately
2. Dissolve 25 mg dried aqueous alcoholic extracts in 2-mL HPLC grade methanol:water (1:1), filter samples through 0.45 μm Millipore micro filter and inject 10 μL each sample into the HPLC system (*see* **Note 9**).
3. Prepare stock solutions of pure reference compounds in HPLC grade methanol and stored in dark at 4°C. From the stock solutions, prepare solution for calibration curve (17.5–35.0 μg/mL) for each reference compound by diluting with HPLC grade methanol. These working solutions of all the reference compounds are mixed together in equal volumes for further analysis (*see* **Note 6**).
4. The compounds exhibit linear response in the concentration range of 17.5–35.0 μg/mL and the calibration curves are prepared.
5. Working solutions after mixing are injected in different concentrations. The calibration curves were obtained for amarogentin; r^2 = 0.998432 amaroswerin; $r^{2\circ}$=°0.997883
6. Calibration curves are determined on the basis of six different concentrations of each standard in the mixture (*see* **Note 7**).

4. Notes

1. In this study on micropropagation of *Swertia chirata* nodal segments, with 1–2 nodes, were best suited material to initiate axillary shoot formation.
2. Length of shoots, 4- to –6-cm long, is critical for root initiation.
3. For hardening to get better survival rate initial low temperature, 15–17°C, is critical with the increase in initial temperature survival rate will also decrease.
4. Growtech vessel is supported with a plastic mesh at the base which provides support to proliferating shoot cultures.
5. Pre-piking is done to check the recovery of the compound from the extract by adding standard solution of amarogentin, amaroswerin, and magniferin. Solution should be homogenized and recovery checked.
6. 17.5–35.0 μg/ml, for each reference compound by diluting with HPLC grade methanol. These working solutions of all the reference compounds are mixed together in equal volumes for preparation of calibration curve using multipoint method. 10–100 μL mixture is diluted to 2 mL and a linear calibration curve is prepared to check the accuracy of the method.
7. Minimum three injections of extract solution should be injected to determine mean value.
8. During ^{1}H and ^{13}C NMR spectra selective ^{1}H decoupling technique is used to find out directly bonded and carbon signals.
9. For quantification by HPLC pump LC-10ATVP, column oven CTO-10ASVP, diode array detector SPD-M10ATVP, and system controller SCL-5.40 should be used. Shimadu Class VP software 6.10 is used for data processing.

References

1. Joshi, P. and Dhawan, V. (2005). *Swertia chirayiata* – an overview. *Curr. Sci.* **89**, 635–640
2. Bhatia, A., Karan, M. and Visisht, K. (2003). Morphological and chemotypic comparison of certain Indian species of *Swertia. J. Med. Arom. Plant Sci.* **25**, 336–343
3. Conservation on International Trade in Endangered Species of Wild Fauna and Flora (CITES) Official document appendices I, II & III *http:www.cites.org/eng/append/latest_appendices.shtml*
4. Edwin, L. and Chungath, J.I. (1988). Studies on *Swertia chirata. Indian Drugs.* **25**, 143–146
5. Shrestha, J.N. and Joshi, S.D.(1992).Tissue culture techniques for medicinal important herbs *Orchis incarnata and Swertia chirata. Banko Janakari. Journal For. Inf. Nepal* **3**, 24–26.
6. Raina, R., Johri, A.K. and Srivastava, I.J. (1994). Seed germination studies in *Swertia chirata* L. *Seed Res.* **22**, 62–63.
7. Murashige, T. and Skoog, F. (1962). A revised medium for rapid growth and bioassay with tabacco tissue cultures. *Physiol. Plant.* **15**, 473–497
8. Wawrosh, C., Maskay, N. and Kopp, B. (1999). Micropropagation of threatened Napalese

medicinal plant *Swertia chirata*. Buch-Ham ex wall. *Plant Cell Reports* **18**, 997–1001

9. Doyle, J.J. and Doyle, J.L. (1990). Isolation of plant DNA from fresh tissue. *Focus* **12**, 13–15.

10. Chakarvarty,A.K.,Mukhopadhyay,S.,Moitra,S.K. and Das B (1994). (-) Syringaresinol, ahepatoprotechtive agent and other constituents from *Swertia chirata*. *Indian J. Chem.* **33B**, 405–408

Chapter 13

Protocols for Establishment of an In Vitro Collection of Medicinal Plants in the Genus *Scutellaria*

Ian B. Cole, Faisal T. Farooq, and Susan J. Murch

Summary

The study of medicinal plants has many unique challenges and special considerations. These plants are studied for their specific chemistry, or pharmacologic activity. Plants are highly sensitive to their environment and respond through changes in their chemistry. To date, one of the biggest problems in the study of medicinal plants has been the acquisition of consistent, positively identified material for chemical analysis. Successful protocols for the collection, identification, and establishment of medicinal plants species in tissue culture is invaluable for future studies. This protocol outlines methods to establish *Scutellaria baicalenisis*, and *Scutellaria lateriflora* from commercial seed sources, and collection and establishment of *Scutellaria racemosa* from wild populations.

Key words: Medicinal plants, *Scutellaria*, Medicinal biotechnology, In vitro propagation, Germplasm collection

1. Introduction

The genus *Scutellaria* is a diverse and widespread genus within the family *Lamiaceae*, the mint family, with more than 350 species worldwide, covering geographic regions from Alaska to South America, from Siberia to Japan, and throughout Europe, the Middle East, and Asia *(1)*. Species within the genus *Scutellaria* have been widely used by many cultures to treat a variety of different ailments.

Scutellaria baicalensis (Huang-qin) remains one of the most commonly prescribed herbs in both traditional Chinese medicine and in Japanese Kampo medicine *(2, 3)*. More than 275 scientific

S. Mohan Jain and Praveen K. Saxena (eds.), *Methods in Molecular Biology, Protocols for* In Vitro *Cultures and Secondary Metabolite Analysis of Aromatic and Medicinal Plants, vol. 547*
© Humana Press, a part of Springer Science+Business Media, LLC 2009
DOI: 10.1007/978-1-60327-287-2_13

reports have appeared since 2000, outlining the medicinal effectiveness of the extracts of Huang-qin in the prevention and treatment of many disorders including prostate cancer *(4)*, hepatic cancer *(5)*, HIV *(6)*, and neurodegeneration *(7)*. Animal studies of *S. baicalensis* supplementation to the diet indicated protection against aflatoxin –B1-induced liver mutagenesis *(8)*, inhibition of liver fibrosis *(9)*, inhibition of hemin-nitrite-H_2O_2 induced liver damage *(10)*, and reduction of symptoms of Type 1 allergic reactions *(11)*.

Scutellaria lateriflora, also known as "mad dog" or "skullcap," is a native North American species that was used by the Cherokee, Applachian, and Iriquois as a treatment for anxiety, psychosis, neurologic disease, and feminine difficulties *(12, 13)*. Currently, skullcap is sold in North America as a tea; tonic or capsules made from dried aerial parts, and is used to treat epilepsy, St Vitus's dance, insomnia, anxiety, neuralgia, and withdrawal from tranquilizers or barbiturates *(14, 15)*. Flavonoids found in *S. lateriflora* also inhibited [3H]-LSD binding to 5-HT7 (sero-tonin) receptors *(15)* and reduced anxiety levels in rats *(16)*.

Scutellaria racemosa, is a native of Central and South America that has recently become an invasive species in the southern United States *(17, 18)*. Ethnobotanic evidence suggests that the Cauca peoples of Columbia and Ecuador use specifically selected ecotypes of *S. racemosa* in a ceremonial or narcotic preparation. Bianchi et al. *(19)* demonstrated that extracts of *S. racemosa* had neuroprotective activity in stressed animal models but very little research has been done with this species, in part because of a lack of available plant material for study.

One of the major requirements for research is to identify specific medicinally active phytochemicals from *Scutellaria* species is the development of optimized protocols for growth, production, harvest, handling, and for preventing the loss of wild germplasm *(20, 21)*. Controlled environment production systems have the potential to provide a continuous supply of consistent plant material, free from pathogens or abiotic contamination [*(22)*, (*see* **Note 1**)]. Previous reports have provided detailed methods for in vitro establishment and production of *S. baicalensis(3, 23, 24)*. However, the potential of many species including *S. lateriflora*, *S. racemosa* and their ecotypes remains unexploited. In many cases, species of *Scutellaria* are identified only by wild collection and herbaria vouchers. For example, TROPICOS, the database of the Missouri Botanical Garden, lists approximately 800 field collections of *Scutellaria* identified by dried shoot and floral tissues, with GPS coordinates locating populations and corresponding maps that locate the site of collection. Recently, these resources were used to identify ten new species of *Scutellaria* in Mesoamerica, an indication of the rich genetic resources remaining undiscovered in the genus *(25)*.

In this chapter we describe standardized, efficient protocols for *in situ* field collections, species identification, in vitro establishment, regeneration, and controlled environment production of axenic cultures of three *Scutellaria* species namely *S. baicalensis, S. lateriflora and S. racemosa.*

2. Materials

2.1. Plant Material

S. baicalensis and *S. lateriflora* seeds were obtained from Richter's Herb's, Goodwood, ON (*see* **Note 2**); *S. racemosa* material was collected from wild populations in Florida. The location of populations of desired plant species was facilitated by herbarium databases such as Missouri Botanical Garden's W3TROPICOS database (http://www.mobot.mobot.org/W3T/Search/vast.html), Florida State University (http://www.herbarium.bio.fsu.edu/search-specimens.php), Fairchild Tropical Garden (http://www.virtualherbarium.org/vh/db/index.htm), The New York Botanical Garden (http://sciweb.nybg.org/science2/virtualherbarium.asp) and from the University of South Florida's Atlas of Florida Vascular Plants database (http://www.plantatlas.usf.edu/) *(26)*. In the case of *S. racemosa*, the databases have records of about 27 populations, which have been collected in North America between 1974 and 2002. We chose to sample populations in Florida for several reasons including: access to populations on public lands, ease of collection, persistence of the population since first description and geographic distribution. *S. racemosa* was collected from three sites in Florida in January of 2005 **(Fig. 2b)**.

2.2. Field Collection Supplies

Conditions in the field are highly dynamic and change frequently. Check the weather of the area in which you will be working and dress appropriately. No matter what the weather predictions are; bring a raincoat. Additionally it is important to understand risks from wild animals, spiders, insects, or snakes. Immunizations are a good idea if you are working in an area with the risk of malaria or hepatitis. Once in the field you will need to have all of the necessary supplies for collecting material and getting it safely back to the laboratory. An assortment of Ziploc bags, plastic 25- or 50-mL tubes, and paper towels or newspaper should be in your pack as well as a weather-proof field notebook, camera, and several pencils. To physically harvest plant material, it is important to have a large knife or machete, pruners or secateurs, and a small pocket-knife for delicate cuts (*see* **Note 3**).

2.3. Building a Plant Press

Plant presses are typically constructed of two pieces of slatted wood 12″ × 18″ (~30 × 45 cm) filled with pieces of corrugated cardboard and blotter paper or newspaper to provide air ventilation and absorb moisture. The plant press is secured with webbed straps [*(27)*; **Fig. 2d**].

2.4. Surface Sterilization Solutions

Solutions used in surface sterilization of seeds and tissues include commercially available bleach (~5.25% sodium hypochlorite) are diluted with water, and made into 10 and 20% solutions. Other chemicals include: 90% ethanol, the detergent Tween-20 (Phytotechnology Laboratories Inc.; Lexana, KS); and Plant Preservation Mixture (PPM) which was purchased from Plant Cell Technology Inc., (Washington DC).

2.5. Tissue Culture Media and Culture Vessels

All media are prepared in 1 L media bottles with screw-on caps. The ingredients of media are 30 g/L sucrose, 4.33 g/L MS salts, and 1 mL/L Gamborg's B5 vitamin solution (Sigma; Canada). The pH of the media is adjusted to 5.75 using 1 N NaOH, and 1 N HCl solutions. The plant growth regulators indoleacetic acid and indolebutyric acid are added to media prior to autoclaving. To solidify media, 7 g/L agar is added (Laboratory Grade Agar; Fisher Scientific, Mississauga, ON). All media are autoclaved at 121°C and 20 lb pressure for 25–30 min.

Seeds are germinated in sterile clear Petri plates (Fisher-Scientific,;Canada). Plant material is grown in Phytotech P700 culture boxes (Phytotechnology Laboratories Lexana, KS).

3. Methods

3.1. Seed Germination

1. Seeds are surface sterilized by immersion in 95% ethanol for 30 s, and then in 20% bleach containing 2 drops per 100 mL Tween-20 for 18 min. Seeds are then rinsed at least three times with sterile distilled water (*see* **Note 4**).
2. Seeds are germinated in Petri dishes containing 25 mL 0.8% agar containing 4 mL PPM.
3. Seed germination occurs after incubation in a dark growth chamber at 24°C for 14 d.

3.2. Identification and Collection of Scutellaria Racemosa (Pers.) from Wild Populations for In Vitro Culture

3.2.1. Preparations for Field Collections

1. All field collections should be performed with appropriate permissions and in compliance with the Convention on Biological Diversity (CBD) and the Convention on International Trade in Endangered Species of Wild Fauna and Flora (CITES). In the case of *S. racemosa*, the plant is an obligate weed in Florida, it is not an endangered species and therefore not covered by the CITES treaty and there are no restrictions on collection from public lands. However, collection of any plant material from private lands should always be avoided unless permission to collect is requested and granted.

2. It is necessary to obtain appropriate permits for transport of any field collections of plant material from the site of collection to the research laboratory. Plant import permits are issued by the Plant Protection agency of each country. For details of this process and application forms, please see APHIS/USDA (http://www.aphis.usda.gov/plant_health/permits/index.shtml) or the Canadian Food Inspection Agency, Plant Protection Branch (http://www.inspection.gc.ca/english/for/pdf/c5256e.pdf). For wild collections of *S. racemosa* in Florida, USA and subsequent transport to the research laboratory in Canada, permits were required to allow the collection and transportation of seeds and plant meristems (*see* **Note 3**).
3. There are a number of effective resources that are helpful in identifying unique characteristics of the plant. Such resources include herbaria specimens and published monographs that are available as published books or journals including *Systematic Botany Monographs* published by the American Society of Plant Taxonomists (http://herbarium.lsa.umich.edu/SBMweb/index.html), and the *Kew Bulletin* (http://www.kew.org/publications/kewbulletin.html). An example of a herbarium voucher *for S. racemosa* is seen in **Fig. 1**. *Scutellaria* is highly heterophyllus genus however; *S. racemosa* is easily identifiable in the field by the consistently hastate leaves and flower corollas less than 7 mm long [*(1)*; **Fig. 2a**].

3.2.2. Collection of Scutellaria in the Field

1. The field notebook is essential for accurate records of plant collection, preparation of herbarium vouchers and creating a record of plant locations. It is important to use a notebook with cloth, waterproof, all-weather writing paper and to write field notes in pencil that will not run when wet. The following information is crucial to the plant collection process:
   (a) Collector(s) full name(s).
   (b) Date of collection.
   (c) Detailed location: country, state, county or province, roads, road junctions, mile markers, distance to cities or towns, elevation, and GPS coordinates.
   (d) Habitat: type of plant community and other plants growing in the area.
   (e) Plant habit: the form of the plant (herb, vine, tree shrub) and its height
   (f) Frequency: is the plant rare, occasional, frequent, or common?
   (g) Plant description: record any characteristics which may be lost upon drying such as aroma, flower colour, fruit colour, or leaf orientation.
   (h) Unique collector number: assign each collection an individual number.

Fig. 1. In vitro propagation and production of *Scutellaria* species. (**a**) Field collected meristem explants. (**b**) Development of single shoots from the auxillary meristem of stem explants. (**c**) De novo shoot organogenesis on stem explants of *Scutellaria*. (**d**) Axenic cultures of *Scutellaria baicalensis* maintained by repeated meristem culture. (**e**) Axenic cultures of *Scutellaria lateriflora* maintained by repeated meristem culture. (**f**) Axenic cultures of *Scutellaria racemosa* maintained by repeated meristem culture. Note: Flowers on apical meristems of *S. racemosa* after 6–8 wk in culture. (**g**) Large-scale production of *Scutellaria* in Liquid Lab™ temporary immersion bioreactors (**h**) Acclimatization of in vitro grown plantlets to ex vitro conditions. (**i**) Development of plants under standard greenhouse conditions.

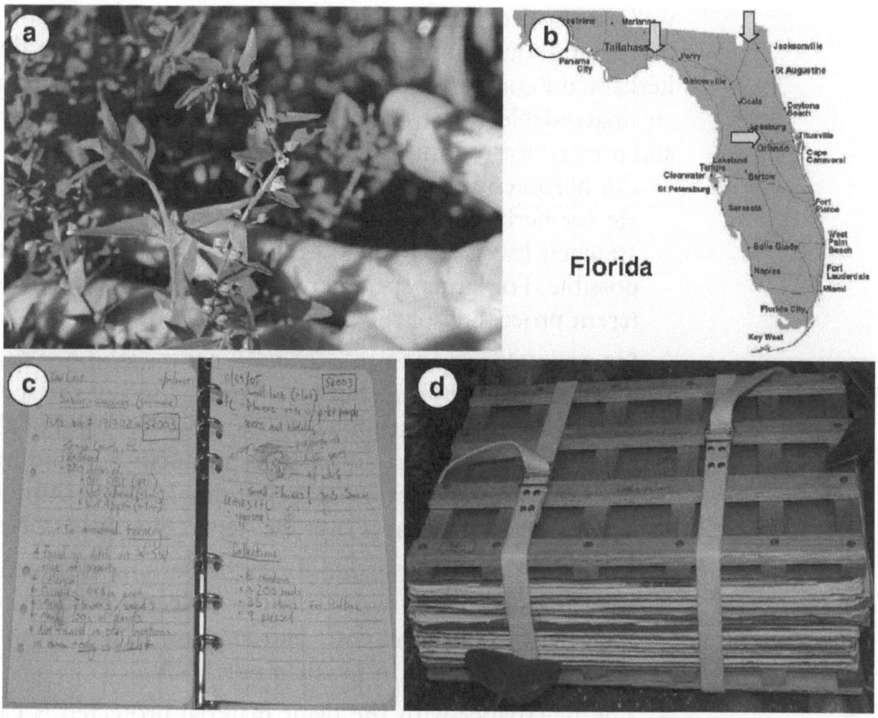

Fig. 2. Field collection of *Scutellaria racemosa*. (**a**) Identification of plants in the field. (**b**) Collection locations for the 2005 populations. (**c**) Field notebook. (**d**) Plant press for herbarium specimens.

2. Field notes from the collection of *S. racemosa* are shown in **Fig. 2c**. In brief, field collection of *S. racemosa* consisted of collection of seeds and meristems. Initially, all plant material was photographed to provide a visual record. Seeds were collected from fully mature, dehisced flower heads. About 100–200 seeds per flower stalk should be collected into plastic 50-mL tubes. Seeds should be collected from at least three different plants at random and details of the collection should be recorded in the field notebook. For meristem collections, healthy stems, about 15- to –35-cm long are excised from the growing plants using secateurs or a knife. It is important to select healthy material from as many different plants in the population as possible however, this may be difficult if populations are in decline or dormant.
3. Following harvest, plant material should be transported quickly. For *S. racemosa*, the most effective means of transport was immersion of the cut surface in a container of clean, tepid water. Store the container with cuttings out of the direct sun in a cool place and transport to tissue culture facility within 24 h.

3.2.3. Making Herbarium Vouchers

Herbarium vouchers create a permanent record of your collections and should be stored at institutes in many different locations (*see* **Note 5**). In order to comply with the requirements of different herbaria, all specimens must be prepared according to well-established international guidelines. It is important to file herbarium vouchers of field collections at four or five different herbaria for comparison and to avoid loss if one herbarium suffers an unavoidable catastrophe. Specimens must be properly dried and pressed for mounting as described below (*see* **Note 5**).

1. For herbaceous species, stems about 31-cm long are appropriate for herbarium vouchers. Stems should be complete with attached leaves and roots, as well as flowers and/or fruits if possible. For larger specimens, aquatic plants or cacti etc., different processing of the tissues may be required.
2. Stems are arranged onto a 27 × 38-cm piece of newspaper in a lifelike manner (i.e., shoots toward the top of the sheet and roots toward the bottom which allows observation of all morphological structures, especially reproductive, including both sides of leaves and several stages of development of shoots, or flowers). Fold newspaper over the stems **(Fig. 2d)** (*see* **Note 6**).
3. A label should be prepared for all herbarium specimens. The label is usually 4.5 -wide with the collector number, collector's name, collection date, and location of collection corresponding to records in the field notebook.
4. The newspaper with the plant material protected is carefully folded and placed in press between sheets of blotter paper and cardboard. Straps are tightened on plant press and the press is dried using one of several different techniques. For most applications with small herbaceous plants, dry under light using three 60–120 W incandescent light bulbs for 24–36 h and transport to herbarium for mounting (*see* **Note 7**).
5. Replicate herbarium voucher specimens should be donated to herbaria in at least four different locations in order to ensure that the samples will be preserved in perpetuity (*see* **Note 8**).

3.3. Creation of an In Vitro Germplasm Collection of Scutellaria Racemosa from Field Collections

1. Stems of *S. racemosa* collected in the field are transported to the tissue culture laboratory within 24 h of collection (*see* **Note 5**).
2. Stems are cut into 4- to 5-cm pieces and surface sterilized with a 20% solution of bleach and two drops of Tween-20 for 30 min, washed with sterile distilled water 3 times and cultured onto basal culture media (MSO) for shipping to final destination (*see* **Note 6**).
3. Subculture of the field collected plant material onto fresh MSO is required within 3 d of shipping. Meristem sections (0.5–1.0 cm) are excised from the field collected shoots and subcultured onto fresh MSO medium in Phytotech P700 Culture Boxes. All cultures are incubated in a growth chamber with a 16-h photoperiod under cool-white light, 22–45 $\mu mol/m^2/s$ at 25°C. Shoot apices were subcultured every 3

wk (**Fig. 1f**). It is interesting to note that several cultures of *Scutellaria* produced profusions of shoots after 4–6 wk of culture under these conditions (**Fig. 1f**) (*see* **Note 6**).

3.3.1. Maintenance of the In Vitro Collection of Scutellaria

1. Once a collection of germplasm of *Scutellaria* species has been established, the plants provide tissue for optimization and phytochemical quantification experiments. All cultures must be transferred to MSO medium devoid of any growth regulators prior to bioassays, optimization studies or chemical analysis (*see* **Note 9** and **Note 10**).
2. Plants kept on MSO medium with a 16-h photoperiod will need to be subcultured every 6 wks to maintain the collection.
3. More rapid growth of *Scutellaria* plantlets in culture is achieved by supplementation of the MSO medium with 1.0 μM kinetin.
4. Rooting of the in vitro plantlets is facilitated by supplementation of the medium with 2.5 μM indoleacetic acid (IAA) or 0.5 μM indolebutyric acid (IBA).

4. Notes

1. In vitro-grown plant material has many advantages, especially for importing and exporting plants across borders. Axenic, sterile shoots are free from many of the strict regulations surrounding the shipping of cuttings, bare roots, seeds, or other plant parts.
2. In the commercial marketplace, adulteration of *Scutellaria lateriflora* seeds with *Teucrium*, (germander) has been reported and selection of a reputable seed source is crucial to the success of both experiments and commercialization efforts. Further, the germination of some seeds under greenhouse conditions followed by comparison of the plants with herbarium vouchers is required for positive identification of the species.
3. This is a very basic set of supplies for the field. It is necessary to make many decisions on what to bring, based on where the work will take place.
4. *Scutellaria* seeds responded very well to this basic surface sterilization protocol. However, some species require more extensive methods to remove pathogens from the surface of the seeds.
5. Another important use of herbarium vouchers is the ability to study and observe the plants of interest to assist in identifying plants in the field. This section is a brief description of

herbarium voucher mounting and preparation. Such work is considered to be an art. For further information please refer to *(27)*.

6. Ideally, plant material would be cultured immediately upon harvest. However, because of the nature of field collection, this is not always possible. Often material must be transported prior to culture. It is important to keep plant material damp either in a vessel of water, or wrapped in wet paper towels or newspaper. It is also important to keep plant tissues safe from extreme environmental conditions such as direct sunlight or snow. These extreme conditions can also affect the material while it is stored in a backpack, or clothing.
7. The value of a herbarium voucher is determined by the care in which the plant material is dried, pressed, and mounted. If the stem tissue is dried in a manner which shows many morphologic characteristics, the voucher will be invaluable for later studies. For detailed information on this process please see reference *(27)*.
8. There are many different drying methods and drying-oven designs. Some of which are very simple, such as the field method suggested above, whereas other oven designs can be very complicated. It is important to use the best design for the project. If many vouchers will need to be prepared, a larger oven may be necessary.
9. In these studies, *Scutellaria* was subcultured by excising stem segments, with 1–2 nodes, about 5-mm in length. These nodal segments were then aseptically placed on fresh media in a laminar flow hood.
10. For phytochemical studies, it is common to need large quantities of plant material for controls and all of the treatments in a given experiment. We found that subculturing *Scutellaria* into 750-mL culture vessels allowed the plantlets to grow larger and supplied much more material for extractions and chemical analysis. It is very important to culture material onto media free from plant growth regulators (auxins and cytokinins) for several weeks prior to phytochemical analysis. Some of these compounds can affect the chemical profile of samples, and could skew the data.

References

1. Paton, A. (1990). A global investigation of *Scutellaria* (Labiatae). *Kew Bull.* **45**, 399–450.
2. Watanabe, H., Kobayashi, T., Meibou, T., Sekiguchi, Y., Uchida, K., Aoki, T., and Cyong, J.C. (2002). Effects of Kampo herbal medicine on plasma melatonin in murine sarcoma S 180 both *in vitro and in vivo. Biol Pharm. Bull.* **29**, 1132–1137.
3. Murch, S.J., Rupasinghe, H.P.V., Goodenowe, D., and Saxena, P.K. (2004). A metobolomic

analysis of medicinal diversity in Huang-qin (*Scutellaria baicalensis* Georgi.) genotypes discovery of novel compounds. *Plant Cell Rep.* **23**, 419–425.

4. Hsieh, T.C., Lu, X., Chea, J., and Wu, J.M. (2002). Prevention and management of prostate cancer using PC-SPES: a scientific perspective. *J Nutrition.* **132**, 3513–3517.
5. Chang, W.T., Chem, C.H., and Lu, F.J. (2002). Different effects of baicalein, baicalin and wogonin on mitochondrial function, glutathione content and cell cycle progression in human hepatoma cell lines. *Planta Medica.* **68**, 128–132.
6. Li, B.Q., Fu, T., Yan, Y.D., Baylor, N.W., Ruscetti, F.W., and Kung, H. (1993). Inhibition of HIV infection by baicalin-a flavonoid compound purified from Chinese herbal medicine. *Cell Mol Biol Res.* **339**, 119–124.
7. Li, F.Q., Wang, T., Pei, Z., Liu, B., and Hong, J.S. (2004). Inhibition of microglial activation by the herbal flavonoid baicalein attenuates inflammation-mediated degeneration of dopaminergic neurons. *J Neur Trans.* **112**, 331–347.
8. de Boer, J.G., Quiney, B., Walter, P.B., Thomas, C., Hodgson, K., Murch, S.J., and Saxena, P.K. (2005.) Protection against aflatoxin-B1-induced liver mutagenesis by *Scutellaria baicalensis Mut. Res.* 578, 1–22.
9. Nan, J.X., Park, E.J., Kim, Y.C., Ko, G., and Sohn, D.H. (2002). *Scutellaria baicalensis* inhibits liver fibrosis induced by bile duct litigation on carbon tetrachloride in rats. *J Pharm Pharmaco.* **54**, 555–563.
10. Zhao, Y., Li, H., Gao, Z., Gong, Y., and Xu, H. (2000). Effects of flavonoids extracted from *Scutellaria baicalensis* Georgi. On hemin-nitrite-H_2O_2 induced liver injury. *Eu J Pharm.* **536**, 192–199.
11. Lim, B.O. (2003). Effects of wogonin, wogonoside and 3,5,7,2′,6′-pentahydroxyflavone on chemical mediator production in peritoneal excudate cells and immunoglobin E of rat mesenteric lymph node lymphocytes. *J Ethnopharm.* **84**, 23–29.
12. Moerman, D.E. (1998). Native American Ethnobotany. Timber Press, Portland, OR.
13. Johnson, T. (1999). CRC Ethnobotany Reference. CRC, Boca Raton, FL.
14. Foster, S. and Tyler, V.E. (1999). Tyler's honest herbal. New York, Haworth Herbal Press, NJ.
15. Gafner, S.C., Bergeron, L.L., Batcha, J., Reich, J., Arnason, J.T., Burdette, J.E., Pezzuto, J.M., and Angerhofer, C.K. (2003). Inhibition of [3H]-LSD binding to 5-HT7 receptors by flavonoids from *Scutellaria lateriflora. J Nat Prod.* **66**, 535–537.
16. Awad, R., Arnason, J.T., Trudeau, V., Bergeron, C., Budzinski, J.W., Foster, B.C., and Merali, Z. (2003). Phytochemical and Biological analysis of Skullcap (*Scutellaria laterifolia* L.): a medicinal plant with anxiolytic properties. *Phytomed.* **10**, 640–649.
17. Kral, R. (1973). Some notes on the flora of the southern states, particularly Alabama and Tennessee. *Rhodora.* **75**, 366–410.
18. Krings, A. and Neal, J.C. (2001). South American Skullcap (*Scutellaria racemosa*: Lamiaceae) in the Southeastern United States. *SIDA.* **19**, 1171–1179.
19. Bianchi, A. (2006). Attivita antidepressiva di due specie di Scutellariae Colombiane. VI Convegno AMIAR. Torino, Italy.
20. Murch, S.J., KrishnaRaj, S., and Saxena P.K. (2000). Phytopharmaceuticals: Problems, Limitations and Solutions. *Sci Rev Alt Med.* **4**, 33–38.
21. Saxena, P.K., Cole, I.B., and Murch, S.J. (2007). Approaches to quality plant based medicine: Significance of chemical profiling. In: Applications of plant metabolic engineering. (Verpoote, R., ed.), Springer, NY, pp. 303–323.
22. Murch, S.J., KrishnaRaj, S., and Saxena, P.K. (2000) Phytopharmaceuticals:Mass-Production, Standardization and Conservation. *Sci Rev Alt Med.* **4**, 39–43.
23. Zobayed, S.M.A., Murch, S.J., Rupasinghe, H.P.V., de Boer, J.G., Glickman, B.W., and Saxena, P.K. (2004). Optimized system for biomass production, chemical characterization and evaluation of chemo-preventive properties of *Scutellaria baicalensis* Georgi. *Plant Sci.* **167**, 439–446.
24. Cole, I.B., Saxena, P.K., and Murch, S.J. (2007). Medicinal biotechnology in the genus *Scutellaria. In Vitro Cell. Dev. Biol. Plant.* **43**, 318–327.
25. Pool, A. (2006). New species of Scutellaria (Lamiaceae) from Mesoamerica. *Novon.* **16**, 388–403.
26. Wunderlin, R. P., and B. F. Hansen. (2004). Atlas of Florida Vascular Plants (http://www.plantatlas.usf.edu/). (Landry, S. M. and Campbell, K. N., ed.), Florida Center for Community Design and Research. Institute for Systematic Botany, University of South Florida, Tampa, FL.
27. Bridson D. and Forman, L. (2004). *The Herbarium Handbook.* The Royal Botanic Gardens, Kew Press, Richmond, BC.

Chapter 14

Protocols for In Vitro Culture and Phytochemical Analysis of *Phyllanthus* Species (Euphorbiaceae)

Elizabete Catapan, Fábio Netto Moreno, Márcio Luís Busi da Silva, Michel Fleith Otuki, Rivaldo Niero, Valdir Cechinel Filho, Rosendo Augusto Yunes, and Ana Maria Viana

Summary

We developed reproducible protocols for micropropagation, callus culture, and root culture of the medicinal plant *Phyllanthus urinaria*, *P. niruri*, *P. tenellus*, *P. corcovadensis*, *P. caroliniensis*, *P. stipulatus*, and *P. fraternus* by using single node explants. Genotype-linked differences are visible among the Phyllanthus species concerning shoot culture, callus culture, and root culture growth. The protocols developed for phytochemical screening of callus and root extracts of *P. urinaria*, *P. caroliniensis*, *P. stipulatus*, and *P. fraternus* have shown the production of sterols and triterpenes. Both compounds are known to account for the antinociceptive activity of the methanolic extracts as glochidone and stigmasterol have strong activity against neurogenic and inflammatory pain. Similarly, methanolic callus extracts of *P. tenellus*, *P. niruri* and *P. corcovadensis* have potent analgesic properties, however phenolics are major compounds isolated from these species. The optimized micropropagation, callus culture, and root culture protocols offer the possibility to use cell/root culture techniques for vegetative propagation and secondary metabolite studies.

Key words: Genus Phyllanthus sp, Micropropagation, Callus, Root culture, Sterols, Triterpenes, Analgesia

1. Introduction

The genus *Phyllanthus* (*Euphorbiaceae*) has between 550 and 750 species and infusion of leaves, stems, and roots of several species are used in the Brazilian folk medicine for treating disturbances of the kidney, urinary tract and bladder infections, intestinal infections, diabetes, and hepatitis B virus *(1,2)*. Several compounds, such as alkaloids, flavonoids, lignans, phenols, and terpenes with

S. Mohan Jain and Praveen K. Saxena (eds.), *Methods in Molecular Biology, Protocols for* In Vitro *Cultures and Secondary Metabolite Analysis of Aromatic and Medicinal Plants, vol. 547*
© Humana Press, a part of Springer Science+Business Media, LLC 2009
DOI: 10.1007/978-1-60327-287-2_14

potential beneficial therapeutic action against hepatitis B, nefrolitiase, and painful disorders have been isolated and identified from a great variety of *Phyllanthus* species *(1)*. The main constraint for plant production of many *Phyllanthus* species is the low seed germination rate and slow plant growth *(3)*. Therefore, there is an urgent need to develop micropropagation, callus, and root culture systems to provide plant material to foster studies on plant production, phytochemical, and pharmacologic analysis and to investigate the application of biotechnologic approaches for germplasm conservation. According to the pharmacologic importance we have optimized in vitro culture systems of *P. tenellus*, *P. corcovadensis*, *P. niruri*, *P. fraternus*, *P. caroliniensis*, *P. urinaria*, and *P. stipulatus(4–8)*. Both in vitro and ex vitro morphogenic responses of different species are very much genotype-dependent. The pharmacologic screening carried out with callus extracts of these species demonstrates potent analgesic properties against neurogenic and inflammatory pain *(5,9)*. Sterols and triterpenes such as stigmasterol and glochidone are responsible for the antinociceptive effect *(10)*. The protocols described in this chapter are based on efficient and reproducible methods for shoot initiation, rooting, acclimatization, callus culture, and root culture of selected Phyllanthus species. Protocols for evaluation of their chemical composition are also described *(11,12)*.

2. Materials

2.1. Plant Material

Collect *Phyllanthus* sp seeds from stock mother plants grown in the gardens of the Federal University of Santa Catarina (UFSC, Florianópolis, SC, Brasil). They were classified by Dr. Leila da Graça Amaral (Departamento de Botânica, UFSC) and a voucher of each specimen was deposited in the herbarium FLOR (Departamento de Botânica, UFSC).

2.2. Culture Media

1. Murashige and Skoog (MS) *(13)* powder basal medium (Sigma; St. Louis, MO).
2. Anderson Rhododendron *(14)* powder basal salt mixture (Sigma).
3. 20 g/L Sucrose grade I crystalline (Sigma).
4. 2 g/L Phytagel (Sigma).
5. Plant growth regulators for plant cell culture (Sigma): 6-benzylaminopurine (BAP), kinetin, 2iP [6-(γ,γ-dimethylallylamino)-purine], indole-3-butyric acid (IBA), NAA (αα-naphthaleneacetic acid), 2, 4-dichlorophenoxyacetic acid (2,4-D), indole-3-acetic acid (IAA)). Prepare standard aqueous solutions with two drops of 1 M NaOH.

2.3. Culture Condition

1. Seal the vessels with a 76 × 76 mm transparent polypropylene film and maintain at 25°C and 70% relative humidity for a 16-h photoperiod and photosynthetic photon flux of 20–25 μmoL/m^2/s supplied by fluorescent light tubes. In all protocols use these culture conditions.

2.4. Phytochemical Analysis

1. Methanol, Chloroform PA (Merck).
2. 20×20-cm TLC silica precoated aluminium plates 20 μM thick (Merck).
3. 90:10 *(v/v)* Chloroform:methanol solvent system.
4. Ferric chloride, magnesium hydrochloride, and anisaldehyde-sulfuric acid reagents.
5. Glochidone, glochidonol, glochidiol, and stigmasterol extracted and isolated from *Phyllanthus* plants are used as standards.
6. High-performance liquid chromatograohy (HPLC) and gel chromatography (GC) equipment.

3. Methods

3.1. Establishment of Shoot Cultures

3.1.1. Preparation of Culture Medium

1. Prepare standard aqueous solutions of plant growth regulators (*see* **Subheading 2.2, item 5**) by adding two drops of 1 M NaOH.
2. Prepare appropriate culture medium by adding all ingredients (*see* **Tables 1** and **2**) together with plant growth regulators.
3. Adjust medium pH to 5.8 with 1 M NaOH before adding Phytagel. Dispense into 25 × 150-mm glass tubes (10 mL/tube) before autoclaving at 121°C for 18 min.

3.1.2. Seed Surface Sterilization and Germination

1. Surface sterilize seeds of selected *Phyllanthus* species for 20–30 min in commercial bleach, 2.5% active chlorine. Add with 2–3 drops of Tween-20 for 30 min.
2. Rinse three times for 10 min in sterile distilled water and culture on MS basal medium supplemented with 2% *(w/v)* sucrose and 0.2% *(w/v)* Phytagel. Adjust the pH to 5.8 before adding Phytagel and dispense the medium into 25 × 150-mm glass culture tubes (10 mL/tube) before autoclaving at 121°C for 18 min.
3. Seal the vessels with a 76 × 76-mm transparent polypropylene film and place the cultures at 25°C at 70% relative humidity for a 16-h photoperiod and photosynthetic photon flux of 20–25 μmol/m^2/s supplied by fluorescent light tubes.

Table 1
Composition of Culture Media used for Micropropagation of Selected Species of *Phyllanthus*

Species	Shoot multiplication	Microshoot Rooting	References
P. tenellus	MS or AR, 1.25 μM IBA	1.25–2.5 μM MS NAA	Unpublished
P. corcovadensis	MS or AR, 1.25 μM 2iP	5.0 μM NAA MS	Unpublished
P. niruri	MS or AR, 1.25 μM BA	MS without PGRs	Unpublished
P.urinaria	MS or AR, 5.0 μM kinetin	1.25–5.0 μM MS NAA	*(8)*
P. caroliniensis	MS or AR, 2.5–5.0 μM BA or 1.25–5.0 μM kinetin	1.25 μM NAA MS or 1.25–5.0 μM IAA	*(6)*
P. stipulatus	MS or AR, 2.5–5.0 μM IBA	MS without PGRs	*(7)*
P. fraternus	MS without PGRs	MS without PGRs	*(4)*

Note: Basic MS medium containing appropriate concentrations of plant growth regulators (PGRs) were used for different objectives. The photosynthetic photon flux at culture level was 20–25/μmol/m²/s

Use these culture conditions in all the protocols mentioned below unless otherwise stated.

3.1.3. Shoot Initiation and Multiplication

1. Remove the single node explants, 0.8- to 1.2-cm long, from 40-d-old seedlings aseptically grown in the light, photosynthetic photon flux of 20–25 μmol/m²/s supplied by fluorescent light tubes.
2. Place the explants vertically on the MS basal medium supplemented with 2% *(w/v)* sucrose, 0.2% *(w/v)* Phytagel, and plant growth regulators specified in **Table 1**.
3. Incubate cultures at 25°C at 70% relative humidity for a 16-h photoperiod and photosynthetic photon flux of 20–25 μmol/m²/s supplied by fluorescent light tubes.
4. Evaluate the percentage of shoot formation, flowering, and number of shoots initiated per explant, after 50 d. Example results of shoot proliferation are shown in **Fig. 1**.

3.1.4. Rooting

1. Remove microcuttings from 40-d-old clusters of shoots produced in the multiplication medium.
2. Inoculate explants on the MS medium supplemented with 2% *(w/v)* sucrose, 0.2% *(w/v)* Phytagel, and adequate amount of plant growth regulators (*see* **Table 1**).
3. Incubate cultures at 25°C at 70% relative humidity fot a 16-h photoperiod and photosynthetic photon flux of 20–25 μmol/m²/s supplied by fluorescent light tubes.
4. Evaluate adventitious root regeneration after 5 weeks of culture.

Table 2
Composition of Culture Media used for Callus Induction and Growth in Selected Species of *Phyllanthus*

Species	Explant position	PGRs	Culture condition	References
P. tenellus	Horizontal	20–40 μM IBA or IAA	Dark	Unpublished
P. corcovadensis	Horizontal	10–20 μM IBA	Dark	Unpublished
P. niruri	Horizontal	10–20 μM IBA	Dark	Unpublished
P. urinaria	Vertical Horizontal	5.0 2,4-D μM or NAA 5.0 μM IBA	Light Dark	*(8)*
P. caroliniensis	Vertical Horizontal	5.0 μM 2,4-D 40.0 μM IBA	Light Dark	*(6)*
P. stipulatus	Vertical Horizontal	5.0 μM NAA 2.5–5.0 μM BA, 2iP, NAA, or 2, 4-D	Light Dark	*(7)*
P. fraternus	Vertical	5.0 μM 2,4-d or IAA	Light	*(4)*

Note: Basic MS medium containing 20 g/L sucrose, 2 g/L phytagel, and plant growth regulators (PGRs) in appropriate concentrations were used. The photosynthetic photon flux at culture level was 20–25 μmol/m^2/s

3.1.5. Acclimatization

1. Remove 30-d-old rooted regenerated plants from the test tubes, wash roots with running water to remove Phytagel. Care should be taken to prevent damage to roots.
2. Trim the main shoot having two axillary buds per plantlet.
3. Transfer the rooted plants to trays containing steam-sterilized medium grade river sand covered with a polyvinyl chloride (PVC) transparent film.
4. After 1 wk, gradually remove the PVC film from the seed trays and expose them to the relative humidity, average 70%, of the culture room.
5. After 20 d, transfer plants to 250-mL plastic pots containing a 1:1 mixture of sand and soil. At all stages of acclimatization keep them in the culture room, under identical culture conditions to those used for the tissue culture studies. After 90 d acclimatization transfer the plants to the greenhouse. Example results of acclimatized microplants are shown in **Fig. 1**.

3.2. Callus Induction

1. Inoculate single node explants (0.8–1.2 cm long) obtained from 40-day-old aseptically grown seedlings on MS medium supplemented with 2% (w/v) sucrose, 0.2% (w/v) Phytagel and the

Fig. 1. Multiple shoot formation from single node explants of *Phyllanthus caroliniensis* (**a**) and *Phyllanthus urinaria* (**b**) after 45 d. Acclimatized micropropagated plants of *Phyllanthus fraternus* (**c**), *Phylanthus stipulatus* (**d**), *P. caroliniensis* (**e**) and *P. urinaria* (**f**) after 4 mo (**c**, **d**) and 2 mo (**e**, **f**) acclimatization.

adequate level of plant growth regulators (*see* **Table 2**). Place explants either vertically or horizontally into the medium.

2. Incubate cultures at 25°C in either light under 16 h photoperiod, photon flux 20–25 μmol/m^2/s, supplied by fluorescent light tubes or in the dark.
3. After 8 weeks evaluate the callus fresh mass. Example results of callus produced are shown in **Fig. 2**.

3.3. Root Culture

1. Excise root tip segments (1- to 3-cm long) from —30- to 40-d-old microplants aseptically grown on MS basal medium supplemented with 2% *(w/v)* sucrose, 0.2% *(w/v)* Phytagel.
2. Inoculate root segments in 125-mL Erlenmeyer flasks containing 20 mL MS liquid medium supplemented with 2% *(w/v)* sucrose and 1.1 μM NAA.

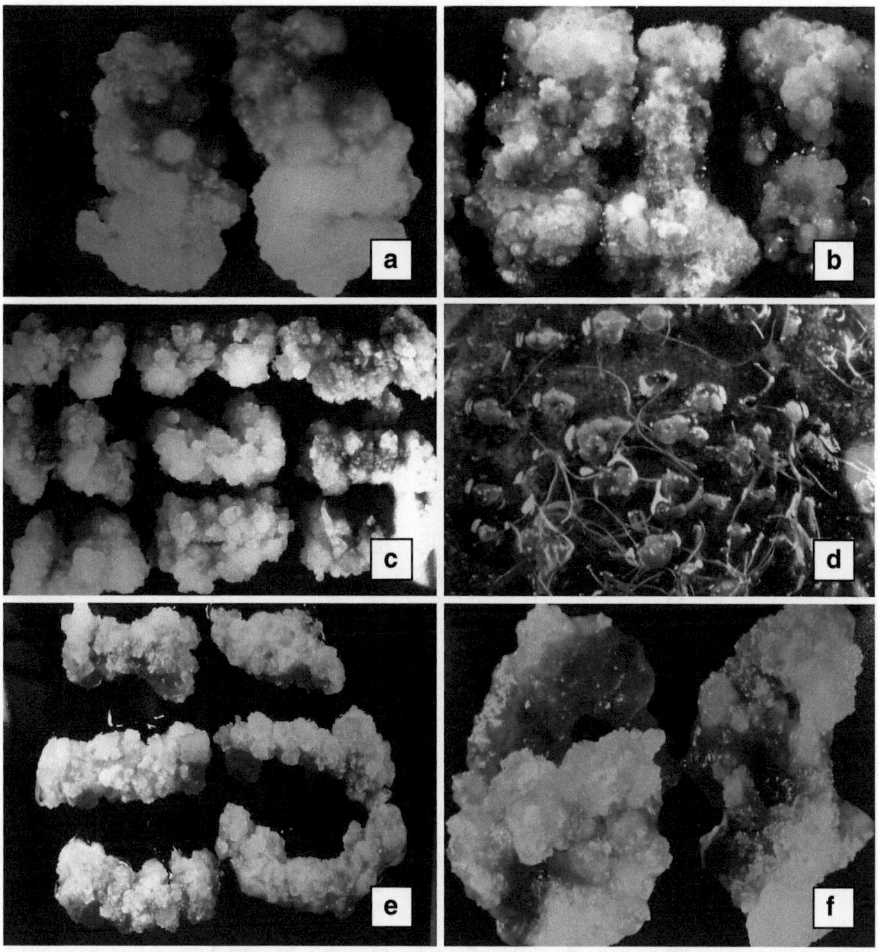

Fig. 2. Callus produced from shoot segments of *Phyllanthus caroliniensis* (**a**), *Phyllanthus urinaria* (**b**), *Phyllanthus stipulatus* (**c**, **d**), *Phyllanthus fraternus* (**e**), and *Phyllanthus tenellus* (**f**) after 8 weeks. Note rhyzogenesis in callus of *Phyllanthus stipulatus* when transferred to MS liquid medium (**d**).

3. Incubate cultures on a rotatory shaker at 100 rpm in the dark at 25°C.
4. After 60 d evaluate the root culture fresh mass.

3.4. Phytochemical Analysis

1. Lyophilize and oven dry 5–72 g callus or 14–29 g roots at 40–50°C for 48 h and macerate with 95% methanol at room temperature for approximately 5 d.
2. Carry out the partioning of the methanolic extracts with ethyl acetate and the pharmacological tests with both extracts.
3. Evaporate the extracts under reduced pressure to yield a residue of approximately 1% original weight of callus tissue.
4. Analyze the chromatographic profile of all methanolic extracts by TLC using Merck silica precoated aluminium plates 20 μM thick with 90:10 *(v/v)* chloroform:methanol solvent system.

Table 3
Chemical Composition and Antinociceptive Action in Mice of Callus Methanolic Extracts of Selected Species of *Phyllanthus*

Species	PGRs	Compounds	Antinociceptive action against formalin-induced pain	References
P. tenellus	20–40 μM IAA	Phenolics	First and second phases	*(9)*
P. corcovadensis	20–40 μM IBA	Phenolics	First and second phases	*(9)*
P. niruri	20–40 μM IBA	Phenolics	First and second phases	*(9)*
P. caroliniensis	5.0 μM 2, 4-D	Glochidonol glochidone	First phase	*(5)*
P. stipulatus	5.0 μM 2, 4-D	Glochidonol glochidone stigmasterol	First phase	*(5)*
P. urinaria	40.0 μM BA (9) + IBA	Glochidonol stigmasterol	First and second phases	*(5)*
P. fraternus	5.0 μM 2, 4-D	Glochidonol glochidone stigmasterol	First phase	*(5)*

Note: Basic MS medium containing 20 g/L sucrose, 2 g/L phytagel, and plant growth regulators (PGRs) in appropriate concentrations were used for callus culture

5. To detect the presence of phenolics in the methanolic extracts use Ultra Violet, the ferric chloride test, and the magnesium hydrochloride reduction test to visualize alkaloid use the Dragendorff and Wagner reagents and to detect steroids or terpenoids use the anisaldehyde-sulfuric acid reagent.
6. Confirm the presence of glochidone, glochidonol, glochidiol, and stigmasterol, using co-injection with authentic samples by co-TLC and co-GC.
7. Carry out the GC (Shimadzu, model GC 14-B) analysis using a 30-m capillary column of phase type LM-1 (polymethylsiloxane) for GC separation of the pre-concentrated compounds.
8. Set the temperature program for the analyses as follows: hold the initial temperature of 40°C for 3 min and then increase to 240°C at 12°C/min, and from 240 to 310°C at 5°C/min. Set the injector and detector temperatures at 280°C and 320°C, respectively. The run time is 20 min. Use helium as a carrier gas at a flow rate of 1–2 mL/min. Example results of isolated compounds are shown in **Table 3**.

4. Notes

1. *Phyllanthus* seeds are very small therefore we recommend packing them in empty tea bags to facilitate seed sterilization.
2. The media and plant growth regulator solutions are prepared in double distilled water. Adjustment of pH is done before the addition of phytagel as it will degrade the electrode of the pH-meter because of its stickiness.
3. All the stock solutions can be stored at 4°C until use. Clean and autoclaved flasks are used to prevent contamination.
4. To optimize the protocols on shoot proliferation and rooting of selected *Phyllanthus* species the separate effects of BA, 2iP, kinetin, NAA, 2, 4-d, IAA, and IBA are tested and the best culture conditions for the different steps of in vitro propagation are shown (*see* **Table 1**).
5. MS and Anderson Rhododendron media promotes significant shoot culture growth in terms of numbers of shoots and nodes produced per explant in all selected Phyllanthus species. Growth of axillary buds is achieved in most of the treatments and as a consequence shoots are produced in 90–100% of the explants for all species. Cytokinins are effective in stimulating shoot multiplication from single node explants of *P. corcovadensis*, *P. niruri*, *P. urinaria*, and *P. caroliniensis* but not of *P. stipulatus* and *P. tenellus*. The multiplication rate

for *P. tenellus*, *P. corcovadensis*, *P. stipulatus*, and *P. fraternus* is 8–9 propagules per culture cycle and for *P.niruri*, *P. urinaria*, and *P. caroliniensis* is 20–21 propagules per culture cycle.

6. Rooting is achieved in 93–100% of the microshoots without plant growth regulators, especially in *P. niruri*, *P. stipulatus*, and *P. fraternus*, although in all selected *Phyllanthus* species the addition of NAA significantly increases the number of roots per explant. after 16 d. Approximately the same length of time, 16 d, is required for all the other selected species to achieve maximum in vitro rooting rates, however for *P. caroliniensis* 48 d are necessary. For all selected *Phyllanthus* species regenerated plants are successfully acclimatized without mist-house/shade-house facilities. One hundred percent of the plantlets survived hardening on river sand and transfer to potting medium consisting of garden soil:sand mixture (1:1). All selected *Phyllanthus* species flowers after ex vitro transfer *(4)*.
7. To test the effect of PGRs on callus initiation on selected Phyllanthus species the separate effects of cytokinins BA, kinetin, 2iP, and auxins 2,4-d, IBA, NAA, IAA are used and the best culture conditions for protocols on callus growth are shown in Table 2. Callus growth is affected by explant position in the culture medium, light/dark culture condition.
8. The growth rate of the root culture systems developed for *P. urinaria* is superior to those observed for root cultures of *P. caroliniensis*, *P. stipulatus*, and *P. fraternus* in the same culture conditions, when maximum root fresh weight is obtained after 45 d of culture.
9. The main classes of compounds confirmed by GC in callus extracts of *P. urinaria*, *P. caroliniensis*, *P. stipulatus*, and *P. fraternus* are shown in **Table 3**. Preliminary phytochemical analysis of callus extracts of *P. tenellus*, *P. corcovadensis*, and *P. niruri* reveals the absence of alkaloids and flavonoids. The infrared and ^{1}H-NMR analysis suggests that phenolic compounds are the main constituent present in callus extracts of theses species. The TLC test for root extracts reveals the presence of glochidone in *P. stipulatus*, *P. fraternus*, and *P. caroliniensis* and of stigmasterol in *P. stipulatus*, *P. urinaria*, and *P. fraternus.*

Acknowledgments

The authors acknowledges the support of the International Foundation for Science (Sweden, grants D/1265-1, D/1265-2 and D/1265-3 to AMV); The British Council (UK); the Conselho Nacional de Desenvolvimento Científico e Tecnológico (CNPq/

Brasil); the Coordenação de Aperfeiçoamento de Pessoal de Nível Superior (PET/CAPES/Ministry of Education/Brasil); and LG Amaral for classification of plant material.

References

1. Calixto J.B., Santos A.R.S., Cechinel-Filho V. and Yunes R.A. (1998). A review of the plants of the genus *Phyllanthus*: Their chemistry, pharmacology and therapeutic potential. *Med. Res. Rev.* **18**, 225–258.
2. Bagalkotkar G., Sagineedu S.R., Saad M.S. and Stanslas J. (2006). Phytochemicals from *Phyllanthus niruri* Linn and their pharmacological properties a review. *J. Pharm. Pharmacol.* **58**, 1559–1570.
3. Unander D.W., Bryan H.H., Lance C.J. and McMillan R.T. (1995). Factors affecting germination and stand establishment of *Phyllanthus amarus* (Euphorbiaceae). *Econ. Bot.* **49**, 49–55.
4. Catapan E. (1999). *In vitro* culture and phytochemical analysis of Phyllanthus spp. MSc. Thesis, Federal University of Santa Catarina, Florianopolis, SC, Brazil (in Brazilian Portuguese with English summary).
5. Catapan E., Otuki M.F., Viana A.M., Yunes R.A., Bresciani L.F.V., Santos A.R.S., Calixto J.B. and Cechinel-Filho V. (2000a). Pharmacological activity and chemical composition of callus culture extracts from selected species of *Phyllanthus*. Die Pharmazie **55** (12), 1–2.
6. Catapan E., Otuki M.F. and Viana A.M. (2000b). *In vitro* culture of *Phyllanthus caroliniensis* (Euphorbiaceae). *Plant Cell Tiss. Org. Cult.* **62**, 195–202.
7. Catapan E., Otuki M.F. and Viana A.M. (2001). *In vitro* culture of *Phyllanthus stipulatus* (Euphorbiaceae). *Rev. Bras. Bot.* **24**, 1–8.
8. Catapan E., Silva M.L.B., Moreno F.N. and Viana A.M. (2002). Micropropagation, callus and root culture of *Phyllanthus urinaria* (Euphorbiaceae). *Plant Cell Tiss. Org. Cult.* **70**, 301–309.
9. Santos A.R.S., Cechinel-Filho V., Viana A.M., Moreno F.N., Campos M.M., Yunes R.A. and Calixto J.B. (1994). Analgesic effects of callus culture extracts from selected species of *Phyllanthus* in mice. *J. Pharm. Pharmacol.* **46**, 755–759.
10. Krogh R., Kroth R., Berti C., Madeira A.O., Souza M.M., Cechinel Filho V., Delle-Monache F. and Yunes R.A. (1999). Isolation and identification of compounds with antinociceptive action from *Ipomoea pes-capre* (L.) *R.Br. Pharmazie* **54**, 1–6.
11. Ugaz O.L. (1988) Investigación Fitoquímica. Pontificia Universidad Catolica, Peru, pp. 97–98.
12. Niero R. (1993) Isolation and identification of analgesic compounds of *Phyllanthus corcovadensis* (Euphorbiaceae). MSc. Thesis, Federal University of Santa Catarina, Florianopolis, SC, Brazil (in Brazilian Portuguese with English summary).
13. Murashige T. and Skoog F. (1962) A revised medium for rapid growth and bioassays with tobacco tissue cultures. *Physiol. Plant.* **15**, 473–497.
14. Anderson W.C. (1978) Tissue culture propagation of Rhododendrons. *In vitro* **14**, 334.

Chapter 15

In Vitro Clonal Propagation of *Asparagus racemosus*, a High Value Medicinal Plant

Sanjay Saxena and Nishritha Bopana

Summary

Asparagus racemosus Willd., an important medicinal plant of tropical and subtropical India, is a potent phytoestrogen which is used extensively in the treatment of menopausal symptoms, diarrhea, dyspepsia, and neurodegenerative disorders. The multiple uses of this species have increased its commercial demand resulting in over-exploitation. Consequently, the plant is recognized as being "critically endangered" in its natural habitat. Development of an efficient micropropagation protocol will not only play a significant role in meeting the requirement of planting material for commercial cultivation, but also in aiding the conservation process. This chapter describes the protocol for in vitro propagation of *A. racemosus* by axillary branching method followed by inter simple sequence repeat marker assay to establish the clonal fidelity of the regenerants.

Key words: *Asparagus racemosus*, Axillary branching, Clonal fidelity, ISSR, PCR

1. Introduction

Asparagus racemosus Willd. (Asparagaceae), commonly called Shatavari, is a woody climber that grows throughout the tropical and subtropical parts of India upto an altitude of 1500 m. Shatavari is an important medicinal plant *(1)*, well known for its phytoestrogen properties and extensively used in combating menopausal symptoms and increasing lactation *(2–4)*. It also has antioxidant *(5)* and immunostimulation properties *(6)* and is widely used in Ayurveda for treating dyspepsia *(7)*. The major active constituents imparting medicinal property to the plant are

S. Mohan Jain and Praveen K. Saxena (eds.), *Methods in Molecular Biology, Protocols for* In Vitro *Cultures and Secondary Metabolite Analysis of Aromatic and Medicinal Plants, vol. 547*
© Humana Press, a part of Springer Science+Business Media, LLC 2009
DOI: 10.1007/978-1-60327-287-2_15

steroidal saponins (shatavarins I-IV) that are present in the roots of the plant *(8, 9)*.

Multiple usages, rising demand, and destructive harvesting has led to the shrinking of natural populations of *A. racemosus*, which is now categorized as "critically endangered" in its natural habitat. In nature, the species regenerates through seeds though it can also be propagated vegetatively but this is a slow process that results in smaller number of new propagules *(10)*. As meeting the growing demand of this species poses a major challenge, in vitro techniques for propagation can be of immense value in offsetting the pressure on natural populations and facilitating the conservation efforts. Micropropagation also has the potential application of mass multiplication of the germplasm containing higher amounts of Shatavarin.

To the best of our knowledge, there is only a single report on the micropropagation of *A. racemosus (11)* wherein shoots were regenerated from callus cultures obtained from shoot segments. Organogenesis, however, carries the risk of somaclonal variations. The occurrence of somaclonal variations in the plants can seriously limit the broader utility of micropropagation systems. In the absence of any reports on axillary branching method and the huge market for this plant, the development of an efficient micropropagation protocol for the large scale production of clonally uniform plants is significant. The present communication describes an efficient in vitro propagation protocol for the production of clonally uniform plants of *A. racemosus*.

2. Materials

2.1. Micropropagation

2.1.1. Stock Preparation

1. The formulation of Murashige and Skoog's (MS) *(12)* medium is provided in **Table 1**.
2. Growth regulator stocks: 2–isopentyl adenine (2ip), kinetin, and 1-napthaleneacetic acid (NAA) (Sigma; St. Louis, MO).
3. Growth adjuvants: Malt extract (Qualigen; India), adenine sulphate (Sigma).
4. Weighing balances (Mettler PE3600 Delta Range and Afcoset electronic balance ER-182A) Microwave (Panasonic); pH meter (Control Dynamics pH meter); autoclave.
5. Culture vessels: test tubes (Borosil; India), glass jars (Hindustan Glass and Industries Ltd; India), cotton plugs, polypropylene caps plugged with cotton.
6. For the preparation of stock solutions, analytical grade chemicals and reverse osmosis water are used. Prepare separate stock

Table 1
Composition of MS medium

Constituents	Amount (mg/L medium)
Major salts	
NH_4NO_3	1650
KNO_3	1900
$CaCl_2 \cdot 2H_2O$	440
$MgSO_4 \cdot 7H_2O$	370
$KH_2PO_4 \cdot$	170
Minor salts	
H_3BO_3	6.2
$MnSO_4$	22.3
$ZnSO_4 \cdot 7H_2O$	8.6
KI	0.83
$Na_2MoO_4 \cdot 2H_2O$	0.25
$CuSO_4 \cdot 5H_2O$	0.025
$CoCl_2 \cdot 6H_2O$	0.025
Iron	
$Na_2EDTA \cdot 2H_2O$	37.3
$FeSO_4 \cdot 7H_2O$	27.8
Organics	
Glycine	2.0
Pyridoxine	0.5
Nicotinic acid	0.5
Thiamine HCl	0.1
Myoinositol	100
Sucrose	3%
Concentration of gelling agent	
Agar	0.8%

solutions for macronutrients, micronutrients, iron, vitamins, and *myo*-inositol by dissolving the weighed constituent chemicals individually in 500-mL or 1-L conical flasks.

7. For the preparation of iron stock, weigh $Na_2EDTA{\cdot}2H_2O$ and $FeSO4{\cdot}7H_2O$ and dissolve them separately in warm water. Then mix the two solutions with the help of a magnetic stirrer.

2.1.2. Inoculation

1. Laminar airflow cabinet (Narang Scientific Works; India), glass bead sterilizer (Dent-eq, BS 1000) (Bangalore; India).
2. Forceps, scalpels, sterile blades, burner, autoclaved plates, 70% ethanol, mercuric chloride (Sigma).

2.1.3. Culture Conditions

1. Culture room conditions: 26 ± 2°C; 12-h photoperiod; light intensity of approximately 47.29 $\mu mol/m^2/s$ provided by cool, white fluorescent tubes (40 W) (Phillips; India).

2.2. ISSR Assay

2.2.1. Leaf Material

1. Collect leaves from randomly selected tissue-cultured plantlets (at 4th, 8th, 12th, 16th, and 20th passages) and from hardened plants (polyhouse stage and field level) for analysis.
2. In addition, collect leaf material from related species to serve as outliers in the study.
3. Label and lyophilize the leaf material in the "Virtis freezemobile G" lyophilizer for 48 h at –70°C. Store the lyophilized leaf material in air-tight zip lock bags at –20°C until future use.

2.2.2. Reagents

1. DNA extraction buffer: 2% CTAB, 1.4 M NaCl, 20 mM ethylenediamine tetraacetic acid (EDTA), and 100 mM Tris, pH 8.0; Make upto 2 L; store at room temperature.
2. 20X TBE buffer: 0.089 M Tris, 0.089 M boric acid, and 0.002 M EDTA. Make up to 2 L and store at room temperature.
3. Washing solution: 70% ethanol and 10 mM ammonium acetate in 1 L of sterile distilled water.
4. Loading dye: 0.25% bromophenol blue, 0.25% xylene-cyanol, and 30% glycerol in electrophoresis buffer or water.

2.2.3. PCR Amplification for ISSR Assay

1. Genomic DNA: Diluted to a concentration of 25 ng/μL.
2. Polymerase chain reaction (PCR) mix: Sterile water, 10X buffer, 50 mM $MgCl_2$, dNTPs, ISSR primers (University of British Columbia) UBC 812, 814, 818, 840, 842, 843, and 848 and *Taq* polymerase (prepared in a 2-mL Eppendorf tube on ice).
3. Equipment: "Bio-Rad DNA Engine" (Peltier Thermal cycler); Gel electrophoresis unit (Bio-Rad Sub-Cell® GT), power-pack (Gene Power Supply GPS 200/400; Pharmacia), gel documentation system AlphaImager® EC (Alpha Innotech Corporation, ChemiImager™ Ready).
4. For gel electrophoresis: Agarose (United States Biochemicals, Ohio), ethidium bromide (Sigma), 5X dye (bromophenol blue), λ1-kbp ladder (Gibco BRL).

3. Methods

3.1. Stock Solutions for Media Preparation

1. To avoid precipitation, store all stock solutions at 4°C and for the same reason, stock solutions of plant growth regulators should be prepared afresh each time and in small quantities.
2. To prepare cytokinin stocks such as that of 2ip and kinetin, weigh the hormone and dissolve it in a few drops of NaOH (1 N). Make-up the desired volume with reverse osmosis water. Auxin stocks are prepared by dissolving weighed quantities in minimum amount of ethanol followed by making up the volume with reverse osmosis water.

3.2. Media Preparation

1. Measure the stock solutions according to their dilution factors and add to 0.8% molten agar. Then make-up the desired volume with reverse osmosis water.
2. Prior to dispensing media into culture vessels, adjust the pH of the medium to 5.8 with the help of 1 N NaOH and 1N HCl.
3. Depending on the culture vessel, pour 20 mL culture medium into test-tubes and 80 mL into 400-mL ordinary glass jars. Use cotton plugs to cap the test tubes while the jars can be covered with transparent polypropylene lids bearing a hole plugged with non-absorbent cotton wrapped in muslin cloth.
4. Autoclave the media at 121°C and 1.05 kg/cm^2 pressure for 20 min.

3.3. Inoculation

1. Sterilize instruments (forceps, scalpels) by controlled flaming with 95% ethanol inside the laminar airflow cabinet.
2. During the course of the inoculations however, the instruments are sterilized by inserting them inside a glass bead sterilizer maintained at 250°C for 30 s.

3.4. Initiation of Aseptic Cultures of Asparagus racemosus

1. Collect single node segments from *A. racemosus* plants maintained in the greenhouse. Give the explants a quick rinse in 70% ethanol for 20–25 s. Wash with 4 drops of Teepol in 100 mL water for 10 min followed by a thorough washing under running tap water for 15 min.
2. Surface sterilize the nodal segments by treating them with 0.1% solution of mercuric chloride for 10 min in the laminar air flow chamber. Then wash the explants thrice with sterile distilled water to remove traces of the sterilant, trim the cut ends of the segments and culture in basal MS medium (3% sucrose and 0.8% agar) in 25 × 100-mm glass tubes.

3.5. Multiplication

1. After 3 wk of culture on initiation medium, carefully excise the in vitro formed shoots from the nodal segment.

2. Culture in semi-solid MS medium containing 3.69 μM 2ip, 3% sucrose, and 0.8% agar for further multiplication of shoots (*see* **Notes 1** and **2**) (**Fig. 1a**).

3.6. Rooting

1. At the end of the shoot multiplication cycle, transfer the clusters (6–8 shoots in each cluster) to rooting media (half strength MS medium (i.e., major salts reduced to half with 1.61 μM NAA, 0.46 μM kinetin, 98.91 μM adenine sulfate, 500 mg/L malt extract, 198.25 μM phloroglucinol, and 3% sucrose) in 400-mL glass jars containing 80 mL of medium.
2. Incubate the jars in dark for 5 d followed by transfer to light for 20 d.

3.7. Hardening and Transplantation

1. After 25 d remove the plantlets from the rooting medium (**Fig. 1b**) and wash them thoroughly under running tap water to remove all traces of medium attached to the roots.
2. Transplant the rooted plants to 16 cm × 8-cm polythene bags containing soil and agropeat in a ratio of 2:1 *(v/v)*.
3. Rear the plants in a greenhouse maintained at 28 ± 2°C. Initially the plants should be placed close to the cooling pads (relative humidity: 80–85%), then gradually shifted away from the pads toward the exhaust fans over a period of 1 wk.
4. Thereafter, shift the plants to a polyhouse (a tunnel-like structure made up of polythene sheet with air coolers at one end and exhaust fans at the other) where, depending on the weather conditions prevailing outside, the plants can be nurtured until they are ready for transfer to shade area and nursery (**Fig. 1c**).

3.8. ISSR Assay

3.8.1. Genomic DNA Extraction

Total DNA is extracted following a modified Cetyl Trimethyl Ammonium Bromide (CTAB) DNA extraction procedure based on the Doyle and Doyle *(13)* protocol.

1. Grind the lyophilized leaf material (*see* **Note 3**) to a fine powder using a pestle and mortar with the help of glass beads as an abrasive agent to aid in the grinding process. Transfer 0.5 g of the powder to 2-mL eppendorf tubes.
2. Add 1 mL of pre-warmed (65°C) CTAB extraction buffer with 2 μL β-mercaptoethanol to the crushed plant material and mix thoroughly. Incubate the samples at 65°C in a water bath for 45 min with intermittent mixing every 10 min.
3. Remove the samples and let them attain room temperature. Add an equal volume of chloroform: isoamyl alcohol (24:1) to the DNA samples and mix well to ensure emulsification of the phases.
4. Centrifuge the samples at 9,660 × g for 10 min at 20°C and transfer the upper aqueous phase to a clean tube. Add chilled isopropanol (0.6 vol) to precipitate the DNA.

Fig. 1. Micropropagation of *A. racemosus*: (**a**) multiplication, (**b**) rooted cluster and (**c**) hardening in pots.

5. Mix the sample well and incubate at room temperature for 20 min. Then pellet the precipitated DNA by centrifuging at 4°C and 9,600 × g for 15 min.
6. Decant the supernatant and remove the residual CTAB by adding 500 μL of washing solution and gently agitating the pellet followed by centrifuging at 4°C and 9,600 × g for 10 min.
7. Decant the supernatant and dry the pellet. To this add 50 μL TE buffer with 2 μL RNAse A and incubate at 65°C for 10 min.
8. Store the extracted DNA at –20°C.

3.8.2. DNA Quantification

1. For DNA quantification, weigh 0.8 g agarose and melt in 100 mL 0.5X TBE buffer.
2. Allow the molten agarose to cool to about 50–60°C and add ethidium bromide to a final concentration of 0.5 μg/mL to enable visualization of DNA under ultraviolet light.
3. Arrange the gel mould with the comb placed in a manner such that it does not touch the base of the mould. Pour the molten agarose into the mould ensuring even gel thickness and avoiding the appearance of bubbles.

4. Once the agarose has solidified, remove the comb carefully and place the gel in the electrophoresis tank filled with TBE buffer.
5. Add 5 μL 2X dye to 5 μL genomic DNA and load into the wells. Carry out electrophoresis at 30 mA (60 V) for 2–3 h. Uncut lambda DNA (25 ng/μL) is used as a standard molecular weight marker to quantify genomic DNA.
6. Visualize and photograph the gel using the gel documentation system.

3.8.3. PCR Assay

1. PCR cocktail (in a 2-mL eppendorf tube): 10.40 μL sterile water, 2 μL 10X buffer, 1 μL 50 mM $MgCl_2$, 0.5 μL dNTPs, 1 μL primers **(Table 2)** (UBC 812, **Fig 2**; 814; 818; 840, **Fig 3**; 842; 843; and 848) and 0.1 μL *Taq*.
2. Suspend 5 μL aliquot of the DNA sample (25 ng as estimated by the quantification procedure) in 0.5 μL PCR tubes and add 15 μL of the cocktail mix to each tube. Give the mix a momentary spin.
3. Cover the PCR tubes and transfer the samples to "Bio-Rad DNA Engine" programmed for 35 cycles with the following parameters:

3.8.4. Agarose Gel Electrophoresis

1. The gel for separation of PCR products should be prepared by weighing 2 g agarose and melting it in 100 mL TBE buffer.
2. Load the first well of the gel (lane M) with 10 μL of λ 1-kbp ladder.

Table 2
Fingerprint Patterns of Micropropagated *A. racemosus* using 7 ISSR Primers

UBC ISSR primer	Primer sequence[a]	GC content (%)	Total bands amplified
812	$(GA)_8A$	47.06	23
814	$(CT)_8A$	47.06	19
818	$(CA)_8G$	52.97	27
840	$(GA)_8YT$	44.44–50.00	20
842	$(GA)_8YG$	50.00–55.55	33
843	$(CT)_8RA$	44.44–50.00	10
848	$(CA)_8RG$	50.00–55.55	24

[a]Key to base compositions
R = A, G; *Y* = C, T

Cycle	Steps	Number of cycles
Pre-amplification	2 min at 94°C	1
Extension	30 s at 94°C	35
Denaturation	30 s at 42°C	35
Annealing	1 min at 72°C	35
Post-amplification	5 min at 72°C	1
Extension	4°C	

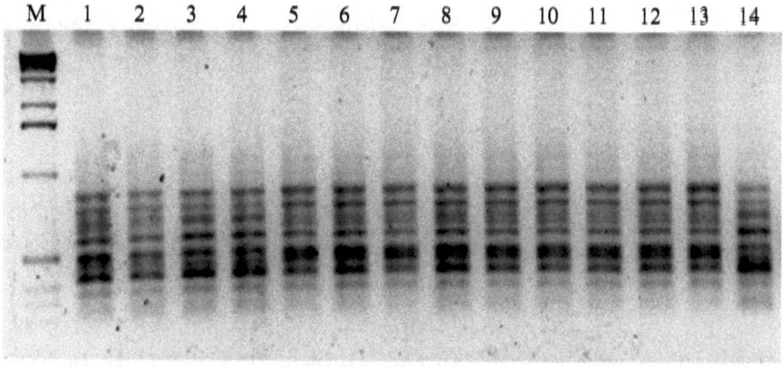

Fig. 2. DNA amplification obtained with primer UBC 812; 1 kbp DNA ladder *(lane M)*; mother plant *(lane 1)*; micropropagated plants at 4th *(lanes 2 and 3)*, 8th *(lanes 4 and 5)*, 12th *(lanes 6 and 7)*, 16th *(lanes 8 and 9)*, and 20th *(lanes 10 and 11)* passages; plant at polyhouse stage *(lanes 12 and 13)*, and plant at field stage *(lane14)*.

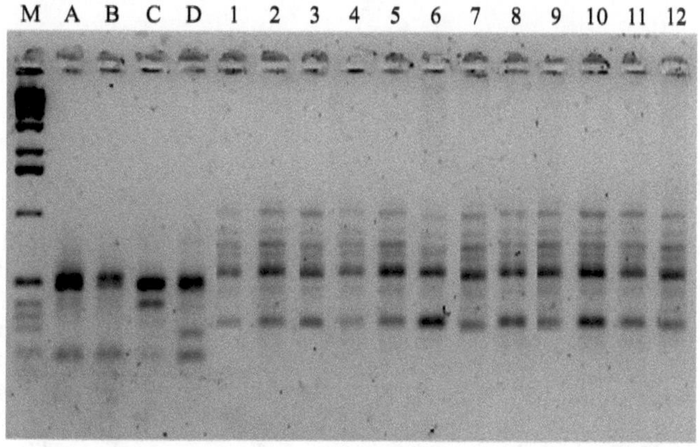

Fig. 3. DNA amplification obtained with primer UBC 840; 1-kbp DNA ladder *(lane M)*; outliers *(lanes **A–D**)*; mother plant *(lane 1)*, and micropropagated plants at 20th passage *(lanes 2–12)*.

3. Add 2.5 μL of 10X loading dye to each of the PCR amplification products and load the entire volume into the well.
4. Complete the assembly of the gel unit and connect to a power supply. Run the gel at 80 V for 3–4 h.

5. Visualize the gel (**Figs. 2** and **3**) and photograph using the gel documentation system AlphaImager® EC.

3.8.5. Scoring of Bands

1. Score the ISSR amplification products for their presence and absence across all the individuals to determine clonal fidelity vis-à-vis the mother plant.

4. Notes

1. Sub-culturing of shoots was done at intervals of 30 d. During sub-culturing, the larger sized shoot clusters obtained at the end of each multiplication cycle were further divided into smaller clusters.
2. Multiplication rates were calculated on the basis of the number of clusters derived from one culture at the end of each passage. As a rule, all the dead and decayed tissue was removed before transfer to fresh medium.
3. Lyophilization enables long term storage and easy handling of the samples. Whereas fresh or frozen material usually gives high yields and a somewhat more intact DNA, lyophilized tissue is easier to handle and can be stored for longer periods of time.

References

1. Bopana, N. and Saxena, S. (2007). *Asparagus racemosus* – Ethnopharmacological evaluation and conservation needs. *Journal of Ethnopharmacology* **110**, 1–15.
2. Mitra, S.K., Gopumadhavan, S., Venkataranganna, M.V., Sarma, D.N.K. and Anturlikar, S.D. (1999). Uterine tonic activity of U-3107 a herbal preparation in rats. *Indian Journal of Pharmacology* **31**, 200–203.
3. Mayo, J.L. (1998). Black cohosh and Chasteberry: Herbs valued by women for centuries. *Clinical Nutrition Insights* **6**(15), 1–4.
4. Sabnis, P.B., Gaitonde, B.H. and Jethmalani, M. (1968). Effect of alcoholic extract of *Asparagus racemosus* on mammary glands of rats. *Indian Journal of Experimental Biology* **6**, 55–57.
5. Kamat, J.P., Boloor, K.K., Devasagayam, T.P.A. and Venkatachalam, S.R. (2000). Antioxidant properties of *Asparagus racemosus* against damage induced by γ-radiation in rat liver mitochondria. *Journal of Ethnopharmacology* **71**(3), 425–435.
6. Rege, N.N., Nazareth, H.M., Issac, A.A., Karandikar, S.M. and Dahanukar, S.A. (1989). Immunotherapeutic modulation of intraperitoneal adhesions by *Asparagus racemosus*. *Journal of Postgraduate Medicine* **35**, 199–203.
7. Dalvi, S.S., Nadkarni, P.M. and Gupta, K.C. (1990). Effect of *Asparagus racemosus* (Shatavari) on gastric emptying time in normal healthy volunteers. *Journal of Postgraduate Medicine* **36**, 91–94.

8. Anonymous. (1987). *Asparagus*. In: Wealth of India. New Delhi: CSIR Publications. pp. 468–472.
9. Thomsen, M. (2002). Shatavari – *Asparagus racemosus*. <http://www.phytomedicine.com.au/files/articles/shatavari.pdf>
10. Tewari, D.N. (2000). Report of the Task Force on Conservation and Sustainable use of medicinal plants. Government of India, Planning commission; pp 90.
11. Kar, D.K. and Sen, S. (1985). Micropropagation of *Asparagus racemosus*. *Plant Cell, Tissue and Organ Culture* **5**, 89–95.
12. Murashige, T. and Skoog, F. (1962). A revised medium for rapid growth and bio-assays with tobacco tissue cultures. *Physiologia Plantarum* **15**, 473–497.
13. Doyle, J.J. and Doyle, J.L. (1990). Isolation of plant DNA from fresh tissue. *Focus* **12**, 13–15.

Chapter 16

Micropropagation of *Penthorum chinense* Through Axillary Buds

Jun Yang and Zheng-song Peng

Summary

Penthorum chinense Pursh is a traditional medicinal herb in China. Micropropagation protocol of this plant has been established. The shoot induction rate is high by culturing nodal explants on Murashige and Skoog (MS) medium supplemented with 6-benzyladenine (BA) as compared with kinetin (Kn). After 6 wks, the highest shoot formation (5.2) is achieved in 59.2% nodal explants cultured on MS medium combined with 2.0 mg/L BA. After 4 wk of subculture on the fresh MS medium, the highest shoot multiplication rate 6.4 is accomplished. MS medium containing 1.0 mg/L BA is most suitable for shoot propagation. Individual well developed shoots (0.8-to 1.0-cm long) are excised and culture on the MS medium containing with 0.5 mg/L indole-3-acetic acid (IAA), for rooting. Three weeks later, 98.8% of shoots had rooted with an average of ten roots per shoot. Plantlets transferred to soil were successfully acclimatized. This protocol will facilitate the conservation and propagation of this important medicinal plant.

Key words: *Penthorum chinense,* Micropropagation, Medicinal plant, Cytokinin, Auxin

1. Introduction

Penthorum chinense Pursh belongs to *Saxifragaceae* family, and is a traditional Chinese medicinal herb, and mainly distributed in Guling, Sichuan Province, P. R. China *(1)*. The plant has the function of humidity-dispel and diuresis, and is used to cure jaundice, edema, and injuries *(1)*. The secondary metabolic products of the plant contain flavonoids, flavones, and alkaloids showing anti-oxidation *(2,3)*, anti-tumor *(4,5)*, and coagulation properties *(6)*. Presently, *P. chinense* is utilized as a component of Chi-

S. Mohan Jain and Praveen K. Saxena (eds.), *Methods in Molecular Biology, Protocols for* In Vitro *Cultures and Secondary Metabolite Analysis of Aromatic and Medicinal Plants, vol. 547*
© Humana Press, a part of Springer Science+Business Media, LLC 2009
DOI: 10.1007/978-1-60327-287-2_16

nese medicine Gansu grain which is effective to cure declining levels of tansaminases, and for protecting the liver and spleen. This medicine can potentially be used to cure chronic hepatitis B and other acute viral hepatitis *(7, 8)*. Furthermore, the seedling of this plant is edible, and is quite popular among resident farmers of Guling, Sichuan Province, P. R. China *(1)*.

Even though *P. chinens,* a wet land plant, is currently being cultivated by resident farmers in Guling, Sichuan Province, it grows very slowly in natural conditions *(1)*. Therefore, it is yet to be adopted for the production of commercial products, restricting its applications. It is necessary to establish a protocol for clonal propagation of *P. chinense*. Micropropagation has proved to be an efficient method for in vitro propagation of medicinal plants, and has successfully been used in numerous medicinal plants *(9,10)*. This chapter describes protocol of in vitro propagation of *P. chinense*, which may be useful for massive cultivation and utilization.

2. Materials

2.1. Plant Material

1. *Penthorum chinense* plants, collected from Guling County, Sichuan Province, P. R. China.

2.2. Plant Growth Regulators

1. 6-benzyladenine (BA) (ABC; Luoyang, China).
2. Kinetin (Kn) (Sigma; St. Louis, MO).
3. Indole-3-acetic acid (IAA) (ABC; Luoyang, China).
4. α-Naphthaleneacetic acid (NAA)) (ABC; Luoyang, China).

2.3. Chemicals for Plant Cell Culture Medium

1. All inorganic chemicals for preparation of MS *(11)* medium (**Table 1**) (Kelong Chemical Co.; Chengdu).
2. *Myo*-inositol, nicotinic acid, and glycine (Shanghai Chemical Co.; Shanghai).
3. Thiamine—HCl and pyridoxine–HCl (Tianjin Fucheng Chemical Co.; Tianjin).
4. Sucrose (Kelong Chemical Co.; Chengdu).
5. Agar powder (Bioprimer Life Tech. Co. Ltd.; Chengdu).

2.4. Chemicals for Treatment

1. $HgCl_2$ (Cuihua Chemical Co.; Chengdu).
2. Ethanol, HCl, and NaOH (Kelong Chemical Co.; Chengdu).

2.5. Instruments

1. Pipets (Eppendorf, Germany).
2. Electronic balance (Shanghai Electronic Analytical Instrument; Shanghai).

Table 1
Composition of MS Medium and Concentration in Stock and Medium Used

Essential element	Concentration in stock solution (mg/L)	Concentration in medium (mg/L)
Macroelements		
NH_4NO_3	33,000	1650
KNO_3	38,000	1900
KH_2PO_4	3400	170
$MgSO_4{\cdot}7H_2O$	7400	370
$CaCl_2{\cdot}2H_2O$	8800	440
Microelements		
KI	166	0.83
H_3BO_3	1240	6.2
$MnSO_4{\cdot}4H_2O$	4460	22.3
$ZnSO_4{\cdot}7H_2O$	1720	8.6
$Na_2MoO_4{\cdot}2H_2O$	50	0.25
$CuSO_4{\cdot}5H_2O$	5	0.025
$CoCl_2{\cdot}6H_2O$	5	0.025
Iron source		
$FeSO_4{\cdot}7H_2O$	5560	27.8
$Na_2EDTA{\cdot}2H_2O$	7460	37.3
Organic supplement		
Myoinositol	20,000	100
Nicotinic acid	100	0.5
Pyridoxine–HCl	100	0.5
Thiamine–HCl	100	0.5
Glycine	400	2
Carbon source		
Sucrose	Added as solid	30,000
Gelling agent		
Agar	Added as solid	6500

3. pH Meter (PB 10; Satorius, Beijing).
4. Laminar flow hood (Sujing Group Co.; cat. no. SW-CJ-1F).
5. Autoclave (Shanghi Huaxian Medical Nuclear Instrumant Co.; cat. no. LS-B50L).
6. Induction cooker (Fushibao; Fushan, Guangdong, China).
7. Water Purifier (Pall; cat. no. PL4235).
8. Cellulose membrane (Satorius; Beijing).
9. Petri dish (Kelong Chemical Co.; Chengdu).

3. Methods

3.1. Plant Material Preparation

1. Use mature plants of *P. chinense* with intact roots.
2. Wrap roots together with soil in a plastic bag to maintain humidity, allowing leaves exposed to air for respiration.
3. Grow the plants in flower pots, and irrigate every day from the top of the plant.

3.2. Preparation of Plant Growth Regulators Stock Solution

3.2.1. NAA Stock Solution (1 mg/mL)

1. Weigh precise 100 mg NAA and dissolve in 1 mL 95% alcohol.
2. Add distilled water to raise volume up to 100 mL.
3. Store it in a refrigerator at 4°C for later use.

3.2.2. IAA Stock Solution (1 mg/mL)

1. Weigh 100 mg IAA, and prepare stock solution similar to NAA.
2. Filter-sterilize IAA stock solution by using 0.22 μm cellulose membrane and store in a sterile brown reagent bottle (*see* **Note 1**).
3. Store it at 4°C for further use.

3.2.3. BA and Kn Stock Solutions (1 mg/mL)

1. Weigh 100 mg each and dissolve separately in several drops of 1 M HCl.
2. Make volume up to 100 mL of each by adding distilled water.
3. Keep them at 4°C for later use.

These stock solutions must be discarded when plant growth regulator precipitate in the solution and prepare fresh stock solutions (*see* **Note 2**).

3.3. Preparation of Stock Solutions of MS Medium (See Table 1)

3.3.1. Macro-Elements Stock Solution (20×)

1. Weigh NH_4NO_3, KNO_3, KH_2PO_4, $MgSO_4 \cdot 7H_2O$, and $CaCl_2 \cdot 2H_2O$ individually, the amounts as listed in **Table 1**.
2. Dissolve them one by one in distilled water, and make volume up to 1000 mL by adding distilled water.
3. Keep stock solution at 4°C in a refrigerator for use.

3.3.2. Micro-Elements Stock Solution (200×)

1. Weigh KI, H_3BO_3, $MnSO_4 \cdot 4H_2O$, $ZnSO_4 \cdot 7H_2O$, $Na_2MoO_4 \cdot 2H_2O$, $CuSO_4 \cdot 5H_2O$ and $CoCl_2 \cdot 6H_2O$ (*see* **Note 3**) individually. Amounts as listed in **Table 1**.

2. Add each one by one to dissolve completely in distilled water, and raise volume up to 1000 mL by adding distilled water.
3. Store at 4°C in a refrigerator for use.

3.3.3. Iron Source Stock Solution (200×)

Iron is actually a microelement in MS medium, but the stock solution is prepared separately (*see* **Note 4**). Iron sulfate is usually used as an iron source.

1. Weigh $FeSO_4 \cdot 7H_2O$ and $Na_2EDTA \cdot 2H_2O$ individually (*see* **Table 1**), and prepare stock solutions according to **Subheading 3.3, step 1**, and **Subheading 3.3, step 2**.
2. Store iron stock solution in brown agent bottle at 4°C.

3.3.4. Organic Supplements Stock Solution (200×)

1. Prepare separately stock solutions of myoinositol, nicotinic acid, pyridoxine–HCl, thiamine–HCl, and glycine (*see* **Table 1**; see previous steps).
2. Keep all stock solutions at 4°C.

High concentrations of elements in above stock solutions can lead to the precipitation of ingredients elements (*see* **Note 5**).

3.4. Culture Medium Preparation

All the media used in this experiment are MS medium, containing 3.0% *(w/v)* sucrose and 0.65% *(w/v)* agar for solidification.

3.4.1. Preparation of Medium Supplemented with 6-BA, NAA or Kn

1. Add 50 mL macro-elements, 5 mL microelements, 5 mL iron solution, and 5 mL organic solutions into about 800 mL distilled water.
2. Add 30 g sucrose into the solution.
3. Supplement precise amount of plant growth regulator into medium according to the need of experiment.
4. Adjust the pH of the medium solution to 5.8 by 1 M HCl or 1 M NaOH.
5. Heat the solution and stir it for proper mixing of solutions.
6. Add 6.5 g agar. Heat and stir in order to melt agar.
7. Make volume up to 1000 mL by adding double distilled water.
8. Autoclave at 121°C, 104 kPa for 15 min.
9. Allow medium to cool down and pour quickly into flasks or Petri dishes before solidification.

3.4.2. Preparation of Medium Supplemented with IAA

1. Follow previous steps.
2. Add filter sterilized IAA (*see* **Note 1**) into autoclaved medium and mix it well and pour quickly in flasks and Petri dishes before it starts solidifying.

3.5. Explants Preparation

1. Cut branches of *P. chinense* plants grown in flower pots.
2. Soak branches in commercial soap solution for 10–15 min.
3. Rinse in running tap water to remove soil and other superficial dust.
4. Transfer plant material to the laminar flow hood.
5. Dip it in 75% ethanol for 10 s for surface-sterilization.
6. Rinse with sterile distilled water by shaking three times.
7. Dip in 0.1% (*w/v*) $HgCl_2$ for 9 min.
8. Wash with sterile distilled water five times by shaking (*see* **Note 6**).

3.6. Shoot Initiation

1. Remove leaves and cut branch into nodes as explants, each with one axillary bud *(12)*.
2. Culture explants on solid MS medium supplemented with BA or Kn (0.1–2.0 mg/L) for shoot induction *(13,14)* (*see* **Note 7**).
3. Incubate cultures at 25 ± 1°C, 12-h photoperiod at an irradiance of about 100 µmol/m²/s provided by cool-white fluorescent tubes.
4. Culture 40 explants in each treatment. Repeat three times.
5. Use control MS medium devoid of plant growth regulators.
6. Record the frequency, length, and number of shoots per explant after 6 wk of culture.
7. Record roots, if spontaneous roots formed in the shoot.
8. Analysis the data statistically by the Turkey's test (*see* **Note 8**).
9. Results are mean ± SE of about 40 nodes per treatment in 3 repeated experiments (*see* **Table 2** and **Fig. 1a–c**).

3.7. Shoot Multiplication

1. Cut shoots proliferated from initial cultures into nodes (each with one bud), after 6 wk of culture.
2. Inoculate explants proportionally on fresh shoot initiation MS medium (*see* **Table 3**).
3. Maintain cultures at 25 ± 1°C under 12-h photoperiod 100 µmol/m²/s.
4. Follow other steps similar to the previous steps.
5. Results are mean ± SE of about 40 nodes per treatment in three repeated experiments (*see* **Table 3** and **Fig. 1d–g**).

3.8. Root Induction

1. Excise shoot clusters to separate in single shoots.
2. Choose shoots 0.8- to 1.0-cm long (*see* **Note 9**), and insert vertically into MS medium containing IAA or NAA *(16,17)* (*see* **Note 10**).

Table 2
Effects of Different Concentrations of Cytokinins (BA, Kn) Added to the MS Medium on Shoot Initiation from Nodal Explants of *P. chinense* After 6 wk of Culture

Cytokinins (mg/L)		Percent nodal segment forming shoots (%)	Number of shoots per axillary bud	Length per shoot (cm)	Percent shoot forming roots (%)
BA	0	54.1d	1.5 ± 0.5e	1.6 ± 0.8a	60.1a
	0.1	55.3d	1.6 ± 0.4d	1.6 ± 0.2a	0.0b
	0.5	74.8b	2.9 ± 2.8c	0.7 ± 0.3b	0.0b
	1.0	83.3a	4.8 ± 0.2b	0.4 ± 0.2c	0.0b
	2.0	59.2c	5.2 ± 0.2a	0.3 ± 0.2d	0.0b
Kt	0	54.1c	1.5 ± 0.5a	1.6 ± 0.8d	60.1b
	0.1	74.7ab	1.2 ± 0.2c	2.0 ± 0.6b	62.0a
	0.5	76.5a	1.3 ± 0.8b	1.6 ± 0.7d	40.0c
	1.0	77.4a	1.0 ± 0.3d	1.7 ± 1.6c	27.1d
	2.0	67.2b	1.0 ± 0.0d	2.9 ± 1.5a	26.6e

Note: Values represent mean ± SE of about forty nodes per treatment in three repeated experiments. Means within each column followed by the same letter are not significantly different by the Turkey test at 0.05% probability level.

Reproduced with permission from **ref.** *15* from *the Society for* In Vitro *Biology*

3. Incubate cultures at 25 ± 1°C, under light 12-h photoperiod, 100 μmol/m^2/s.
4. Culture 40 shoots in each treatment, and repeat three times.
5. Culture shoots on MS medium free of regulator for control.
6. Record data on rooting frequency, length and number of roots per shoot, after 3 wk.
7. Analysis the data statistically by the Turkey's test.
8. Results are mean ± SE of about forty nodes per treatment in three repeated experiments (*see* **Table 4** and **Fig. 1h**).

3.9. Acclimatization and Transfer

1. Make a tiny hole to the polypropylene closures and incubate for 3 d to acclimatize regenerated plantlets in natural conditions (*see* **Note 11**).
2. Remove closures entirely and incubate for another 3 d of acclimatization.
3. Pull plantlets with well developed roots carefully from conical flasks.
4. Wash roots under running tap water to get rid of agar stuck on roots.
5. Transfer plantlet to flower pots with pre-sterilized soil.

Fig. 1. Micropropagation of *P. chinense* Pursh. (**a**) shoot initiation with 1.0 mg/L BA, after 7 d of culture, bar = 0.2 mm; (**b**) shoot initiation with 1.0 mg/L BA, after 15 d of culture, bar = 0.1 mm; (**c**) shoot initiation with 1.0 mg/L BA, after 6 wk of culture, bar = 0.2 mm; (**d**) shoot multiplication with 1.0 mg/L BA after 4 wk of culture, bar = 0.2 mm; (**e**) shoot multiplication with 2.0 mg/L BA, after 4 wk of culture, bar = 0.2 mm; (**f**) shoot multiplication with 1.0 mg/L Kn, after 4 wk of culture, bar = 0.2 mm; (**g**) shoot multiplication with 2.0 mg/L Kn, after 4 wk of culture, bar = 0.2 mm; (**h**) roots regenerated from microshoots on medium with 0.5 mg/L IAA, after 3 wk of culture, bar = 0.2 mm; (**i**) in vitro derived plantlet grown in a pot, bar = 0.8 mm (Reproduced from **ref.** *15* with permission from the Society for In Vitro Biology)

6. Cover with closures to ensure high humidity of transferred plantlet, leaving a tiny hole for respiration, for 1 wk of acclimatization.
7. Irrigate the plantlets with tap water each day.
8. Remove closures entirely, after 1 wk.
9. The results of plantlet transplanting are shown in **Fig. 1i**.

Table 3
Effects of Different Concentrations of Cytokinins (BA, Kn) Added to the MS Medium on Shoot Proliferation from Nodal Explants of *P. chinense* after 4 wk of Subculture

Cytokinins (mg/L)		Percent nodal segment forming shoots (%)	Number of shoots per axillary bud	Length per shoot (cm)	Percent shoot forming roots (%)
BA	0	81.4d	1.0 ± 0.0e	0.8 ± 0.7b	70.1a
	0.1	82.1c	1.5 ± 0.3d	1.0 ± 0.5a	4.1b
	0.5	82.2c	2.4 ± 0.9c	0.5 ± 0.3d	0.0c
	1.0	89.5b	5.1 ± 1.8b	0.6 ± 0.2c	0.0c
	2.0	91.8a	6.4 ± 0.1a	0.2 ± 0.1e	0.0c
Kt	0	81.4c	1.0 ± 0.0c	0.8 ± 0.7c	70.1a
	0.1	75.0d	1.2 ± 0.2bc	0.8 ± 0.4c	70.1a
	0.5	85.1b	1.4 ± 0.2a	1.1 ± 0.5b	34.8b
	1.0	88.2a	1.3 ± 0.3ab	0.6 ± 0.6d	29.3c
	2.0	81.4c	1.3 ± 0.3ab	1.3 ± 0.2a	4.4d

Note: Values represent mean ± SE of about 40 nodes per treatment in three repeated experiments. Means within each column followed by the same letter are not significantly different by the Turkey test at 0.05% probability level.

Reproduced with permission from **ref.** *15* from the *Society for* In Vitro *Biology*

Table 4
Effects of Different Concentrations of Auxins (NAA, IAA) Added to the MS Medium on Rooting Response of Shoots from 1.0 mg/L BA Cultivated Medium for 3 wk

Auxins (mg/L)		Rooting percent (%)	Number of roots per shoot	Length per root (cm)
NAA	0	98.5a	8.9 ± 3.6b	1.5 ± 0.8a
	0.1	89.7b	6.8 ± 0.5d	1.3 ± 0.2b
	0.5	80.2bc	7.0 ± 2.6c	1.1 ± 0.4c
	1.0	85.4b	13.4 ± 8.0a	0.6 ± 0.1d
	2.0	72.4c	6.6 ± 2.7e	0.8 ± 0.2e
IAA	0	98.5a	8.9 ± 3.6c	1.5 ± 0.8a
	0.1	90.6c	8.6 ± 2.6d	1.2 ± 0.1b
	0.5	98.8a	10.0 ± 1.8a	1.2 ± 0.2b
	1.0	96.0b	9.8 ± 0.7b	1.0 ± 0.1c
	2.0	95.2b	8.3 ± 1.4e	0.8 ± 0.4d

Note: Values represent mean ± SE of about 40 shoots per treatment in three repeated experiments. Means within each column followed by the same letter are not significantly different by the Turkey test at 0.05% probability level.

Reproduced with permission from **ref.** *15* from the *Society for* In Vitro *Biology*

4. Notes

1. IAA is unstable to both heat and light, so it could only be filter-sterilized. While preparing IAA stock solution, filter-sterilize by passing through 0.22 μm cellulose membrane and

store in a sterile brown reagent bottle at 4°C for further use. After cooling down of the autoclaved medium to about 50°C, IAA is added and mix well by shaking. High temperature is harmful and could degrade IAA.

2. Discard immediately precipitated PGR stock solutions and prepare fresh. results will not otherwise be consistent.
3. It is rather difficult to weigh precisely small quantity of $CuSO_4 \cdot 5H_2O$ and $CoCl_2 \cdot 6H_2O$ while preparing microelements stock solution less than 1000 mL. We resolve this problem easily by the following procedure: weigh 20 mg $CuSO_4 \cdot 5H_2O$ and $CoCl_2 \cdot 6H_2O$ respectively. Dissolve into 20 mL by distilled water and make primary stock solutions. Proceed to pipet precise amounts of primary stock solution to microelement stock solution (secondary solution) according the need.
4. Although ethylenediamine tetraacetic acid (EDTA) is not necessary for tissue culture medium, it was traditionally used as a chelator for iron sulfate. Subsequently iron can be released into the medium slowly thus avoid iron precipitation in the medium *(18)*.
5. Very high concentration of stock solution may lead to the precipitation of elements.
6. $HgCl_2$ is highly toxic, so care should be taken to prevent skin exposure while handling it. At least five times the material should be washed with sterile distilled water by shaking in order to eliminate toxic residual of $HgCl_2$ sticking to plant materials.
7. Plant growth regulators are critical components in the medium for plant micropropagation. The ratio of auxin to cytokinin determines the nature of in vitro cultures. A high cytokinin to auxin ratio favors shoot formation, thus cytokinins are usually used either alone or in combination with auxins for shoot induction *(13,14)*.
8. According to Duncans' Multiple Range Test ($p \leq 0.05$), different among the treatment of variance concentrations of BA and Kn were designated by different letters which were followed with means in each column in the table.
9. Optimum frequency and number of multiple regenerated shoots were obtained on MS medium supplemented with 1.0 mg/L BA, but considering the frequency and node length of the initiated shoots, 0.8- to 1.0-cm long regenerated shoots were excised from the shoot clusters for root induction.
10. Auxins, IAA, and NAA have been commonly used for root induction in vitro *(16, 17)*.

11. Direct removal of polypropylene closures would wither regenerated plantlets. Therefore, removing closures gradually to acclimatize in vitro plantlets would minimize the loss of plantlets and enhance the survival rate.

Acknowledgments

This work is financially assisted by Sichuan Provincial Youth Science and Technology Foundation (05ZQ026-017), and Sichuan Provincial Key Subject Program (SZD0420), P. R. China.

References

1. Pan, J. T. (1992). Delectis florae reipubilcae popularis sinicae agendae academiae sinicae edita, In: Flora Peipublicae Popularis Sinices. Science Press, Tomus **34**(2): 2–3.
2. Mahesh, T., Menon, V. P. (2004). Quercetin allievates oxidative stress in streptozotocin-induced diabetic rats. *Phytother Res.* **18**: 123–127.
3. Moon, Y. J., Wang, X., Morris, M. E. (2006). Dietary flavonoids: effects on xenobiotic and carcinogen metabolism. *Toxicol. In Vitro.* **20**: 187–210.
4. Zhang, M. J., Huang, J. (2004). Recent research progress of anti-tumor mechnism matrine. *China J. Chinense Materia Medica* **29**: 115–118.
5. Sigudsson, S., Ogmundsdottir, H. M., Hallgrimsson, J., Gudbjarnason, S. (2005). Antitumour activity of *Angelica archangelica* leaf extract. *In Vivo.***19**: 191–194.
6. Kang, B. H., Ryoo, I. J., Park, D. J., Lee, H. S. (1996). Screening of biologically active compounds from various weeds. Agr. *Chem. Biotech.* **39**: 409–413.
7. He, S. M., Li, M., Wu, Z., Ni, P., Zhu, W. L. (2002). Progress on research of *Penthorum chinense*. *Chinese Tradit Herb. Drugs* **33**: U005-U006.
8. Zhao, J. Q., Yang, M., Zhao, L. S., Zhou, M. M., Wang, X., Zhou, W. S., He, F., Qin, S. (2002). Inhibitory effect of Penthotum chinense pursh on hepatitis B virus *in vitro*. *Chinese J Integ. Trad. Western Med. Liver Dis.* **12**: 26–27.
9. Saini, R., Jaiwal, P. K. (2000). In vitro multiplication of *Peganum harmala*-an important medicinal plant. *Indian J. Exp. Biol.* **38**: 499–503.
10. Nandwani, D., Kumaria, S., Tandon, P. (2001). Micropropagation of *Pinus kesiya* Royle ex Gord (Khasi pine). *Gartenbauwissenschaf.* **66**: 68–71.
11. Murashige, T., Skoog, F. (1962). A revised medium for rapid growth and bio assays with tobacco tissue culture. Physiol. *Planatarum* **15**: 473–497.
12. Haw, A. B., Keng, C. L. (2003). Micropropagation of *Spilanthes acmella* L., a bio-insecticide plant, through proliferation of multiple shoots. *J. Appl. Hort.* **5**: 65–68.
13. Skoog F., Armstrong D. J. (1970). Cytokinin annual review. *Plant Physiol.***21**: 359–384.
14. Gyana, R. R. (2004). Effect of cytokinins and auxins on micropropagation of *Clitoria ternatea* L. *Biol. Lett.* **41**(1): 21–26.
15. Cao H., Yang J., Peng Z. S., Kang C. Y., Chen D. C., Gong Z. C., Tan X. (2007). Micropropagation of Penthorum chinense through axillary bud. *In Vitro Cell. Dev. Biol. -Plant* **43**: 149–153.
16. Anis, M., Faisal, M., Singh, S. K. (2003). Micropropagation of Mulberry (*Morus alba* L.) through *in vitro* culture of shoot tip and nodal explants. *Plant Tissue Cult.* **13**: 47–51.
17. Gao, H. H., Li, W., Yang, J., Guo, G. Q., Zheng, G. C. (2003). Effect of 6-benzyladenine and casein hydrolysate on micropropagation of *Amorpha fruticosa*. *Biologia Plantarum* **47**: 145–148.
18. Heller, R. (1953). Researches on the mineral nutrition of plant tissues. *Ann. Sci. Nat. Bot. Biol. Veg.* 11th Ser. **14**: 1–123.

Part II

Transgenic Approaches

Chapter 17

Spontaneous Plant Regeneration and Production of Secondary Metabolites from Hairy Root Cultures of *Centaurium erythraea* Rafn

Angelina Subotić, Slađana Jevremović, Dragoljub Grubišić, and Teodora Janković

Summary

We have established an efficient protocol for plant regeneration and production of secondary metabolites in hairy root culture of *Centaurium erythraea* Rafn. Because the hairy roots and regenerated plants produce bitter secoiridoid glucosides and xanthones similar to the plants in nature, the use of in vitro cultures as an alternative source of their production is feasible. This chapter describes a protocol for the induction of adventitious shoots and transgenic plants from hairy root cultures of *C. erythraea* and their phytochemical analysis.

Key words: Adventitious shoots, *Centaurium erythraea*, Hairy root, Medicinal plant, Plant regeneration, Secoiridoids, Xanthones

1. Introduction

Centaurium erythraea Rafn. is a winter annual plant and belongs to Gentanaceae family, and grows throughout Atlantic countries of Southern Europe and North Africa. This species was traditionally collected as a major component of pharmaceutical drug "Centaurii herba" designed for medicinal purposes, such as treatment of gastrointestinal tract diseases, fever, anemia, and for increasing appetite *(1)*. *C. erythraea* is found relatively rarely in nature as a consequence of intensive plant collection. Therefore, there is an urgent need to develop an appropriate protocol for

S. Mohan Jain and Praveen K. Saxena (eds.), *Methods in Molecular Biology, Protocols for* In Vitro *Cultures and Secondary Metabolite Analysis of Aromatic and Medicinal Plants, vol. 547*
© Humana Press, a part of Springer Science+Business Media, LLC 2009
DOI: 10.1007/978-1-60327-287-2_17

mass propagation and conservation of this endangered medicinal plant. Induction of organogenesis and/or somatic embryogenesis in *C. erythreae* from different type of explants has been reported *(2, 3)*.

It is well known that many plants infected by *Agrobacterium rhizogenes* harboring *Ri* plasmid form adventitious roots at the infected site. These adventitious roots, referred to as hairy roots, grow vigorously in a culture medium without plant growth regulators *(4)*. These cultures serve as a source material for regenerating transgenic plants. In the number plant species, plant regeneration from hairy roots could be achieved in the presence of phytohormones *(5, 6)*. On the other hand, only few reports describe spontaneous regeneration of shoots from hairy root cultures on hormone-free medium *(7)*. Hairy root cultures have been found suitable systems for the production of secondary metabolites *(8)* and isolation of novel compounds of medicinal value.

Phytochemical investigations of *C. erythraea* revealed the presence of pharmacologically active compounds, bitter secoiridoid glucosides, and xanthones in field-grown *(9, 10)* as well as in aerial parts and roots of in vitro plants *(11, 12)*. As these compounds possess characteristic optical activity in the ultraviolet (UV), the high-performance liquid chromatograohy/diode array detector (HPLC-DAD) detection represents a method of choice for the analysis of the *Centaurium* extracts.

This regeneration method was originally devised for the production of transformed plants of *C. erythraea*. However, this protocol is now routinely used in our laboratory to obtain transgenic plants from Gentianaceae family for various purposes *(13–15)*. Moreover, the use of hormone – free culture medium can make the in vitro production of plants economically viable.

2. Materials

2.1. Plant Material

1. Explants from non-transformed and hairy root cultures of *C. erythraea* (*see* **Note 1**).

2.2. Culture Media

1. The basal Murashige and Skoog (MS) medium (*see* **Note 2**) for the growth of hairy root cultures, induction of adventitious shoots, and rooting of transgenic shoots *(16)*. The chemicals required for preparing stock solutions are listed in **Table 1** (*see* **Note 3**). The pH is adjusted to 5.8 using 1 N KOH. Medium is autoclaved in 500-mL bottles at 121°C for 20 min and stored at room temperature for several weeks. A microwave oven is used to liquefy the media on the day of use.

Table 1
Murashige and Skoog Basal Medium

Components	Final medium (mg/l)	Stock solution	Volume of stock per liter of final medium (mL)
Major inorganic salts			50
NH_4NO_3	825	82.5	
KNO_3	950	95	
$CaCl_2 \times 2\ H_2O$	220	22	
$MgSO_4 \times 7\ H_2O$	185	18.5	
KH_2PO_4	85	8.5	
		g/l	
Minor inorganic salts			0.5
H_3BO_3	6.2	1.24	
$MnSO_4 \times 4\ H_2O$	22.3	3.38	
$ZnSO_4 \times 4\ H_2O$	8.6	1.72	
$CuSO_4 \times 5\ H_2O$	0.025	0.166	
$CoCl_2 \times 6\ H_2O$	0.025	0.05	
$Na_2MoO_4 \times 2\ H_2O$	0.25	0.005	
KJ	0.83	0.005	
		g/L	
Iron components			5
Na_2EDTA	37.3		
$FeSO_4 \times 7\ H_2O$	27.8		
		g/L	
Organic components			0.5
Nicotinic acid	0.5	0.1	
Pyridoxine HCl	0.5	0.1	
Thiamine HCl	0.1	0.02	
Glycine	2.0	0.4	
myo-Inositol	100	20	
Sucrose	30%		
Agar	0.7%		

2. 250-mL Erlenmeyer flasks.
3. 15 mm × 100 mm Petri dishes.

2.3. Growth Conditions

1. All tissue cultures are kept at 25°C ± 1°C for a 16-h photoperiod provided by white fluorescent light 50 μmol m^{-2} s^{-1}.
2. Cultures are then placed in the dark (*see* **Note 4**).

2.4. HPLC Analysis of Secoiridoids and Xanthones

1. Solvents. Acetonitrile and water (*see* **Note 5**).
2. Reverse phase analytical column.

3. Methods

3.1. Explants Preparation and Induction of Adventitious Shoots

1. Roots explants are aseptically prepared from different clones of hairy root cultures (*see* **Note 6**) and roots from non-transformed root culture are used as a control. Ten root tips are excised from 4-wk-old cultures and cut into 15-mm long pieces and place horizontally in a Petri dish containing 10–15 mL MS medium. Care should be taken to avoid damage the root tips with the forceps (*see* **Note 7**). They are cultured under either in complete darkness or light, 16-h light/8-h dark.
2. Small outgrowths are induced at the surface of hairy root explants after 7–10 d of the culture. They have the capacity to form adventitious shoots throughout the whole culture period in all clones. Within 15 d adventitious shoots develop from the hairy root explants (**Fig. 1a, b**). Under dark conditions, both hairy roots and non-transformed roots rarely form adventitious shoots. On the other hand, under light conditions (16-h light/8-h dark) hairy roots produce many adventitious shoots, but non-transformed roots form a few adventitious shoots (**Fig. 2**). Regenerated shoots from hairy roots can be classified in two types—normal and elongated—according to their morphology.
3. At the end of culture, hairy roots form adventitious shoots of normal size and morphology under light (**Fig. 3a**) similar to shoots regenerated in non-transformed cultures (**Fig. 4a**). In the darkness, however, adventitious shoots elongate and most of them remain 5- to 7-mm long at end of culture period (**Fig. 3b**).

3.2. Rooting of Adventitious Shoots

1. The excised shoots are placed on MS medium without plant growth regulator. About 98% of excised shoots develop roots, 25 roots per shoot, within 15–20 d of culture (**Fig. 4b**).

Fig. 1. Formation of adventitious shoots from hairy roots of *Centaurium erythraea* (**a**) Detail of hairy root explants with adventitious shoots, at the early stage of development, after 15 d of culture on MS medium; (**b**) Hairy root with well developed adventitious shoots (bars = 0.5 mm).

2. Remove rooted plantlets from the culture vessels after 3–4 wk.
3. Carefully wash under running tap water to remove the adhering culture medium from the roots, and then transfer to pots containing a mixture of soil in the greenhouse conditions (*see* **Note 8**).

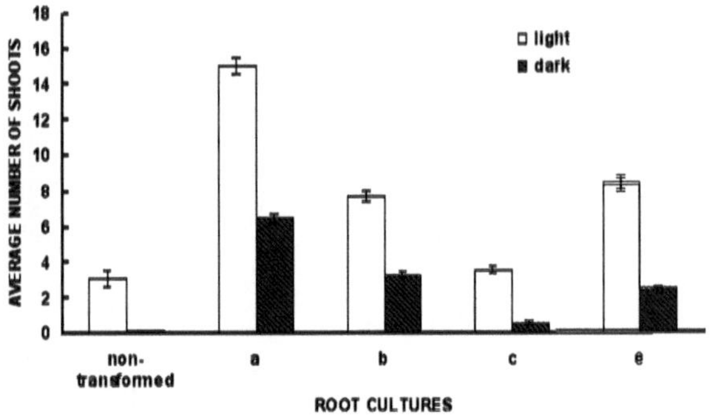

Fig. 2. Effect of light on adventitious shoot formation in root cultures of *C. erythraea* clones a, b, c, d and e. The data were recorded after month of culture. Means ± SE, $n = 50$ root explants. The experiment was repeated three times. (Reproduced from **ref.** *17* with permission from Springer).

3.3. HPLC Analysis

1. Air-dried aerial parts and roots are separately extracted with methanol for 30 min in an ultrasonic bath. The ratio between plant material and solvent is 1:20 *(w/v)*. All extracts are filtered through 0.45 PTFE filter in a HPLC vial.
2. Liquid chromatography is carried out with Agilent HPLC consisting of vacuum degasser, binary pump, autosampler, thermostated column compartment, and DAD. Analyses are performed on reverse phase Hypersil BDS C18 analytical column 125 × 2 mm ID, particle size 5 μm. Mobile phase are (a) H_2O containing 1% H_3PO_4 and (b) Acetonitrile.
3. Extracts are separated by gradient elution according to the following scheme: 98% A 0–2 min, 98–90% A 2–5 min, 90–80% A 5–10 min, 80–60% A 10–13 min, 60% A 13–18 min, 60–45% A 18–20 min, 45–10% A 20–23 min, 10–0% A 23–25 min. Flow was adjusted to 0.5 mL/min with injection volume of 5μl, and detection at 260 and 320 nm. The column temperature is set at 25°C.
4. Secoiridoids and xanthones in extracts are identified by co-injection method, using the standard samples isolated in our laboratory (*see* **Note 9**). Standard solution of secoiridoids and xanthones are prepared by dissolving 5 mg of each compound in a 5-mL methanol into volumetric flask.

4. Notes

1. In vitro-grown seedlings were inoculated with *A. rhizogenes* strains A4M70GUS by puncturing the internodes of stems

Fig. 3. Morphology of *C. erythraea* hairy roots (**a**) under normal light conditions the explants formed adventitious shoots with normal morphology (**b**) under darkness the explants formed elongated adventitious shoot (bars = 1 cm).

Fig. 4. In vitro regeneration of *C. erythraea* (**a**) Non-transformed root cultures with well developed shoots after 4 weeks of culture on ½ MS medium. (**b**) Induction of roots from transformed shoots on ½ MS medium (bars = 1 cm).

with a sterile, hollow needle loaded with bacteria. Hairy roots appear 10 d after inoculation. Apical segments of primary hairy roots were excised and transferred to MS medium for three successive subcultures and subsequently without 250 mg dm^{-3} cefotaxim. Root tips 15 mm in length from the hairy root culture were transferred to MS medium and subcultured once

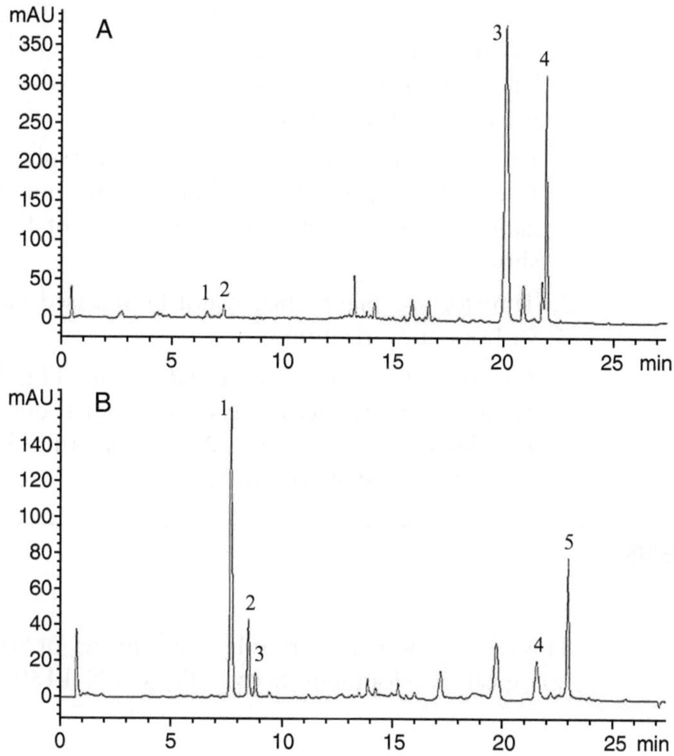

Fig. 5. HPLC profiles of methanol extracts of hairy roots (**a**) and transgenic shoots (**b**) of *C. erythraea* (*see* **Note 9**).

per month. Adventitious shoots were initiated under these conditions. When the shootes are 2- to 3-mm long, excise and transfer to a growth regulator-free MS medium for shoot elongation and root initiation *(17)*.

2. The basal tissue culture medium (MS) is half strength macronutrients.
3. Dissolve all components from the **Table 1** in distilled water. Microelements solutions are stored at 4°C for up to 4 wk. This stock solution is regularly checked and discarded if precipitation occurred. Solutions of microelements and vitamins are stored at –20°C.
4. If adventitious shoots are yellow the light conditions are not optimal. A further potential problem is that the light source produces too much heat so that the culture vessels warm up. In this case a greater distance between the light source and Petri dishes alleviates this problem.
5. All solvents should be of HPLC grade. Phase A water should be prepared daily. Solvents and solutions to be analyzed should be filtered through 0.45 μm PTFE filters before use.

6. Several clones of hairy root are selected for all experiments based on differences in their morphological characteristics. These hairy roots showed normal growth on MS medium without phytohormones.
7. Pay careful attention while cutting the explants of hairy root. Use very sharp scalpel and forceps and not hot after sterilization. Heavily damaged explants do not form adventitious shoots.
8. Transformed plants should not be released into environment without authorization.
9. Assignments in HPLC chromatograms. **Fig. 5a**: (1) swertiamarin; (2) gentiopicrin; (3) eustomin; (4) demethyleustomin. **Fig. 5b**: (1) swertiamarin; (2) gentiopicrin; (3) sweroside; (4) eustomin; (5) demethyleustomin.

Acknowledgments

This research was supported by the Ministry of Science and Technological Development, Serbia (Project N°143026B).

References

1. Pharmacopoeia Yugoslavica IV (1984). Federal Institute of Public Health, Belgrade, Serbia
2. Barešová, H., Herben, T., Kaminek, M., and Krekule, J. (1985). Hormonal control of morphogenesis segments of *Centaurium erythraea* Rafn. *Biol. Plant.* **27**, 286–291.
3. Subotić , A., Janković , T., Jevremović , S., and Grubišić , D. (2006). Plant Tissue Culture and Secondary Metabolites Productions of *Centaurium erythraea* Rafn., a Medical Plant. In: Jaime A. Teixeira da Silva(ed.): Floriculture, Ornamental and Plant Biotechnology: Advances and Topical Issues (1st Edition). Vol. II, Global Science Books, London, UK, 564–570.
4. Meyer, A., Tempe, J., and Constatino, P. (2000). Hairy Root: A Molecular Overview Functional Analysis of *Agrobacterium rhizogenes* T-DNA genes. In: Stacey, G., Keen, N., (ed.): *Plant–Microbe Interactions.* Vol. 5, APS Press, St Paul, MN, 93–139.
5. Choi, P. S., Kim, Y. D., Chung, H. J., Choi, D. W., and Liu, J. R. (2004). Plant regeneration from hairy-root cultures transformed by infection with *Agrobacterium rhizogenes* in *Catharanthus roseus. Plant Cell Rep.* **22**, 828–831.
6. Crane, C., Wright, E., Dixon, A. R., and Wang, Z. (2006). Transgenic *Medicago truncatula* plants obtained from *Agrobacterium tumefaciens*-transformed roots and *Agrobacterium rhizogenes*-transformed hairy roots. *Planta* **223**, 1344 – 1354.
7. Vinterhalter, B., Orbović, V., and Vinterhalter, D. (1999). Transgenic root cultures of *Gentiana punctata* L. *Acta Societatis Botanicorum Poloniae*, **68(4)**, 275–280.
8. Georgiev, M., Pavlov, I., and Bley, T. (2007). Hairy root type plant *in vitro* systems as sources of bioactive substances. *Appl. Microbiol. Biotechnol.* **74**, 1175–1185.
9. van der Sluis, W. G. (1985). Chemotaxonomical Investigation of Genera *Blackstonia* and *Centaurium* (Gentanaceae). *Plant Syst. Evol.* **149**, 253–286.
10. Barešová, H. (1988). *Centaurium erythraea*: Micropropagation and the Production of Secoiridoids. In: Bajaj YPS (ed.): *Biotechnology in Agriculture and Forestry.* Vol. 4. Springer, Berlin, 350–366.
11. Janković, T., Krstić, D., Šavikin-Fodulović, K., Menković, N., and Grubišić , D. (1997). Comparative investigation of secoiridoid compounds of *Centaurium erythraea* grown in

nature and cultured in vitro. *Pharm. Pharmacol. Lett.* 7(**4**), 30–32.

12. Janković, T., Krstić, D., Šavikin-Fodulović, K., Menković, N., and Grubišić, D. (2002). Xanthones and secoiridoids from hairy root cultures of *Centaurium erythraea* and *C. pulchellum*. *Planta Medica* **68**, 944–946.
13. Momčilović, I., Grubišić, D., Kojić, M., and Nešković, M. (1997). *Agrobacterium rhizogenes*-mediated transformation and plant regenration of four *Gentiana* species. *Plant Cell. Tiss. Org. Cult.* **50**, 1–6
14. Vinterhalter, B., Momčilović, I., and Vinterhalter, D. (2000). High biomass producing root cultures of *Gentiana punctata* L. transformed with *Agrobacterium tumefaciens* C58C1 (pArA4b). *Arch. Biol. Sci. Belgrade*, **52(2)**, 85–90.
15. Bjelović, A., Rosić, N., Miljuš-Đukić, J., Ninković, S., and Grubišić, D. (2004). *In vitro* regeneration and transformation of *Blackstonia perfoliata*. *Biol. Plant.* **48(3)**, 333–338.
16. Murashige, T., and Skoog, F. (1962). A revised medium for rapid growth and bioassays with tobacco tissue cultures. *Physiol. Plant.* **15**, 473–497.
17. Subotić, A., Budimir, S., Grubišić , D., and Momćilović, I. (2003). Direct regeneration of shoots from hairy root cultures of *Centaurium erythreae* inoculated with *Agrobacterium rhizogenes*. *Biol. Plant.* **47(4)**, 617–619.

Chapter 18

Transgenic *Hypericum perforatum*

G. Franklin, Margarida M. Oliveira, and Alberto C.P. Dias

Summary

Plant transformation is an important tool with many applications in modern plant biology. Although this technique is primarily used to produce superior crop varieties, it is also being utilized to answer basic questions concerning gene function and regulation in contemporary functional genomics research. In our laboratory, we have established a transformation system for *Hypericum perforatum*. This protocol involves the transfer of foreign DNA into *H. perforatum* organogenic nodule explants *via* particle-bombardment and the regeneration of shoots from the explants under selection pressure. We have successfully used this method to express ß-glucuronidase and hygromycin phosphotransferase genes in *H. perforatum*. Molecular analyses of putative phenotypically normal transgenic plants show stable integration of the transgenes into the plant nuclear genome. Here we describe the procedure for the transformation of *H. perforatum*.

Key words: *Hypericum perforatum*, Organogenic nodular suspension culture, GUS gene, HPT gene, Particle-bombardment, Transgenic plant, Polymerase chain reaction analysis, Southern blot analysis

1. Introduction

Hypericum perforatum L. (St. John's Wort) is an important medicinal plant that has been used since ancient times for the treatment of numerous ailments. Recent clinical studies demonstrate that *H. perforatum* extracts are efficient in the treatment of mild to moderate depression *(1, 2)*. The extract is also reported to possess antiviral *(3)*, anticancer *(4)*, neuroprotective *(5)* and antioxidant *(6)* properties. The pharmaceutical importance of *H. perforatum* extract (secondary metabolites) is the main driving force behind the research that is focused on HP cell cultures *(7–11)*. However, the cell and tissue cultures for large-scale production of secondary metabolites has so far

S. Mohan Jain and Praveen K. Saxena (eds.), *Methods in Molecular Biology, Protocols for* In Vitro *Cultures and Secondary Metabolite Analysis of Aromatic and Medicinal Plants, vol. 547*
© Humana Press, a part of Springer Science+Business Media, LLC 2009
DOI: 10.1007/978-1-60327-287-2_18

achieved only limited success because of the low and unreliable yield of the products. Although significant improvements in product yields have been achieved through conventional biochemical approaches and manipulation of culture and process factors, the reproducibility of results is still a matter of concern *(12)*.

Biosynthesis of therapeutically useful compounds can be effectively improved in medicinal plants by altering the expression of transcription factors or structural genes through metabolic engineering *(13–15)*. As the pharmacologic activities of *H. perforatum* extract are largely attributed to compounds like hypericin and hyperforin that are exclusively produced in this species, improving their production is an important target for genetic manipulation. In spite of the availability of excellent regeneration protocols *(16–19)*, this goal is not realized satisfactorily so far because of the poor knowledge about the biosynthetic pathways involved and also because of the absence of a suitable genetic transformation system for the species.

Particle-bombardment (biolistics) is a versatile technique, by which very different cell types can be transformed *(20, 21)*. Because this technique makes use of physical processes to accelerate DNA directly into intact tissues, it has the advantage of avoiding plant cell defense responses, frequently observed in recalcitrant plants against *A. tumefaciens(22, 23)*. Hence, this technique has been successfully used in the genetic transformation of a wide variety of plant species that remain otherwise recalcitrant to *Agrobacterium*-mediated transformation *(24–26)*. Moreover, this technology can deliver large number of genes into the target cells in a single step *(27)*, which is often necessary for the manipulation of metabolic pathways of plants *(28)*.

Because *H. perforatum* remains highly recalcitrant to *A. tumefaciens* mediated transformation *(22)*, particle-bombardment would be an extremely useful alternative technology in efforts to improve this medicinal plant. Hence, we have developed a particle-bombardment protocol for genetic transformation of *H. perforatum*, which can be applied in the genetic improvement programs of this important medicinal plant. So far, we have used this procedure to introduce the ß-glucuronidase (GUS) gene *(29)* and stilbene synthase gene (G. Franklin and A.C.P. Dias, unpublished data) into the *H. perforatum* nuclear genome.

2. Materials

2.1. Plant Material

H. perforatum seeds are available from many commercial sources including Richters Seeds (Goodwood; ON, Canada). The target tissues (**Fig. 1**) for particle-bombardment mediated

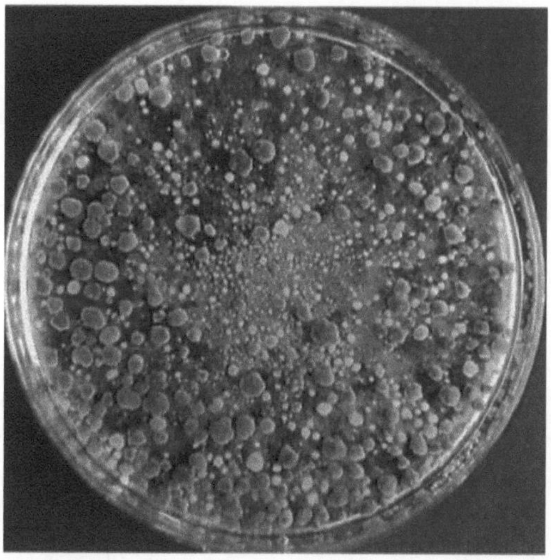

Fig. 1. Organogenic cell suspension of *H. perforatum* variety 'Helos' showing ONS explants.

transformation of *H. perforatum* can be obtained as described in **Subheading 3.4**.

2.2. Plasmid Vector

The binary vector used in the present study (pCAMBIA1301, Cambia, Australia) harbors the hygromycin phosphotransferase (HPT) gene as the selectable marker and the GUS gene disrupted by a catalase intron as reporter gene. Both genes are driven by the CaMV 35S promoter and are cloned in opposite orientation (**Fig. 2**). Plant expression vectors with many other combinations of marker and reporter genes are also available from Cambia and from other sources. We use *E. coli* strain DH5 alpha (Invitrogen, USA) for plasmid multiplication and Wizard® plus midipreps DNA purification system (Promega, USA) for plasmid isolation. Plasmid DNA should be stored at –20°C.

2.3. Plant Culture Media and Components

Plant culture media used in the present study are based on Murashige and Skoog (MS) *(30)* formulation.

1. 10X Murashige and Skoog (MS) basal salt mixture with and without vitamins (Duchefa Biochemie, The Netherlands).
2. 1000X MS vitamin cocktail: Dissolve 200 mg glycine, 50 mg nicotinic acid, 50 mg pyridoxine–HCl and 1 g thiamine–HCl in 100 mL of deionized water.
3. Other additives: myo-inositol, D-Mannitol (Sigma, St. Louis, MO), D-Sorbitol (Sigma), Sucrose (Panreac, Spain) and agar (VWR, Belgium)

2.3.1. Plant Growth Regulator Stock Solutions

1. Plant growth regulators (PGRs) including 6-benzylaminopurine/N6-benzyladenine (BA), Indole-3-butyric acid (IBA), α-napthaleneacetic acid (NAA), and Kinetin (furfurylaminopurine) (Sigma).

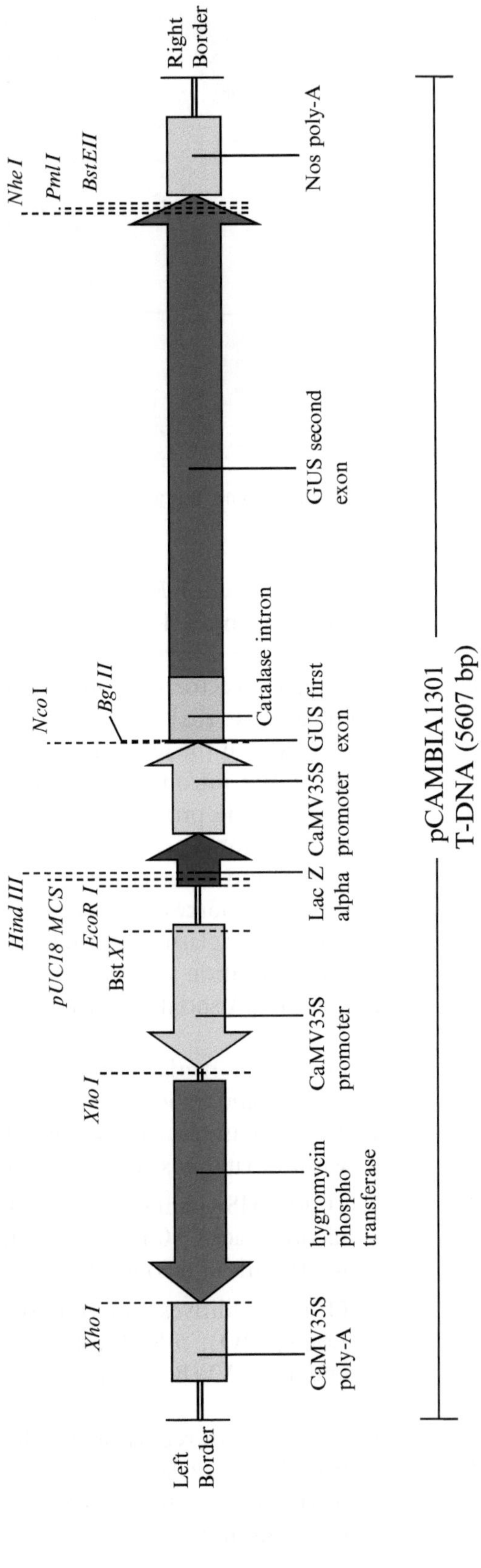

Fig. 2. T-DNA of pCAMBIA1301 showing restriction sites.

2. Stock solutions of the above PGRs can be made by dissolving the required amount in a minimal volume of 1 N NaOH and diluting with distilled water to obtain 1 mg/mL concentration and stored at 4°C.

2.3.2. Antibiotic Stocks

Antibiotics are available from several commercial sources. We have used hygromycin B and ticarcillin clavulanate (timentin) (Duchefa) and kanamycin sulphate (Calbiochem, USA).

1. Prepare stock solutions of desired antibiotic concentrations by dissolving them in distilled water. We generally prepare 10- , 50 and 250-mg/mL stock solutions respectively for hygromycin, kanamycin, and timentin.
2. Filter sterilize kanamycin and timentin using 0.2–0.45 μm pore size syringe filters (Sarstedt, Numbrecht, Germany). Because of its high toxicity, hygromycin B does not need sterilization.
3. Store all antibiotic stocks at –20°C as frozen 1-mL aliquots.

2.4. Particle Bombardment

Particle Delivery System PDS-1000/He, various sizes of microcarrier gold particles, macrocarriers, macrocarrier holders, stopping screens and rupture discs including 1100 psi (Bio-Rad, USA).

1. Maintain and multiply plasmid of interest (pCAMBIA1301) as described in **Subheading 2.2.**
2. Prepare 0.1 M stock solution of spermidine-free base (Sigma) immediately after opening the bottle and store as 100 μL aliquots at –20°C.
3. Prepare fresh $CaCl_2$ solution (2.5 M).

2.5. Analysis of Transgenic Plants

2.5.1. GUS Solution

1. Dissolve 100 mg 5-bromo-4-chloro-3-indolyl β-D-glucuronide (X-GlcA) (Sigma) in 1 mL *N, N*-dimethylformamide or dimethylsulfoxide (Sigma).
2. Add 15 mL 100 mM NaH_2PO_4 buffer, pH 7.0; 2.0 mL 0.5 M ethylenediamine tetraacetic acid (EDTA); and 10 mL 1% *(v/v)* Triton-X 100.
3. Make up the final volume to 100 mL with sterile distilled water.
4. Store the solution as desired aliquots at –20°C.

2.5.2. Polymerase Chain Reaction

1. Isolate genomic DNA from putative transgenic plants. We use DNeasy plant mini kit (Qiagen, Germany).
2. Components for polymerase chain reaction (PCR): gene-specific primers: 5X Taq DNA polymerase, 10 mM dNTP mix, and 25 mM $MgCl_2$ solution.
3. Dissolve primers in distilled water to a final concentration of 5 pmol/μL. We use the following primer sequences to amplify GUS and HPT genes.

   GUS gene forward – 5′GATCGCGAAAACTGTGGAAT3′

   GUS gene reverse – 5′TGAGCGTCGCAGAACATTAC3′

HPT gene forward – 5′ATTTGTGTACGCCCGACAGT3′

HPT gene reverse – 5′GGATATGTCCTGCGGGTAAA3′

4. Store all the components at –20°C until use.
5. Perform PCR analysis of genomic DNA in a thermocycler (Mastercycler gradient®, Eppendorf, Germany).

2.5.3. Southern Blot Analysis

1. Restriction endonuclease *Eco*RI (Fermentas, USA).
2. Hybond nylon membrane (Amersham Biosciences, UK).
3. Ultraviolet (UV) cross-linker (Stratagene, USA).
4. Church buffer: 250-mM sodium phosphate buffer, pH 7.2; 1% *(w/v)* bovine serum albumin (BSA); 7% *(w/v)* sodium dodecyl sulfater (SDS); and 1 mM EDTA.
5. α-[^{32}P] dCTP (Amersham), Prime-a-Gene® labeling kit (Promega).
6. Hybridization oven/shaker (Amersham) and phosphorimager (Bio-Rad).

3. Methods

3.1. Media Preparation

Refer to **Table 1** for the media components.

1. For all media, adjust the pH to 5.8 before autoclaving at 121°C for 15 min.
2. Add antibiotics, if needed, after cooling the media to 50–60°C.
3. Pour 25 mL solid medium in each sterile plastic Petri dish and solidify in a flow hood.

3.2. Sterilization of Instruments

1. Sterilize instruments for explant preparation (forceps, scalpel etc.) by dipping them in 90% *(v/v)* ethanol and flaming. It is important to cool them before use.
2. Alternative methods of sterilization are also possible.

3.3. Seed Germination

1. Take approximately 50–100 *H. perforatum* seeds in an Eppendorf tube containing 1 mL sterile water and add one drop of Tween-20. Keep in dark at 4°C.
2. After 12 h, discard the solution. Decontaminate the seeds with 70% *(v/v)* ethyl alcohol for 60 s and with commercial bleach containing 1.5% *(v/v)* active chlorine for 3 min (*see* **Note 1**).
3. Wash the seeds three times in sterilized distilled water and blot-dry on a sterile filter paper.

Table 1
Media Composition

Component	WA	½ MS	MSB	CIM	ONI	REG	OSM	SEL	RT
MS basal salts (g/L)		2.15	4.3	4.3	4.3	4.3	4.3	4.3	2.15
Sucrose (g/L)		15.00	30.0	30.0	30.0	30.0	30.0	30.0	15.00
Vitamins									
Myo-inositol (mg/L)		50	100	100	100	100	100	100	50
MS vitamins (mL/L)		0.5	1	1	1	1	1	1	0.5
Agar (g/1 L)	8	8	8	8		8		8	8
PGRs									
Kinetin (mg/L)				0.5					
NAA (mg/L)				1	0.5	0.1	0.1	0.1	
BAP (mg/L)						0.1	0.1	0.1	
IBA (mg/L)									0.5
Antibiotics									
Timentin (mg/L)								500	500
Hygromycin (mg/L)								20	20
Osmoticum									
D-Mannitol (g/L)							34		
D-Sorbitol (g/L)							34		

4. Transfer the disinfected seeds onto WA (water–agar) medium for germination.
5. After 10–15 d, transplant the seedlings into Baby Food Jars containing 50 mL ½MS (half strength MS basal) medium for further growth (*see* **Note 2**).

3.4. Establishment of Organogenic Nodular Cell Suspension Culture

1. Transfer the *H. perforatum* seedlings aseptically to a sterile Petri dish containing sterile distilled water (*see* **Note 2**).
2. Excise the leaves and transfer them to CIM (callus induction medium). Green compact callus induction can be seen in 15 d (*see* **Note 3**).
3. Cut the green compact calluses into pieces and transfer to organogenic nodule induction (ONI) medium. Keep the flasks on a rotary shaker at 120 rpm.
4. Organogenic nodules **(Fig. 1)** generally appear in the culture after 4–5 subcultures (*see* **Note 4**).

5. Collect these nodules using steel mesh screen (# 40, Sigma) and culture separately (*see* **Note 5**).
6. This organogenic nodular suspension (ONS) culture can be maintained for many years by subculturing a 10-mL aliquot to 70-mL ONI medium every month (*see* **Note 6**).

3.5. Pretreatment of Organogenic Nodules for Bombardment

1. Harvest ONS from the cultures using a sterile steel mesh screen (# 40, Sigma) and transfer them to OSM (osmotic medium) for osmotic treatment. This treatment should not exceed 4 h (*see* **Note 7**).
2. Transfer 1-mL ONS along with OSM to the centre of a sterile round Whatman No1 filter paper disk (*see* **Note 8**).
3. Following the absorption of the excess OSM, carefully place the paper disc along with the ONS tissues in plastic Petri dishes containing 25-mL solidified OSM.
4. Plates are now ready for bombardment.

3.6. Plasmid DNA Multiplication and Isolation

1. Transform chemically competent *E. coli* with pCAMBIA1301 following the manufacturer's protocol.
2. Transformed *E. coli* can be maintained as glycerol stocks. When necessary, initiate 100 mL broth culture in Luria bertani (LB) medium augmented with 50 mg/L kanamycin.
3. Isolate plasmid DNA from bacterial culture grown overnight (*see* **Note 9**). There are several protocols available for plasmid isolation from *E. coli*. We use Wizard®plus midipreps DNA purification system following manufacturer's instructions.
4. Quantify the plasmid DNA in a spectrophotometer and adjust to 1 μg/μL using TE (*see* **Note 10**) and store at –20°C.

3.7. Preparation of Gold Particles

1. Weigh out 20 mg gold particles, 1 μm size, in a sterile 1.5-mL Eppendorf tube, add 1 mL 70% *(v/v)* ethanol and vortex vigorously for 5 min.
2. Allow the gold particles to settle down by resting the tube for 15 min.
3. Pellet the settled gold particles by spinning for 5 s in a microfuge and discard the supernatant.
4. Add 1 mL sterile distilled water to the pellet and vortex for 1 min.
5. Allow the particles to settle down by resting the tube for 5 min.
6. Pellet the particles by spinning for 5 s in a microfuge and discard the supernatant.
7. Repeat **steps 4–6** two more times.
8. Suspend the gold particles in 350 μL 50% *(v/v)* glycerol and store at 4°C (*see* **Note 11**) until use.

3.8. Coating Gold Particles with Plasmid DNA

1. Pipet out 87.5 μL that is equivalent to 5 mg gold particles from the glycerol stock to a sterile 1.5-mL Eppendorf tube and vortex vigorously.
2. While vortexing, add 5 μL 1 μg/μL plasmid DNA, 87.5 μL 2.5 M $CaCl_2$ and 35 μL 100 mM spermidine sequentially into the tube (*see* **Note 12**).
3. Rest the mixture on ice for 10 min.
4. Centrifuge for 10 s in a microfuge to pellet the DNA coated gold particles and discard the supernatant.
5. Gently resuspend the pellet in 100 μL 100% ethanol and keep on ice (*see* **Note 13**).

3.9. Particle Bombardment

1. Set up the PDS-1000/He Particle Delivery System in a flow hood and arrange the helium gas cylinder and the vacuum pump conveniently to connect with it (*see* **Note 14**). For an interactive guide, refer the manufacturer's website http://www.bio-rad.com
2. Wipe the target shelf, macrocarrier launch assembly and the particle delivery chamber of the apparatus with absolute ethanol.
3. Immerse the macrocarrier holders, macrocarriers, and stopping screens in absolute ethanol and keep rupture discs immersed in isopropanol (see Note 15).
4. Unscrew the rupture disk retaining cap from the gas acceleration tube and place rupture disk (1100 psi) in the recess of the cap.
5. Take the macrocarriers out of ethanol and place on a sterile Whatman no. 1 filter paper to allow ethanol to evaporate and position them into the macrocarrier holders.
6. Briefly vortex the DNA precipitated gold particles for 5 s to disperse any clumps and pipet out 10 μL onto the centre of the each macrocarrier, spread evenly and air-dry (*see* **Note 16**).
7. Arrange a macrocarrier holder containing dry macrocarrier coated with microcarrier particles in the launch assembly in such a way that the microcarriers face the target desk.
8. Place the osmotic plates containing tissues 9 cm below the stopping screen on the target desk (*see* **Note 17**).
9. Close and evacuate the bombardment chamber to 28" of mercury (*see* **Note 18**).
10. Fire the microcarriers onto the target tissue.
11. Perform **steps 4–11** for each plate.
12. Seal the plates containing bombarded tissues with parafilm and incubate in darkness.

3.10. Regeneration of Transgenic Plants

1. Transfer ONS tissues aseptically to REG (regeneration) medium 4 h after bombardment and incubate in dark (*see* **Note 19**).
2. After 2 d, transfer the tissues to selection medium (*see* **Note 20**).
3. Maintain all the cultures in dark until the formation of callus (**Fig. 3a**) or shoot initials (*see* **Note 21**).
4. Transfer ONS with calluses and/or shoot initials (**Fig. 3b**) to MSB medium containing 20 mg/L hygromycin for shoot elongation (*see* **Note 22**).
5. Once the shoots are reached 3- to 5-cm height (**Fig. 3c**) excise them from the explant and transfer to Baby Food Jars containing 50 mL of rooting medium.
6. Rooting can be observed from hygromycin-resistant shoots within 25 d (**Fig. 3d**), while nontransformed (hygromycin-susceptible) shoots will be completely killed (**Fig. 3d**, *arrows*).
7. Fill the jars containing rooted plants with sterile distilled water and leave at room temperature for 2 d (*see* **Note 23**).
8. Remove the plants from the medium without damaging the root system and wash thoroughly in running tap water to remove the traces of rooting medium (*see* **Note 24**).
9. Transfer the plants to plant propagation system (**Fig. 4a**) and cover with polyethylene bags to maintain humidity.
10. Acclimatize the plants by gradually reducing the humidity by making holes in the polyethylene bags and by gradually exposing the plants to sunlight for a period of 1 wk.
11. Transfer the hardened plants to garden pots (**Fig. 4b**) containing soil:compost (1:1), irrigate regularly with tap water and grow under controlled environment. We use growth chamber with 25 μmol/m^2 s 16 h/d incident radiation, 70% humidity and 26°C temperature (*see* **Note 25**).

3.11. Analyses of Transgenic Plants

3.11.1. Enzymatic Histochemical GUS Assay

1. Thaw the frozen aliquots of GUS solution to room temperature.
2. Thoroughly wash the tissue samples in distilled water and put them into tubes containing GUS solution (*see* **Note 26**).
3. Cover the tubes with aluminium foil and incubate at 37°C for 12–24 h.
4. Remove chlorophyll from the tissues with 70% (*v/v*) methanol/ethanol to visualize the result more clearly.
5. Characteristic deep blue coloration is the indication of GUS gene expression in the tissues (**Fig. 5**).

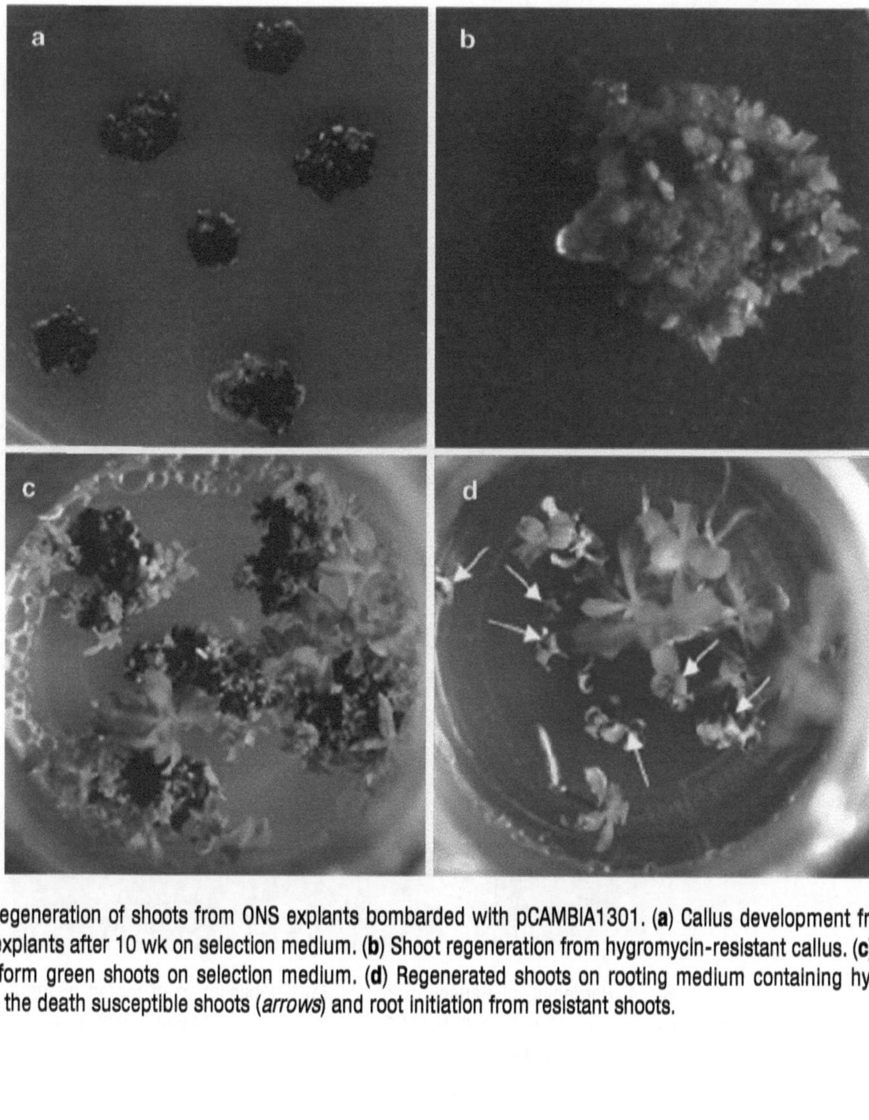

Fig. 3. Regeneration of shoots from ONS explants bombarded with pCAMBIA1301. (**a**) Callus development from bombarded explants after 10 wk on selection medium. (**b**) Shoot regeneration from hygromycin-resistant callus. (**c**) Cultures with uniform green shoots on selection medium. (**d**) Regenerated shoots on rooting medium containing hygromycin showing the death susceptible shoots (*arrows*) and root initiation from resistant shoots.

3.11.2. Polymerase Chain Reaction Analysis

1. Take 50 ng DNA from each sample including the putative transgenic plants, pCAMBIA1301 (positive control) and non-transformed plant (negative control) into sterile PCR tubes. Make up the volume to 10 μL with PCR grade water.
2. Prepare a master mix cocktail sufficient for the desired number of reactions, each with 2 μL forward primer, 2 μL reverse primer, 0.5 μL dNTP mix, 5 μL PCR buffer, 2 μL $MgCl_2$, 0.25 μL Taq DNA polymerase and enough water to bring the volume to 15 μL. Distribute 15 μL to each PCR tubes containing DNA. Mix well by pipetting up and down gently.

Fig. 4. Hardening of transgenic plants. (**a**) Establishment of rooted plants in JIFFY 7 plant propagation system (Lisbon, Portugal). (**b**) A transgenic plant growing in garden pot after hardening.

3. Amplify the specific fragments of transgenes with a hot start at 94°C for 4 min, followed by 30 cycles of denaturation (94°C, 1 min), annealing (55°C, 2 min) and extension (72°C, 2 min), with a final extension of 10 min at 72°C in the thermocycler.
4. Resolve the PCR products in 0.8% *(w/v)* agarose gel electrophoresis. The expected fragment sizes are 1.3 kb for GUS gene and 0.8 kb for HPT gene **(Fig. 6)**.

3.11.3. Southern Blot Hybridization Analysis

1. Digest 20 μg genomic DNA from control and PCR positive plants with *Eco*RI following the manufacturer instructions.
2. Load the restriction-digested DNA and proper molecular weight marker in 1% *(w/v)* agarose gel and run at 30 V overnight (*see* **Note 27**).

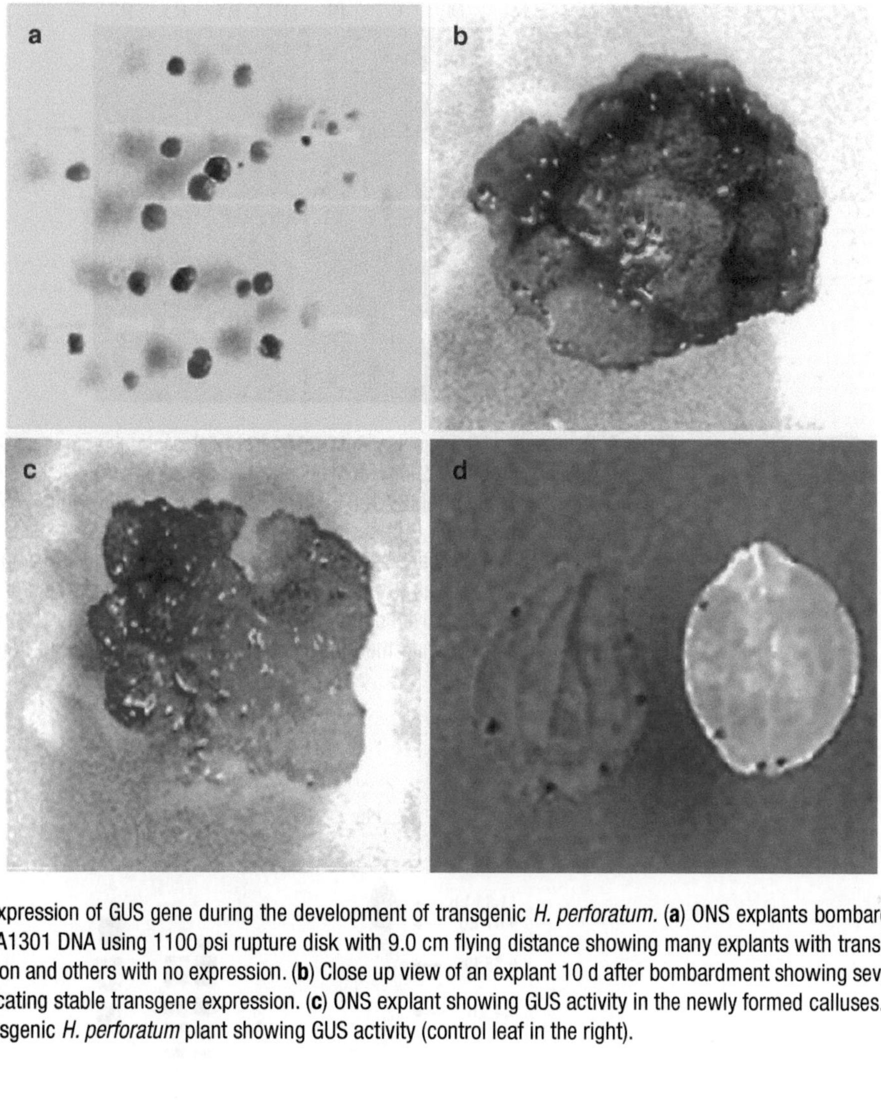

Fig. 5. Expression of GUS gene during the development of transgenic *H. perforatum*. (**a**) ONS explants bombarded with pCAMBIA1301 DNA using 1100 psi rupture disk with 9.0 cm flying distance showing many explants with transient GUS expression and others with no expression. (**b**) Close up view of an explant 10 d after bombardment showing several GUS foci indicating stable transgene expression. (**c**) ONS explant showing GUS activity in the newly formed calluses. (**d**) Leaf of a transgenic *H. perforatum* plant showing GUS activity (control leaf in the right).

3. Transfer the electrophoresed DNA onto a nylon membrane by alkaline capillary blotting.
4. Crosslink the DNA to the membrane using UV Stratalinker 1800, Stratagene under autocrosslink mode (*see* **Note 28**).
5. Prehybridize the membrane for 3 h in church buffer at 55°C.
6. Label a GUS gene specific fragment with α-[^{32}P] dCTP and transfer to the hybridization solution.
7. After 16 h hybridization at 55°C, remove the solution. Wash the membrane at 55°C twice with 2X SSC + 0.1% *(w/v)* SDS (each for 15 min) and with 0.1X SSC + 0.1% *(w/v)* SDS for 5 min.

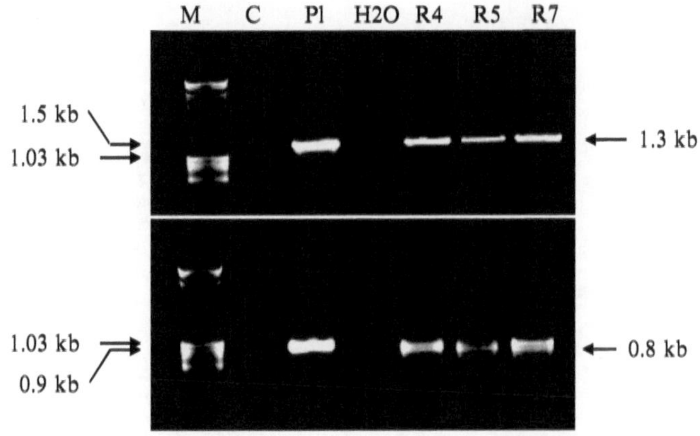

Fig. 6. PCR analyses of transgenic plants. (**a**) Agarose gel electrophorogram of GUS gene PCR amplification products. *Lanes*: *M* Molecular size marker (MassRuler, Fermentas), *C* control plant, *Pl* plasmid pCAMBIA1301 positive control, *H2O* water control, *R4, R5* and *R7* are plants positive in the GUS assay showing the amplification of expected 1.3-kb fragment of GUS gene. (**b**) Agarose gel electrophorogram of HPT gene PCR amplification products. *Lanes*: *M* Molecular size marker (MassRuler, Fermentas), *C* control plant, *Pl* plasmid pCAMBIA1301 positive control, *H2O* water control, *R4, R5* and *R7* are hygromycin resistant plants showing the amplification of the expected 0.8 kb fragment of HPT gene.

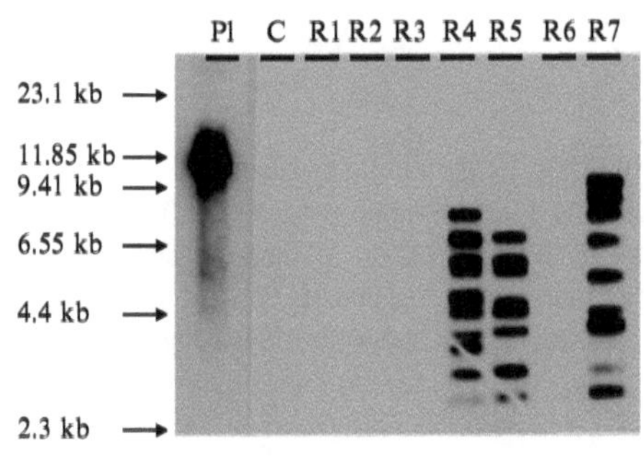

Fig. 7. Southern blot analysis of transgenic plants. Southern blot analysis of DNA isolated from the leaves of seven hygromycin resistant plants. *Lanes*: *Pl* pCAMBIA1301 DNA showing hybridization signal at 11.8 kb, *C* Non-transformed control plant had no hybridization signal, *R1–R7* Seven putative transgenic plants that were rooted on hygromycin, out of them R4, R5, and R7 with hybridization signal are true transgenic plants.

8. Wrap the membrane with Kim wipe and expose the membrane to an imaging screen for 12 h and scan in Phosphorimager (*see* **Note 29**).
9. Lanes with radioactive signals indicates the presence of GUS gene in the corresponding plants (**Fig. 7**).

4. Notes

1. Avoid contact of skin to the bleach. Commercial bleach can vary in active chlorine content. Make sure to check the product label and adjust dilution to obtain a 1.5% *(w/v)* final concentration of active chlorine.
2. Work as quick as possible to avoid desiccation of the plantlets.
3. Callus can be obtained using many other combinations of auxins and cytokinins.
4. We have successfully produced organogenic nodules from the callus that were induced in CIM, but there is no reason to believe that organogenic nodules can not be produced from callus obtained on other media.
5. If the culture flasks left unshaken for few minutes, organogenic nodules settles down and the medium can be simply removed without the need of steel mesh.
6. We use sterile 10.0 mL serological pipes (Sarstedt) in order to manipulate organogenic nodules along with the medium.
7. Because the organogenic nodules are delicate tissues, the period of osmotic treatment can be reduced and should not exceed 4 h. Longer osmotic treatments generally results in drastic reduction of plant regeneration.
8. To transfer 1.0 mL ONS onto filter paper discs, P-1000 micropipet with 1-mL tip with cut at the end can be used.
9. Since bacterial culture is initiated directly from the glycerol stocks, broth may grow slowly. By increasing the inoculum, optimal growth can be achieved overnight.
10. Alternatively, plasmid can be dissolved in sterile water. However TE preserves the plasmid for longer storage.
11. We generally sterilize the gold particles on the day before bombardment, although gold particles can be stored in glycerol for 1 wk.
12. DNA precipitation onto the macrocarrier is a critical step. After the addition of $CaCl_2$, spermidine should be added quickly to the mixture.
13. It is advisable to spread the microcarriers onto the macrocarriers immediately, as ethanol evaporates quickly. Ice incubation only prevents evaporation to a certain extent.
14. We prepare the whole setup a day before bombardment leaving the UV lights on overnight.
15. We keep rupture disks, macrocarriers, macrocarrier holders and stopping screens separately in sterile Petri dishes filled

with the sterilent either isopropanol or ethanol and dry out in the hood on a blotting paper before use.

16. Pipetting of the DNA-microcarrier mix onto the macrocarrier should be performed very rapidly in order to avoid agglomeration of the gold particles.
17. Although transformation of ONS cells can also be achieved with other flying distances, 9-cm flying distance show the best results in our experience. Do not forget to remove the Petri dish lid.
18. The vacuum pump should be powerful enough to reach the 28 mmHg value in 15 s. Regeneration is drastically affected if ONS are left under vacuum for longer period.
19. Post bombardment osmotic treatments exceeding 4 h will lead to the drastic reduction of explant survival.
20. The effect of medium containing hygromycin (SEL) on non-transformed tissues is quite obvious and leads to their death within 20 d.
21. Callus formation from the bombarded explants normally takes about 10 wk of culture on SEL.
22. This step can be avoided for cultures with elongated shoots.
23. Direct exposure of micropropagated *H. perforatum* plants to ambient conditions will lead to quick desiccation. Hence, maintaining the plants in water avoid desiccation and make them hard enough to withstand further acclimatization processes.
24. During this process, the delicate root system should not be damaged. We pass tap water with pressure through the medium which will allow the separation of plants from the medium without damaging the roots.
25. Growth chamber is not required if a greenhouse with appropriate lighting, temperature and humidity is available.
26. Use tissues from non transformed plants as negative control.
27. If not using radioactive labelled molecular weight marker, photographing the gel after electrophoresis will be useful to estimate the molecular weights of the hybridization signals.
28. Membranes can be stored at room temperature for several months before radioactive hybridization.
29. Exposure time should be adjusted based on the signal intensity of the hybridized membrane. As the positive control plasmid exhibits high signal, we cut the corresponding area and expose it separately.

Acknowledgments

We are grateful to Dr. Richard Jefferson for providing the plasmid pCAMBIA1301. This work was supported by Fundação para a Ciência e a Tecnologia (POCTI/AGR/40283/2001). Postdoctoral fellowship (SFRH/BPD/17102/2004) awarded by Fundação para a Ciência e a Tecnologia to G. Franklin is gratefully acknowledged.

References

1. Lecrubier, Y., Clerc, G., Didi, R., and Kieser, M. (2002). Efficacy of St-John's wort extract WS 5570 in major depression: a double-blind, placebo-controlled trial. *Am J Psychiatry.* **159**, 1361–1366
2. Butterweck, V. (2003). Mechanism of action of St. John's wort in depression: what is known? *CNS Drugs.* **17**, 539–562
3. Schinazi, R.F., Chu, C.K., Babu, J.R., Oswald, B., Saalman, V., Cannon, D.L., Ericksson, B.F.H., and Nasr, M. (1990). Anthraquinones as a new class of antiviral agents against human immunodeficiency virus. *Antiviral Res.* **13**, 265–272
4. Agostinis, P., Vantieghem, A., Merlevede, W., and Peter, A.M. (2002). Hypericin in cancer treatment: more light on the way. *Int J Biochem Cell Biol.* **34**, 221–241
5. Silva, B.A., Dias, A.C.P., Ferreres, F., Malva, J.O., and Oliveira, C.R. (2004). Neuroprotective effect of *H. perforatum* extracts on beta-amyloid-induced neurotoxicity. *Neurotox Res.* **6**, 119–130
6. Silva, B.A., Ferreres, F., Malva, J.O., and Dias, A.C.P. (2005). Phytochemical and antioxidant characterization of *Hypericum perforatum* alcoholic extracts. *Food Chem.* **90**, 157–167
7. Kirakosyan, A.B. Vardapetyan, H.R., and Charchoglyan, A.G. (2000). The content of hypericin and pseudohypericin in cell cultures of *Hypericum perforatum* L. (St. John's wort) and production of hypericin. *Russ J Plant Physiol.* **47**, 270–273
8. Smith, M.A.L., Kobayashi, H., Gawienowski, M., and Briskin, D.P. (2002). An in vitro approach to investigate chemical synthesis by three herbal plants. *Plant Cell Tissue Organ Cult.* **70**, 105–111
9. Pasqua, G. Avato, P. Monacelli, B. Santamaria, A.R., and Argentieri, M.P. (2003). Metabolites in cell suspension cultures, calli, and in vitro regenerated organs of *Hypericum perforatum* cv. Topas. *Plant Sci.* **165**, 977–982
10. Dias, A.C.P., and Ferreira, M.F. (2003). Production of phenolics by in vitro cultures of *Hypericum perforatum*: a case study. (Rauter, A.P., Araujo, M.E., Palma, F.B., Justino, J., Santos, S.P., eds.). Kluwer Academic Publishers, Dordrecht, pp. 367–374
11. Dias, A.C.P. (2003). The potential of in vitro cultures of *Hypericum perforatum* and *Hypericum androsaemum* to produce interesting pharmaceutical compounds. (Earnst, E., ed.). Taylor and Francis, UK, pp. 137–154
12. Zhou, L.G., Wu, J.Y. (2006). Development and application of medicinal plant tissue cultures for production of drugs and herbal medicinals in China. *Nat Prod Rep.* **23**, 789–810
13. Hu, W., and Zhong, J. (2001). Effect of bottom clearance on performance of airlift bioreactor in high-density culture of Panax ginseng cells. *J Biosci Bioeng.* **92**, 389–392
14. Kutchan, T.M. (2005). A role for intra- and intercellular translocation in natural product biosynthesis. *Curr Opin Plant Biol.* **8**, 292–300
15. Brown, P. (2004). Transcription factors as tools for metabolic engineering in plants. *Curr Opin Plant Biol.* **7**, 202–209
16. Murch, S.J., Choffe, K.L., Victor, J.M.R., Slimmon, T.Y., Krishnaraj, S., and Saxena, P.K. (2000). Thidiazuron-induced plant regeneration from hypocotyl cultures of St. John's wort (*Hypericum perforatum* cv 'Anthos'). *Plant Cell Rep.* **19**, 576–581
17. Zobayed, S.M.A., and Saxena, P.K. (2003). *In vitro* grown roots: a superior explant for prolific shoot regeneration of St. John's wort (*Hypericum perforatum* L. cv 'New Stem') in a temporary immersion bioreactor. *Plant Sci.* **165**, 463–470
18. Zobayed, S.M.A., Murch, S.J., Rupasinghe, H.P.V., and Saxena, P.K. (2004). In vitro production and chemical characterization of St. John's wort (*Hypericum perforatum* L. cv 'New Stem'). *Plant Sci.* **166**, 333–340

19. Franklin, G., and Dias, A.C.P. (2006). Organogenesis and embryogenesis in several *Hypericum perforatum* genotypes. *In Vitro Cell Dev Biol – Plant.* **42**, 324–330
20. Jiao, S., Cheng, L., Wolff, J. A., and Yang, N.-S. (1993). Particle bombardment-mediated gene transfer and expression in rat brain tissues. *Bio/Technology.* **11**, 497–502
21. Yang, N. S., Burkholder, J., Roberts, B., Martinell, B., and McCabe, D. (1990). In vivo and in vitro gene transfer to mammalian somatic cells by particle bombardment. *Proc Natl Acad Sci U S A.* **87**, 9568–9572
22. Franklin, G., Conceiçao, L.F.R., Kombrink, E., and Dias, A.C.P. (2008). *Hypericum perforatum* plant cells reduce Agrobacterium viability during co-cultivation. *Planta.* **227**, 1401–1408
23. Hansen, G. (2000). Evidence of *Agrobacterium*-induced apoptosis in maize cells. *Mol Plant Microbe Interact.* 13, 649–657
24. Manoharan, M., and Dahleen, L. (2002) Genetic transformation of the commercial barley (*Hordeum vulgare* L.) cultivar Conlon by particle bombardment of callus. *Plant Cell Rep.* **21**, 76–80
25. Vidal, J.R., Kikkert, J.R., Wallace, P.G., and Reisch, B.I. (2003). High-efficiency biolistic co-transformation and regeneration of "Chardonnay" (Vitis vinifera L.) containing *npt-II* and antimicrobial peptide genes. *Plant Cell Rep.* **22**, 252–260
26. Zhang, G., Lu, S., Chen, T.A., Funk, C.R., and Meyer, W.A. (2003). Transformation of triploid bermudagrass (*Cynodon dactylon* x *C. transvaalensis* cv. TifEagle) by means of biolistic bombardment. *Plant Cell Rep.* **21**, 860–864
27. Pink, D., and Puddephat, I. (1999). Deployment of disease resistance genes by plant transformation – a 'mix and match' approach. *Trends Plant Sci.* **4**, 71–75
28. Cappell, T., and Christou, P. (2004). Progress in plant metabolic engineering. *Curr. Opin. Biotechnol.* **15**, 148–154
29. Franklin, G., Oliveira, M., and Dias, A.C.P. (2007). Production of transgenic *Hypericum perforatum* plants via particle bombardment-mediated transformation of novel organogenic cell suspension cultures. *Plant Sci.* **172**, 1193–1203
30. Murashige, T., and Skoog, F. (1962). A revised medium for rapid growth and bioassays with tobacco tissue cultures. *Physiol Plant.* **15**, 473–497

Chapter 19

Agrobacterium-Mediated Transformation of *Ruta graveolens* L.

Karine Lièvre, Thi Lê Minh Tran, Sébastien Doerper, Alain Hehn, Paul Lacoste, Brigitte Thomasset, Frédéric Bourgaud, and Eric Gontier

Summary

Agrobacterium tumefaciens is used to develop a genetic transformation method for a medicinal plant *Ruta graveolens*. The direct plant regeneration strategy is preferred to callus line establishment. In vitro seedlings, 2- –to 3-wk-old, are used to excise hypocotyls and co-cultivated for 3 d with *A. tumefaciens* strain C58C1Rif^R containing plasmid pTDE4 harbouring neomycin phosphotransferase (npt II, kanamycin resistance) and β-glucuronidase encoding genes. The Southern blot analysis has shown that 78% kanamycin resistant plants contain gene encoding β-glucuronidase. The GUS histochemical assay shows that 67% transgenic plants exhibit the corresponding enzymatic activity. Routine transformation efficiency of *R. graveolens* L. is 11% and could reach up to 22%. Transgenic plants are grown in the greenhouse within 4 months after the initial seedlings.

Key words: *Agrobacterium tumefaciens*, *Ruta graveolens*, Hypocotyl, Direct regeneration, Transgenic plant, Gus, nptII

1. Introduction

Furocoumarins are secondary compounds mainly distributed in four families of higher plants: Apiaceae (celery), Moraceae (ficus), Rutaceae (citrus), and Fabaceae (coronilla) *(1)*. Some of these plants, like Apiaceae (carrot, celery, parsley) or Rutaceae (Citrus), are of agronomic interest. They have been largely studied because of consumer's health hazard caused by the toxicity of furocoumarins that are present in the products of these crops. Furocoumarins are also

S. Mohan Jain and Praveen K. Saxena (ed.), *Methods in Molecular Biology, Protocols for* In Vitro *Cultures and Secondary Metabolite Analysis of Aromatic and Medicinal Plants, vol. 547*
© Humana Press, a part of Springer Science+Business Media, LLC 2009
DOI: 10.1007/978-1-60327-287-2_19

pharmacologically active molecules used for skin disease treatments such as vitiligo and psoriasis. In general, the plants producing furocoumarins are also of fundamental interest because they constitute a model of co-evolution with herbivorous insects. It is known today that an association of linear and angular furocoumarins, as found in several Apiaceaes, is impossible to detoxify by insects of the Papilio species whereas they are able to attack plants producing only linear furocoumarins *(2)*. Therefore, it is a big challenge to understand the biosynthesis of this pathway which led us to develop a metabolic engineering strategy.

For a decade, ecophysiologic approaches have been developed in our laboratory for studying biosynthesis of these molecules in *Ruta graveolens (3, 4)*. Previous studies have shown that modification of culture conditions and elicitation have little effects on furocoumarins profiles. Thus our thinking has been that the genetic transformation could be ideal to understand the furocoumarin pathway, enzymes, and genes involved in their biosynthesis. We have developed a new tool to generate transgenic plants *(5)*. *Ruta graveolens* is susceptible to *Agrobacterium* wild strains, and there is a possibility of regeneration of shoot teratomas and hairy roots from *R. graveolens (6, 7)*. The status of transformants could not be confirmed neither genetically nor biochemically.

This chapter describes reliable *Agrobacterium*-mediated transformation method for genetic transformation of *R. graveolens* L.

2. Materials

2.1. Plant and Explants Culture Media

1. Murashige and Skoog *(8)* medium (MS). For 1 L MS medium, mix 100 mL 10X macroelement stock solution, 10 mL 100X microelement stock solution, 10 mL 100X Fe-ethylenediamine tetraacetic acid (EDTA) stock solution, 10 mL 100X vitamin stock solution, and add water to reach the final volume. Adjust to pH 5.8 with NaOH. For solid medium, add 8 g/L Difco agar and autoclave (20 min at 120°C, 10^5 Pa) (*see* **Notes 1–3**).
2. MS (30). MS medium containing 30 g/L sucrose. Adjust to pH 5.8 with NaOH.
3. MS (30) BAP 0.1. MS (30) supplemented with 0.1 g/L growth regulator BAP (Benzylaminopurine). Adjust to pH 5.8 with NaOH when required.
4. MS (30) BAP 0.1 + 250 μM acetosyringone. Add 250 μM acetosyringone after autoclaving MS (30) BAP 0.1.

2.2. Plant Culture

1. *Ruta graveolens* L. seeds (Samen und Pflanzen GmbH, Darmstadt, Germany).

2. Keep all cultures, sealed with gas porous tape, in growth chambers at 22°C with a 16 h photoperiod at 160 mmol/m^2/s light irradiance.
3. Cultures are maintained in 15 mm × 57 mm Petri dishes or in glass vessels (70 mm × 55 mm, Sigma V0830 or 100 mm × 55 mm) (Sigma; cat no. V0633) closed with magenta cap (Sigma; cat. no. B8648) containing 10, 25, or 40 mL medium, respectively.

2.3. Bacteria Culture Media

1. LB medium. 10 g/L tryptone; 5 g/L yeast extract; 10 g/L NaCl; adjust to pH 7.2 with NaOH. Use 15 g/L Difco agar to solidify medium after autoclaving (20 min at 120°C, 10^5 Pa). Add the corresponding antibiotics (*see* **Table 1** and **Note 4**).
2. YEB medium. 5 g/L beef extract; 1 g/L yeast extract; 5 g/L peptone; 0.5 g/L $MgSO_4$; adjust to pH 7.2 with NaOH. Add 15 g/L Difco agar to solidify medium. Supplement 5 g/L 0.2 µm filtered sucrose after autoclaving (20 min at 120°C, 10^5 Pa). Add the corresponding antibiotics (*see* **Table 1** and **Note 4**).
3. YEB Mg^{2+} medium. YEB medium containing 0.5 g/L $MgSO_4$.

Table 1
List of Bacteria, Plasmids, and Corresponding Culture Conditions

Bacteria (plasmid)	Temperature, medium + antibiotics
E. coli HB 101 (pRK2013)	37°C, LB +25 µg/mL kanamycin
E. coli MC 1061 (pTDE4)	37°C, LB +100 µg/mL carbenicillin +20 µg/mL streptomycin
A. tumefaciens C58C1Rif[R] (pGV2260)	26°C, YEB +100 µg/mL rifampicin +100 µg/mL carbenicillin
Triparental mating:	26°C, YEB
A. tumefaciens C58C1Rif[R] (pGV2260, pTDE4)	+100 µg/mL rifampicin +100 µg/mL carbenicillin +300 µg/mL streptomycin +100 µg/mL spectinomycin

4. Autoclaved Lambda buffer. 10 mM Tris–HCl; 10 mM MgSO4. Adjust pH to 7.2 with NaOH.

2.4. Bacteria Strains and Plasmids

1. *Agrobacterium tumefaciens* C58C1Rif^R (resistant to rifampicin) harbours pGV2260 plasmid. This plasmid carries the *vir* genes necessary to infect the plant plus a gene encoding carbenicillin resistance.
2. *Escherichia coli* MC1061 (resistant to streptomycin) contains the pTDE4 plasmid (kindly provided by B. Thomasset, *see* **Note 5**). This binary plasmid confers resistance to kanamycin. Its T-DNA includes a neomycin phosphotransferase (nptII) marker gene driven by the nopaline synthase (*nos*) promoter, allows kanamycin selection, and the β-glucuronidase (*gus*) reporter gene, driven by the cauliflower mosaic virus (CaMV 35S) promoter. This latter gene allows a GUS histochemical assay.
3. *Escherichia coli* HB101 harbors pRK2013 plasmid. This plasmid helps *E. coli* MC1061 to transfer its plasmid pTDE4 to *A. tumefaciens.*

2.5. Bacteria Culture and Storage

2.5.1. Bacterial Culture

Bacteria are grown up-side-down in sterile Petri dishes containing medium solidified with 15 g/L Difco agar. *E. coli* are grown for 16 h in Luria Bertani (LB) medium at 37°C, and *A. tumefaciens* are grown for 3 d in YEB medium at 26–28°C. In order to avoid contamination, the appropriate concentrations of antibiotics are used according to bacteria and plasmid resistances (*see* **Table 1** and **Note 5**).

2.5.2. Short-Term Storage

Bacteria from a fresh plate culture can be kept in its Petri dish and store at 4°C for 1–2 mo.

2.5.3. Long-Term Storage

Fresh culture of bacteria is used to inoculate 12-mL tubes containing 2 mL appropriate liquid medium (*see* **Table 1**). Bacteria are grown overnight at 26 or 37°C under a 200 rpm agitation. Add 750 μL fresh liquid medium to 1.5-mL microtubes containing 750 μL sterile glycerol solution 40% *(v/v)* in LB. Homogeneous bacteria culture can be stored at 80°C for 1 yr.

2.6. Transfer of T-DNA from E. coli MC1061 to A. Tumefaciens by Triparental Mating

1. Grow 2 mL fresh cultures of *E. coli* MC1061 and HB101for 16 h at 37°C
2. Grow 2 mL fresh culture of *A. tumefaciens* C58C1Rif^R for 16 h at 26°C
3. Mix 100 μL of each of three cultures in a sterile microtube
4. Grow the mixture of bacteria overnight in a Petri dish containing YEB Mg^{2+} at 26°C to get *A. tumefaciens* C58C1Rif^R (pGV2260, pTDE4).

5. Add 2 mL sterile Lambda buffer to the culture
6. Use a sterile glass stick to get the bacteria in the buffer
7. Pipette the buffer containing the bacteria and inoculate a Petri dishes with YEB solid medium and antibiotics (*see* **Table 1**)
8. Grow *A. tumefaciens* C58C1Rif^R (pGV2260, pTDE4) for 2 d at 26°C

2.7. Stock Solutions

2.7.1. MS Medium Stock Solutions for Plant Cultures (See Note 3)

1. 10X MS medium macro-element stock solution. KNO_3 19 g/L; NH_4NO_3 16.5 g/L; $CaCl_2 \cdot H_2O$ 4.4 g/L; $MgSO_4 \cdot 7H_2O$ 3.7 g/L; KH_2PO_4 1.7 g/L.
2. 10X MS medium micro-element stock solution. $MnSO_4 \cdot H_2O$ 1.69 g/L; $ZnSO_4 \cdot 7H_2O$ 860 mg/L; H_3BO_3 620 mg/L; KI 83 mg/L; $Na_2MoO_4 \cdot 2H_2O$ 25 mg/L; $CuSO_4 \cdot 5H_2O$ 2.5 mg/L; $CoCl_2 \cdot 6H_2O$ 2.5 mg/L.
3. 100X MS medium Fe-EDTA stock solution. $FeSO_4 \cdot 7H_2O$ 2.78 g/L, $Na_2EDTA \cdot 2H_2O$ 3.72 g/L.
4. 100X MS medium vitamin stock solution. Inositol 10 g/L; nicotinic acid 50 mg/L, pyridoxine HCl 50 mg/L; thiamine HCl 10 mg/L.

2.7.2. Antibiotic and Bacteriostatic Stock Solutions

Except for rifampicin, all the antibiotics and cefotaxime (bacteriostatic) solutions are filtered through 0.22 μm filters and transfer in 1.5-mL aliquots for storage at –20°C.

1. Carbenicillin: 50 mg/mL in water
2. Kanamycin: 50 mg/mL in water
3. Streptomycin: 50 mg/mL in water
4. Spectinomycin: 50 mg/mL in water
5. Cefotaxime: 50 mg/mL in water
6. Rifampicin: 50 mg/mL in DMSO

2.7.3. Molecular Biology Stock Solutions

1. 1X TE . 10 mM Tris–HCl, 1 mM EDTA, pH 8.0. Mix, autoclave and store at room temperature.
2. 50X TAE pH 8.0 for 1 L. 242 g Tris Base, 57.1 mL glacial acetic acid, 37 g Na_2EDTA, add water to the final volume. Mix and store at room temperature
3. 10X DNA loading buffer . 1X TAE , 50% glycerol, 0.2 M EDTA, 0.1% bromophenol blue, 0.1% sodium dodecyl sulfate (SDS). Mix and store at -20°C
4. Sodium acetate 3 M for 100 mL. 24.61 g sodium acetate in water, and adjust pH to 5.6 with glacial acetic acid (>10 mL)
5. Depurination solution. HCl 0.25 M
6. Denaturant solution. NaCl 1.5 M, NaOH 0.5 M
7. Neutralization solution. Tris—HCl 0.5 M, NaCl 3 M pH 7.5.

8. SDS stock solution. 10% *(w/v)* in water. Use 0.45-μm filters and store at room temperature.
9. *N*-lauroylsarcosine stock solution. 10% *(w/v)* in water. Use 0.45-μm filters and store at room temperature.
10. 20X SSC stock solution. 3 M NaCl; 300 mM sodium citrate, pH 7.0; autoclave.
11. 2X Wash solution . SSC; 0.1% SDS.
12. 0.5X Wash solution . 0.5X SSC; 0.1% SDS.
13. Standard hybridization buffer. 5 g Ficoll 400 (Pharmacia), 5 g polyvinylpyrrolidone (Serva; cat. no. 33422), 5 g BSA (Sigma; cat. no. B4287). Heat while you stir. Use 0.45-μm filters and store at 4°C

2.7.4. Other Stock Solutions

1. Acetosyringone. 250 μM in dimethyldulfoxide (DMSO). Use 0.22-μm filters to get a sterile solution and transfer in 1.5-mL aliquots. Store at −20°C
2. Benzyl-amino-purine (BAP). 1 mg/mL in 96% ethanol. Store at 4°C for up to 1 yr.
3. X-Gluc solution. 100 mM 3-bromo-4-chloro-3-indolyl glucuronide in DMSO. Store at −20°C. The stock solution should be kept in the dark because of its light sensitivity
4. X-Gluc buffer. 100 mM Tris–HCl, 50 mM NaCl, pH 7.0.

3. Methods

Hypocotyls of *R. graveolens* are co-cultured for 3 d with *A. tumefaciens* in order to transfer the T-DNA from pTDE4 plasmid. The first selection is based on kanamycin resistance because the T-DNA confers resistance in transgenic plants. Other tests are carried out to detect escapees. T-DNA will also transfer the *GUS* sequence encoding β-glucuronidase activity, polymerase chain reaction (PCR) with our specific primers amplifies a 550-bp sequence; and the gene expression is confirmed with GUS histochemical assay showing blue color in genetic transformants. However, these previous tests can be positive even if the T-DNA is not integrated in the host genome (transient expression). To confirm the stability of the transformants, the Southern blotting is done. The host DNA is digested by restriction enzymes and hybridized with a specific probe designed for the *gus* sequence. This approach also provides additional information on the number of recognized sequences reflects the number of transgenes integrated in the host genome.

3.1. Agrobacterium Tumefaciens Inoculum Preparation

1. Use freshly grown Agrobacteria C58C1Rif^{R} (pGV2260, pTDE4)
2. Inoculate 2-mL YEB liquid medium + 100 μg/mL rifampicine + 25 μg/mL kanamycin
3. Grow for 16 h at 26°C under 200 rpm agitation on a shaker
4. Collect bacteria cells by centrifugation 3500*g* for 15 min
5. Wash twice in 4-mL lambda buffer centrifugation 3500g for 15 min each time
6. Resuspend in 100 mL MS(30)BAP0.1

3.2. In Vitro Culture of Ruta Graveolens L. (See Fig. 1)

1. Seeds of *R. graveolens* L. are scarified min in 95% sulfuric acid for 10 min.
2. Rinse three times with tap water
3. All the following steps are carried out under a laminar flow hood.
4. Sterilize scarified seeds for 5 min in 7% *(w/v)* calcium hypochlorite
5. Rinse three times for 5 min in sterile distilled water
6. Put seeds in sterile 0.1% *(w/v)* agar solution, 100 mL for more than 4000 seeds
7. Enlarge the diameter 3 mm of sterile pipet tips, size 5000 μL. Keep them sterile.
8. Pick up seeds suspended in agar solution with sterile pipet tips
9. Put about 30 seeds on MS (30) medium containing 8 g/L Difco-agar in Petri dishes
10. After 2–4 wk, seeds germinate and hypocotyls are ready to use for genetic transformation

3.3. Inoculation and Co-culture of Ruta Graveolens Explants with Agrobacterium tumefaciens

All the following steps are carried out under a laminar flow hood

1. Take ten plantlets 2- to 3-wk-old (*see* **Note 6**).
2. Put them on a sterile paper and use a sterile scalpel to cut 1 cm long hypocotyl segments.
3. Put the hypocotyls in Petri dishes with solid MS (30) and cover them to prevent drying. 200–250 hypocotyls can be cut in 60–90 min (*see* **Note 7**).
4. Put 200–250 hypocotyls in a flask with 100 mL prepared *Agrobacterium* suspension.
5. Leave hypocotyls for 15 min in the medium, agitate from time to time.

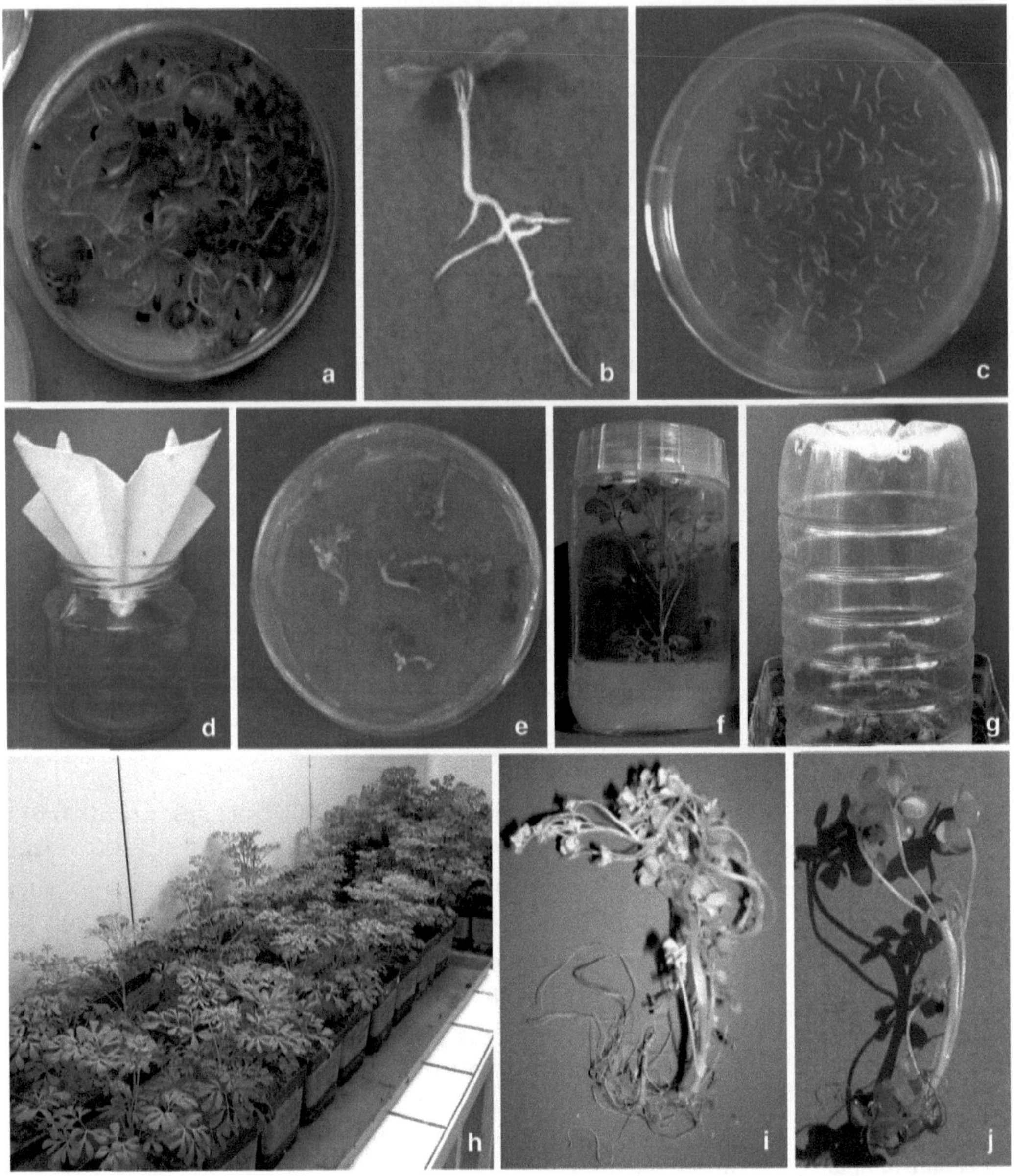

Fig. 1. Different steps for *Ruta graveolens* genetic transformation and characterisation. Seeds are grown in vitro to get sterile plantlets (**a**). When they are 2–4 wk old (**b**), hypocotyls are cut (**c**) and inoculated with *Agrobacterium tumefaciens*. After 3 d of coculture, bacteria are eliminated by washes with cefotaxime and pouring hypocotyls onto sterile papers (**d**). Kanamycin resistant cells can develop on selective medium containing 300 mg/L cefotaxime and 50 mg/L kanamycin (**e**). A second selective step is carried out in glass vessel. As the plant grows, it is transferred to larger vessel for rooting (**f**). Individual closed chambers (**g**) are used to keep high hygrometry when plants are transferred to a greenhouse (**h**). GUS histochemical assay allows identification of plants expressing β-glucuronidase: only transgenic plants turn blue (**i**). After chlorophyll removal, wild-type plants become white (**j**).

6. Remove excessive *Agrobacteria* by brief blotting hypocotyls with a sterile filter paper.
7. Put five hypocotyls in each Petri dishes containing solid MS (30) BAP 0.1 + 250 μM acetosyringone. Seal them with gas porous tape.
8. The co-culture of hypocotyls and *Agrobacteria* continues for 3 d in a growth chamber.
9. Then, eliminate *Agrobacteria* by washing first in sterile water for 15 min.
10. Briefly blot the hypocotyls on sterile paper to absorb most of the bacteria.
11. Wash hypocotyls with liquid MS (30) 0.1 mg/L BAP and 300 mg/L cefotaxime, for 15 min.
12. Briefly blot the hypocotyls on a sterile paper and transfer them on the selective medium.

3.4. Selection of Plants Resistant to Kanamycin

1. Put five hypocotyls in each Petri dish containing selective medium MS (30) 0.1 mg/L BAP, 300 mg/L cefotaxime, and 50 g/L kanamycin.
2. Grow them in phytotron chamber until the transformed plants are developed (3–6 wk)
3. Transfer 5 mm tall plantlets in glass vessels (70 mm × 55 mm), closed with magenta caps, containing MS (30) 0.1 mg/L BAP, 300 mg/L cefotaxime, and 50 g/L kanamycin. Remove the basal (2 mm) part of the plantlets to get rid of *Agrobacteria* (*see* **Note 8**).
4. After 3 wk, transfer plantlets to glass vessels (100 mm × 55 mm) close with magenta cap with MS (30) for rooting in 6–8 wk.
5. Plants are transferred to the greenhouse (*see* **Note 9**): high hygrometry is kept for 1–2 d by placing plants in individual closed chambers that are eliminated at day 5.

3.5. Detection of Transgenic Plants by Histochemical GUS Assay (9)

1. Harvest a few leaves or entire plants and use them fresh.
2. Dilute X-Gluc stock solution in X-Gluc buffer to make the final concentration of 1 mM substrate.
3. Immerse plant parts into X-Gluc solution.
4. Incubation at 37°C for 12–24 h in the dark. Only plant cells expressing β–glucuronidase activity turn blue.
5. To see the result more easily, eliminate chlorophyll by washing with ethanol. The number of washing is dependent on plant parts and age.

3.6. Characterization of Transgenic Plants by Molecular Analyses

3.6.1. DNA Purification and Quantification

1. Extract DNA with Qiagen 'DNeasy plant maxi kit' (*see* **Note 10**) and store at –20°C.
2. To quantify DNA, compare signal intensities on gel electrophoresis with known amount of DNA, 5, 10, and 20 ng/µL (unsonicated Calf Thymus DNA)

3.6.2. PCR Analysis

1. Primer sets for 552 bp *gus* sequence detection: 5′-ATG GTC CGT CCT GTA GAA ACC CCA ACC CGT GAA ATC-3′ and 5′-CGC AGC GTA ATG CTC TAC ACC ACG CCG AAC ACC TGG GTG GAC GAT A-3′.
2. PCR amplification conditions. 95°C for 5 min followed by 35 cycles – 95°C for 60 s, 65°C for 30 s, and 72°C for 40 s and a final step at 72°C for 10 min
3. PCR mix composition for one sample. 50 ng DNA, 0.4 mM of each primer, 0.2 mM dNTP, 1× PCR buffer (provided by the manufacturer), 1.5 mM $MgCl_2$. 0.5 µL Taq DNA polymerase (Qbiogen), water to obtain the final volume 50 µL
4. PCR products analysis is performed by electrophoresis on a 1% *(w/v)* TAE agarose gel under 100 V tension for 20 min. 1 Kb PLUS DNA Ladder is used to determine their length. An example of the results produced is shown in **Fig. 2**.

3.6.3. Southern Blot Analysis (See Note 11 and Fig. 3)

Probe Labelling

1. Use plasmid pTDE4 for PCR reaction as described earlier to obtain the probe.
2. Label the probe with ^{32}P, use Ready-To-Go™ Labeling Beads (-dCTP) according to the manufacturer's recommendation (RPN1675K, Amersham-Bioscience) and use Sephadex column G50 to purify the probe.

DNA Digestion

1. Use 1 g of each DNA sample
2. Add 150 units of *Eco*RI, 150 units of *Bam*HI and 70 units of *Bgl*II (*see* **Note 12**)
3. Add enzymes to 1X reaction buffer according to the manufacturer's recommendation, and 100 µg bovine serum albumin (BSA).
4. Add water to raise the final volume 400 µL.
5. Mix and incubate for 20 h at 37°C.
6. Precipitate the digested DNA with 0.8 volume isopropanol + 0.3 M of sodium acetate, pH 5.6.

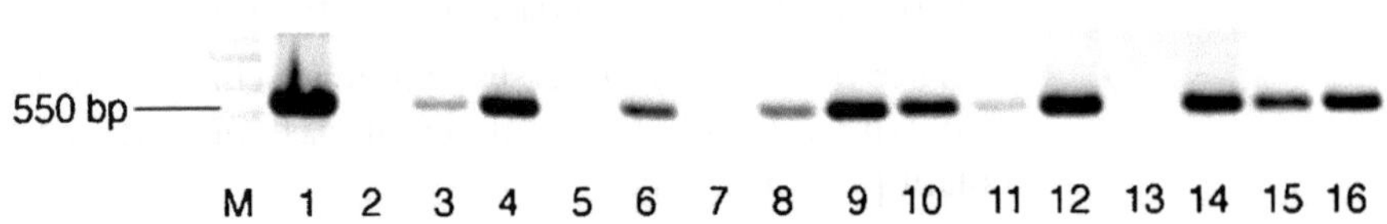

Fig. 2. PCR analysis of putative transformed plants. Genomic DNAs were amplified with primers corresponding to a 550 pb GUS gene sequence. *Lane M,* molecular weight ladder. *Lane 1,* positive control (pTDE4). *Lane 2,* negative control (wild type plant). *Lanes 3–4* and *6–16,* transformed plants. *Lane 5,* untransformed plant.

7. Cool at –20°C for 1 h and spin for 20 min at 4°C, 15,000*g*.
8. Wash pellets with 80% ethanol, and air dry before resuspending them in 20 μL water.

Digested DNA Electrophoresis and Gel Treatment

1. Add loading buffer to the digested DNA, and heat samples for 10 min at 65°C. Add 1 ng plasmid pTDE4 as a positive control.
2. DNA electrophoresis is performed on a 0.8% *(w/v)* TAE 1× agarose gel under 40 V tension for 3–4 h.

   The following steps are carried out without agitation.
3. Rinse the agarose gel with water for 2 min.
4. Soak the agarose gel for 5 min in HCl 0.25 M (*see* **Note 13**) for depurination and rinse with water for 2 min.
5. Soak the agarose gel 45 min in NaCl 1.5 M and NaOH 0.5 M for denaturation and rinse with water for 2 min.
6. Soak the agarose gel two times for 30 min each in Tris–HCl 0.5 M, NaCl 3 M, pH 7.5, for neutralization and rinse with water for 2 min.

DNA Blotting by Capillary Transfer (*See* **Note 14**)

1. In a glass vessel with 1–1.5 cm high of 20X SSC, place a wick higher than the liquid level.
2. From the bottom to the top, place one over the previous ones as follows.
3. 1 sheet of Whatman paper on the wick, large enough to absorb 20X SSC.
4. The agarose gel containing the DNA.
5. One positively charged nylon membrane Hybond N+ (RPN203B, Amersham-Bioscience). The membrane is the same size as the gel and is previously wet with 6X SSC/
6. Two sheets of Whatman paper wet with 6X SSC.
7. Dry paper about 7 cm thick.
8. Add a 1 kg mass on the top and allow digested DNA transfer overnight.
9. Briefly wash the membrane with SSC 6× and air dry for 1 h at room temperature.
10. Fix DNA on the membrane with UV (0.120 J) with a crosslinker (BLX254, Bioblock scientific).
11. Briefly wash the membrane with water.
12. Store the membrane in 2X SSC, or use it for hybridization.

Hybridization

All the following steps are carried out in tubes in a hybridization oven:

1. For pre-hybridization. Denaturate 0.5 ml salmon sperm DNA (1 mg/mL) by heating for 5 min at 100°C and put immediately on ice.

2. Immerge the membrane in 25 mL standard hybridization buffer containing the salmon sperm DNA, and incubate at 68°C for 2 h (*see* **Note 15**)
3. Denaturate the probe by heating for 5 min at 100°C and put it immediately on ice
4. Dilute 20 μL probe in a standard hybridization buffer containing salmon sperm DNA.
5. Immerge the membrane with the probe and incubate at 65°C for 12 h.
6. Remove probe, rinse 2 × 5 min with washing solution 2XSSC at room temperature for 2 min, then with 2X SSC for 30 min at 55°C. Additional steps with washing solution SSC 2× can be carried out according to the signal intensity.

Film Development

1. A film (Hyperfilm™ MP Amersham) is exposed to the membrane in a cassette.
2. Adapt the length of exposure to the ^{32}P probe activity and develop the film. An example of the results produced is shown in **Fig. 3**.

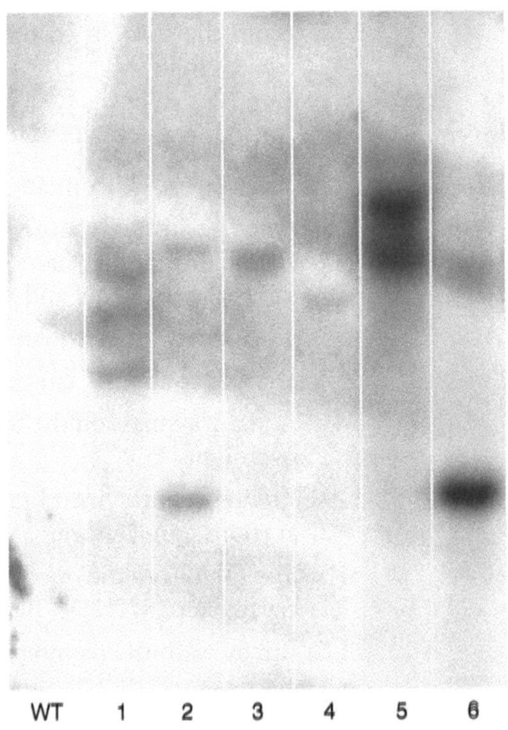

Fig. 3. Southern blot analysis of putative transformed plants. Genomic DNAs were digested with a mix of three enzymes (*Eco*RI, *Bam*HI and *Bgl*I) and probed using a ^{32}P labeled GUS probe, prepared from the plasmid pTDE4. *Lane wt*, wild-type control plant, *lanes 1–6*, transformed plants.

4. Notes

1. Solutions should be prepared with distilled water for in vitro culture, and with water suitable for molecular biology. This is referred to as "water" in this text.
2. The media can be stored at −20°C for 3 mo if necessary. BAP don't degrade during autoclaving, therefore it is added to the media before autoclaving, 20 min, 120°C, 10^5 Pa. To get solid medium use 8 g/L Difco agar. Adjustment of pH is performed prior to the addition of agar in the medium. The stickiness of agar degrades the electrode of the pH-meter.
3. All the stock solutions can be stored at 4°C until use. Clean and autoclaved bottles are used to prevent contamination. Prepare only small amounts of vitamin stock solution as it is rapidly contaminated.
4. Antibiotics are temperature sensitive, thus appropriate concentrations are added after autoclave. The temperature of the medium must be around 60–65°C.
5. The plasmid pTDE4 was developed by M. Van Montagu group and the Plant Genetic System, Ghent, Belgium.
6. Plantlets and hypocotyls are very sensitive to temperature variation and they dehydrate rapidly in laminar hood. It is appropriate to take a few of them at a time, put regularly the hypocotyls in a Petri dish containing MS medium and cover immediately to protect them.
7. About 200–250 hypocotyls are ideal to obtain high transformation rate. Because hypocotyls are very sensitive to dehydratation and temperature, keep them bright green otherwise they will turn either white or brown in 3–4 days. As a result, there won't be any transgenic cell lines. Therefore, it is highly essential to keep close Petri dishes.
8. Cefotaxime is a bactericide and stops bacterial growth. The growth begins again as soon as bactericide is either degraded or stopped adding to the culture medium.
9. When transgenic plants are transferred to the greenhouse within 1–2 mo after root initiation, they develop better and rapidly. Any further delay (5–6 mo), plants develop long stems and few leaves, and are difficult to adapt greenhouse conditions.
10. Several procedures were tested for *R. graveolens* DNA isolation. High amounts of secondary metabolites seem to interfere with these procedures. The best results are obtained with the greenhouse-grown plants as compared to in vitro cultures. This is due to lower sugar concentration than in vitro culture.

11. Southern blotting is designed to locate a particular sequence of DNA within a complex mixture. For example, it can be used to locate a particular gene within an entire genome. The amount of DNA needed is dependent on the size and specific activity of the probe. Short probes tend to be more specific.
12. Digest DNA with chosen enzymes without restriction site in the gene, so the number of insertion is equal to the number of fragments obtained.
13. Depurination with HCl takes the purines out, cutting the DNA into smaller fragments (fragments greater than 15 kb are hard to transfer to the blotting membrane). Be aware however, that the procedure may also be hampered by fragments that are too small. Make sure you neutralize the acid before continuing with the experiment.
14. Capillary action transfer draws the buffer up by capillary action through the gel into the membrane, which will bind ssDNA.
15. A pre-hybridization step is required before hybridization to block nonspecific sites.

Acknowledgments

The authors would like to thank Andrew Saunders for help in improving the English quality of this manuscript.

References

1. Bourgaud. F, Hehn A., Larbat R., Doerper S.,Gontier E., Kellner S. and Matern U. (2006). Biosynthesis of coumarins in plants: a major pathway still to be unravelled for cytochrome P450. *Phytochemistry Reviews*, **5**: 293–308.
2. Berenbaum M. R. (2002). Postgenomic chemical ecology: from genetic code to ecological interactions. *Journal of Chemical Ecology* **28**: 873–896.
3. Massot B., Milesi S., Gontier E., Bourgaud F. and Guckert A. (2000). Optimized culture conditions for the production of furanocoumarins by micropropagated shoots of Ruta graveolens. *Plant Cell, Tissue and Organ Culture* **62**: 11–19.
4. Milési S., Massot B., Gontier E., Bourgaud F. and Guckert A. (2001). *Ruta graveolens* L.: a promising species for the production of furanocoumarins. *Plant Science* **161**: 189–199
5. Lièvre K., Hehn A., Tran T. L. M., Thomasset B., Bourgaud F. and Gontier E. (2005) Genetic transformation of the medicinal plant *Ruta graveolens* L. by an *Agrobacterium tumefaciens*-mediated method. *Plant Science*, **168**: 883–888.
6. Eilert, U. (1989). Elicitor Induction of Secondary Metabolism in Dedifferentiated *in Vitro* Systems of *Ruta graveolens* in W. G. W. Kurz, ed. Primary and Secondary Metabolism of Plant.
7. Porter, J. R. (1991). Host range and implications of plant infection by *Agrobacterium rhizogenes. Critical Reviews inP lant Sciences* **10**: 387–421.
8. Murashige, T. and Skoog F. (1962). A revised medium for rapid growth and bio assay with tobacco tissues cultures. *Physiologica Plantarum* **15**: 473–497.
9. Jefferson RA, Burgess SM, Hirsh D (1986). β-glucuronidase from *Escherichia coli* as a gene-fusion marker. *Proc Natl Acad Sci* **83**: 8447–8451

Chapter 20

Gene Expression Profiling in *Taxus baccata* L. Seedlings and Cell Cultures

Katarína Bruňáková and Ján Košuth

Summary

Limited native resources of paclitaxel from *Taxus* trees initiated the research to produce this compound by biotechnology. In vitro plant cell culture systems have been used for large-scale production of paclitaxel and related taxanes. In the past decade, several genes involved in the taxane biosynthetic pathway have already been sequenced and cloned. This protocol details how to derive cell cultures of *Taxus baccata* L. from young stems of mature trees and from all parts of in vitro- grown seedlings such as root segments, hypocotyls, and cotyledons. The time-course of expression of two genes – *dbat* and *dbtnbt* – coding for two enzymes of the later steps of paclitaxel biosynthesis and the intracellular taxane accumulation has been investigated through a 64-day subculture interval of *T. baccata* cell cultures, during germination, and in early stages of seedling development. The expression level is measured by using quantitative real-time reverse transcriptase polymerase chain reaction. The intracellular content of baccatin III and paclitaxel is quantified by high-performance liquid chromatography HPLC.

We have shown that although the increase in transcriptional activity of *dbat* and *dbtnbt* positively correlate with callus growth, the intracellular accumulation of paclitaxel varies during subculture with the maximum between the late linear and stationary phase. The expression of both genes peaks on day 8 of germination, followed by a decrease in the post-germination phase and during seedling growth. The increase of the steady-state mRNA level of both genes is followed by corresponding metabolite accumulation with a delay of approximately 14–28 d.

Key words: Taxus baccata, Cell culture, Gene expression, *dbat*, *dbtnbt*, Paclitaxel, Baccatin III, HPLC, Quantitative real-time RT-PCR

1. Introduction

A complex diterpenoid paclitaxel (TAXOL®) and its derivatives are plant secondary metabolites with a unique tumor-suppressing mechanism – they prevent mitosis by stabilizing microtubules

S. Mohan Jain and Praveen K. Saxena (eds.), *Methods in Molecular Biology, Protocols for* In Vitro *Culture and Secondary Metabolite Analysis of Aromatic and Medicinal Plants, vol. 547*
© Humana Press, a part of Springer Science+Business Media, LLC 2009
DOI: 10.1007/978-1-60327-287-2_20

against depolymerization *(1)*. Commercially, paclitaxel currently is the most important taxane used in treating a variety of cancers, such as breast, ovarian, lung, head, neck cancers, AIDS-related Kaposi's sarcoma, *etc.* *(2–4)*. The low amount of paclitaxel in the natural source – the bark of *Taxus* spp. and the slow growth of yew trees has led to the search for new alternative approaches: such as isolation from plantation-grown *Taxus* spp. and semisynthesis from precursors (e.g, baccatin III) *(5)*, total chemical synthesis *(6, 7)*, or using *Taxus* spp. in vitro cultures.

Since the first report of Christen research group on the production of paclitaxel by *Taxus* cell cultures *(8)*, many strategies have been used to improve the yield of this compound, such as the addition of sugars, precursors, elicitors, manipulation of gas composition, osmotic pressure, and using of two-stage cultivation or bioreactors *(9)*. In spite of that *Taxus* cell cultures are still limited for large-scale commercial use because of low and unstable production and high cost. One of the most promising ways to increase paclitaxel yield appears to be the molecular engineering and genetic manipulation of the key steps of taxane biosynthesis.

With the exception of a few undefined steps, the entire paclitaxel biosynthetic pathway was revealed and 13 genes for key enzymes were cloned and characterized *(10–12)*. The assumption is that the regulation of the taxane biosynthetic pathway occurs at mRNA level and there is a correlation between steady-state transcript abundance and respective taxane accumulation *(13)*. The detailed knowledge of the course of gene expression in relation to accumulation of subsequent metabolite during the growth cycle of *Taxus* cell culture would give an opportunity to manipulate taxane production.

This chapter describes a protocol for profiling of expression of two genes— *dbat* and *dbtnbt*—which code for two enzymes of the later steps in paclitaxel biosynthesis during callus culture growth and seedling development: the *dbat* gene coding for 10-deacetylbaccatin-III-10β-*O*-acetyltransferase (DBAT) (which catalyzes the conversion of 10-DAB III to baccatin III the last intermediate without phenylisoserine side-chain) and the *dbtnbt* gene which codes for 3′-*N*-debenzoyl-2′-deoxytaxol *N*-benzoyltransferase (DBTNBT) and leads to the final product of the pathway, paclitaxel. Accurate quantification of the studied gene transcripts and the respective products/intermediates is performed by quantitative real time reverse transcriptase-polymerase chain reaction (qRT-PCR) and high-performance liquid chromatography (HPLC), respectively. We have shown that although the increase in transcriptional activity of *dbat* and *dbtnbt* positively correlated with callus growth, the intracellular accumulation of paclitaxel varied during subculture with the maximum between the late linear and stationary phase *(14)*. The expression of both genes peaked on the day 8 of germination followed by a decrease in post-germination phase and seedling growth *(15)*. The increase in the steady-state mRNA level of both genes is followed by corresponding metabolite accumulation with a delay of approximately 14–28 d *(14, 15)*.

2. Materials

2.1. Establishment of *T. Baccata* In Vitro Cultures

2.1.1. Sterilization Procedure

For the sterilization procedure mild soap solution with 2–3 drops of 5% Tween-20 *(w/v)* sodium hypochloride, 70% *(v/v)* ethanol, and 0.2% *(w/v)* $HgCl_2$ are needed (*see* **Note 1**).

2.1.2. Culture Media

1. RM^+ medium containing mineral salts *(16)*, vitamins *(17)*, 2% *(w/v)* sucrose, 100 mg/L *myo*-inositol, 2 mg/L glycine, 1 g/L casein hydrolysate, 1 g/L yeast extract, and 5 g/L activated charcoal *(18)* is used for germination of *T. baccata* embryos.
2. Gamborg's B5 medium *(17)* containing 2% *(w/v)* sucrose, supplemented with 1.5% *(w/v)* polyvinyl pyrrolidone (soluble PVP 10; Mw = 10,000) (*see* **Note 2**) and plant-growth regulators (3 mg/L 2,4-D and 0.5 mg/L kinetin) are used for initiation of callus culture (*see* **Notes 3** and **4**).

2.2. Taxane Analysis

2.2.1. Extraction Procedure

The solvents used in the extraction procedure include water, methanol, hexane and dichloromethane.

2.2.2. Analytical HPLC Method

1. HPLC conditions: column: a stainless steel column (250×4.6 mm ID); stationary phase: Kromasil 100–7 μm, C18 (EKA Chemicals AB); mobile phase: a mixture of acetonitrile and water (65:35 A:B *[v/v]*); flow rate: 0.7 mL/min; sample size: 20 μL.
2. Standards: the stock solutions contain 1 mg paclitaxel (Sigma; St. Louis, MO) or baccatin III (Sigma) dissolved in 5 mL HPLC grade methanol.

2.3. Gene Expression Analysis

2.3.1. RNA Isolation and Reverse Transcription

1. RNeasy Plant Mini kit (Qiagen), RLT buffer supplied in the Qiagen Kit is the lysis buffer of choice, β-mercaptoethanol, absolute ethanol.
2. RNase-free DNase I Set (Qiagen).
3. Quant-iT™ RiboGreen® RNA Kit (Molecular Probes); TE buffer: 10 mM Tris–Cl pH = 7.5, 1 mM ethylenediamine tetraacetic acid (EDTA).
4. 100 mM oligoT primer (1:1:1 mix of primers – $T_{12}GN$, $T_{12}CN$, $T_{12}AN$); M-MLV reverse transcriptase (Invitrogen).

2.3.2. Quantitative Real Time RT-PCR

1. The gene specific primers for amplification:
   430-bp-long cDNA fragment of *dbat*:
   dbat-F 5′CCA AGC CAG CCA TCG CCC AAA G 3′.
   dbat-R 5′ GGC AGA AAC TCA CCC CCA CAA CAA A 3′,
   199bp-long cDNA fragment of *dbtnbt*:
   dbtnbt-F 5′ TTG CCG TTG GAG TGA CTT TGC 3′.
   dbtnbt-R 5′ AGC ATT GGA GGT GGG CAT ATC G 3′.
2. Desgin PCR primers using GenTool Lite 1.0 software based on the published cDNA sequences coding for DBAT from *T. baccata* (GenBank accession no.AF193765), and DBT-NBT from *T. cuspidata* (GenBank accession no. AF466397).
3. iQ™ SYBR Green Supermix (0.2 mM dNTP; 3 mM $MgCl_2$) (BioRad).

2.3.3. Gel Electrophoresis

1. Agarose.
2. 1 mg/mL Ethidium bromide (EB).
3. TAE buffer: 40 mM Tris acetate and 1 mM EDTA. pH 8.3.

3. Methods

3.1. Establishment of T. Baccata L. In Vitro *Cultures*

3.1.1. Sterilization Procedure

1. Excise explants from young stems, 1- to 2-years old, flexible, bearing dark green needles of mature *T. baccata* trees.
2. Strip stems of needles and wash with running tap water, and mild soap solution with 2–3 drops of Tween-20 in order to remove dust and solid particles.
3. Stems are surface sterilized by immersing in 5% *(w/v)* NaClO solution for overnight (12–16 h), dry and immerse in 5% *(v/v)* ethanol for 5 min, then in 0.2% *(w/v)* $HgCl_2$ for 20 min. Rinse three times with sterile distilled water (*see* **Note 5**).
4. The seeds free of red arils are surface disinfected by soaking in 70% *(v/v)* ethanol for 5 min, then in 1% $HgCl_2$ *(w/v)* for 15 min. Rinse three times with sterile distilled water.

3.1.2. Media Preparation

1. RM^+ medium for embryo culture is prepared with double distilled water, solidified with 0.8% *(w/v)* agar, adjusted to pH 5.7 with NaOH and HCl before autoclaving at 121°C and 105 kPa for 15 min.
2. Gamborg's B5 medium for callus culture is prepared with double distilled water, solidified with 0.6% *(w/v)* agar, adjusted to pH5.6 with NaOH and HCl prior to sterilization by autoclaving at 121°C and 105 kPa for 15 min.

3.1.3. Initiation of Embryo Culture

1. For initiation of germination use mature seeds removed from fruits with fully developed red arils.
2. After removing the red arils the seeds are washed with running tap water for 7 d to break the seed dormancy *(18)*.
3. Embryos are aseptically excised from surface sterilized and longitudinally halved seeds with a surgical blade and transfer on 5 mL fresh RM^+ medium in glass tubes.
4. Germinating embryos (**Fig. 1a**) are incubated at 22±2°C, on a 16-h photoperiod 72 μmolxm^{-2} × s^{-1}.
5. Transfers to fresh medium are usually done every 28 d.

3.1.4. Callus Cultures

1. Following sterilization excise the exposed cut areas of the stems damaged by sterilizing agents.
2. Stems of outdoor trees are aseptically chopped into approximately 10-mm long segments and immersed in the medium with one of transversally sectioned ends (*see* **Note 6**). Hypocotyls, root segments, and cotyledons of in vitro seedlings are placed on the surface of the B5 medium (*see* **Note** 7).
3. The explants are grown in glass tubes, each containing 5 mL B5 medium. The cultures are incubated at 22±2°C in the dark.
4. After 4–6 wk of culture the primary callus is isolated, and use as an inoculum, 0.2–0.4 g/5 mL medium in a glass tube. An example of 4-wk-old primary callus is shown in **Fig. 1b**. Transfers to the fresh medium should be done every 28–35 d depending on growth rate of calli.
5. After several cycles of propagation, calli should be transferred to 100-mL Erlenmeyer flasks with 20 mL solid B5 medium. The appropriate inoculum's size is in the range of 1–2 g. **Figure 1c, d** show an example of *T. baccata* callus cultures.
6. The regular subculture interval is 28–35 d depending on the cell line.

3.1.5. Cell Suspension Cultures

1. For the initiation of cell suspension cultures use rapid growing friable calli derived from stem or root segments and hypocotyls or cotyledons.
2. Place 1–2 g in 100-mL Erlenmeyer flask filled with 25 mL liquid B5 medium.
3. The suspension culture is cultivated in the dark at 22 ± 2°C on a gyratory shaker at 90 rpm.
4. In first weeks in culture the media can be replaced at 2-wk intervals to increase the cell density.
5. After several cycles the cell density is high enough to start inoculation in new cultivation vessels. The proportion

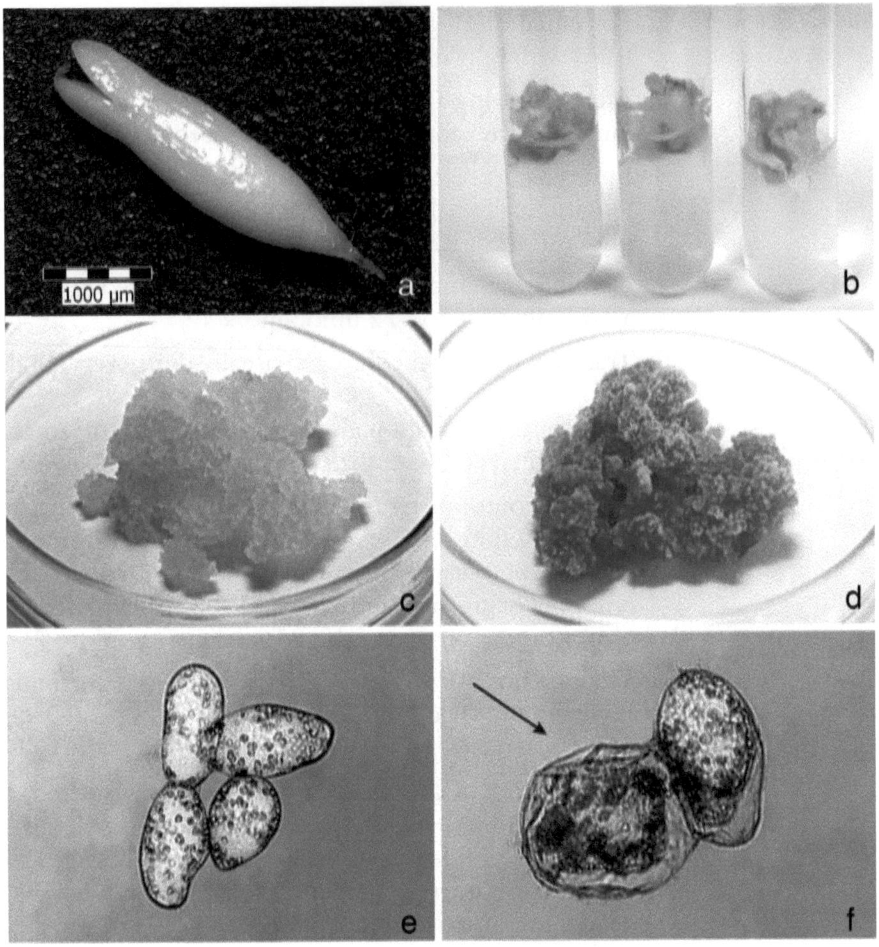

Fig. 1. In vitro cultures of *T. baccata* L.; (**a**) germinating 8-d-old zygotic embryo; (**b**) callogenesis of stem segments, 4-wk-old primary callus; (**c**, **d**) two different hypocotyl's-derived callus cultures after 1 yr of cultivation; (**e**, **f**) the cells of a suspension culture; the trypane-blue positive cell (*arrow*) is considered not viable.

of inoculum to fresh liquid medium should be in range of approximately 1:5–1:6 *(v/v)*.

6. The transfers to the fresh liquid medium should be done at the end of exponentially growth phase.
7. The viability of cells can be checked under microscope after viable staining with trypane blue: 10 µL cell suspension is mixed with 10 µL 0.5% *(w/v)* trypane blue solution. The trypane-blue positive cells are non-viable (**Fig. 1e, f**). The regular control of cell viability and culture sterility is recommended.
8. The regular subculture interval is 15–20 d depending on the cell line.

3.2. Taxane Analysis

3.2.1. Extraction Procedure

1. At least 1 g lyophilized cell culture or 10 mg lyophilized seedlings homogenized in a mortar with a small amount of sterile sea sand is extracted with 50 mL methanol in 100-mL Erlenmeyer flask. The extraction is performed in sonicator bath for 30 min at room temperature.
2. The extract is filtered through filter paper into a 250-mL round-bottomed flask and the volume of filtrate is reduced by evaporation under vacuum till about 10 mL using a rotatory evaporator at 70°C. The volume of extract is then adjusted to 12.5 mL with methanol:water (1:1) *(v/v)*.
3. The water-diluted extract is transferred to a separator funnel and 25 mL hexane is added. The mixture is shaken thoroughly and separate into two layers. The individual phases are separated and the hexane fraction (the upper layer) is discarded. The purification step should be repeated 2–3 times until the hexane fraction is clear (*see* **Note 8**).
4. The water-methanol fraction (the bottom layer) is extracted with 25 mL dichloromethane. After vigorous shaking allow the two phases to separate into distinct phases (*see* **Note 9**). Separate the phases; the bottom one (dichloromethane phase) contains extracted taxanes. This step should be repeated 2–3 times to maximize the taxane recovery.
5. All dichloromethane fractions are combined into a clean 100-mL Erlenmeyer flask and the solid drying agent (magnesium sulphate or sodium sulphate) is added. The mixture is swirled and allowed to stand until the solution is clear.
6. The solution is filtered through filter paper into dry 100-mL round-bottomed flask and the extract is evaporated to dryness on the rotatory evaporator with bath temperature maintained at 40°C.
7. The taxanes excreted into the culture media are extracted with equal amounts of dichloromethane (1:1) *(v/v)*. The mixture is thoroughly shaken in flask shakers overnight followed by sonication for 30 min at room temperature and partitioned in separation funnel. The dichloromethane fraction (the bottom layer) is filtered and dried under vacuum as described earlier.
8. The dried extracts can be stored in the refrigerator at 4°C. Prior to the HPLC analysis dried extracts are dissolved in an exact volume of HPLC grade methanol (0.5–1.0 mL) and proceed to analysis.

3.2.2. Analytical HPLC Method

1. Paclitaxel and baccatin III standards are dissolved in HPLC grade methanol to prepare stock solutions, 0.2 mg/mL. Store the solutions in a refrigerator.

2. Paclitaxel and baccatin III stock solutions are diluted with HPLC grade methanol in range of 0.001–0.1 mg/mL to prepare calibration curve.
3. For HPLC analysis 20 μL standard curve dilutions and experimental samples are injected into the column. For identification of paclitaxel and baccatin III peaks the co-chromatography with the appropriate dilution of standard is used. Run the experimental samples and the samples of calibration curve.
4. According to the standards retention times the paclitaxel- and baccatin III-corresponding peaks in samples are identified. The amounts of paclitaxel and baccatin III in mg/mL in samples are calculated according to calibration curve of the respective standard.

3.3. Gene Expression Analysis

3.3.1. RNA Isolation and Reverse Transcription

1. The RNeasy Plant Mini kit is very convenient – fast, easy and high-throughput for isolation of total RNA from *T. baccata* seedlings and calli. For total RNA isolation use maximum of 100–120 mg frozen callus or 50 mg seedlings and ground in liquid nitrogen (LN). RTL buffer is the buffer of choice for disruption of *T. baccata* cell and extraction of RNA. According to the supplier instructions 2% β-mercaptoethanol *(v/v)* should be added to the RLT buffer before use. The incubation of samples with RLT buffer at 56°C for 5–10 min during extraction helps to disrupt the material (*see* **Note 10**).
2. Treat the RNA samples with DNase I to remove the co-purified DNA. Digestion with RNase free DNase I (Qiagen) eliminates the residual genomic DNA without negative effect on RNA integrity. The digestion step is performed during RNA purification on RNeasy mini column supplied in RNeasy Plant Mini kit according to the supplier's instructions.
3. Check the integrity of the purified RNA by gel electrophoresis in 0.5 μg/mL EB stain 1% agarose gel prepared in 1X TAE buffer. Non-degraded RNA is indicated by distinct 28S and 18S rRNA bands.
4. Calculate the concentration of RNA in the samples after measuring absorbance at 260 nm wavelength. Absorbance of RNA solution diluted in water equal to one unit ($A_{260} = 1$) means concentration of 40 μg/mL RNA.
5. Because the spectrophotometric quantification of nucleic acids is not precise enough the accurate concentration of RNA should be determined by fluorescent RNA-binding dye RiboGreen (*see* **Note 11**). 1:1000 *(v/v)* dilution of the RiboGreen dye allows for accurate quantification of diluted solutions of RNAs in the range of 50–150 ng/mL RNA (the linear range of quantification). First, prepare the samples of standard curve in the linear range of quantification by making

several dilutions of the ribosomal RNA standard, 100 μg/mL, in 1X TE buffer. The 1X TE buffer is used also for dilution of aliquots of the isolated RNA samples. According to the spectrophotometric readings of concetrations the dilution of each RNA sample in linear range of quantification is prepared. Mix equal amounts of diluted RNA samples and RiboGreen working solution in microplate and measure the fluorescence in microplate reader fitted with 485 nm (excitation) and 520 nm (emission) filters. The accurate concentrations of total RNA are then extrapolated from calibration curve.

6. Equal amounts of total RNA quantified by RiboGreen (max. 10 μg) are reverse transcribed to the first strand cDNA. Reverse transcription should be performed at 37°C in 20 μL volume using 10 mM anchored oligoT primer and 200 U M-MLV reverse transcriptase, according to the manufacturer's instruction.

3.3.2 .Quantitative Real Time RT-PCR

1. Run qRT-PCR in PCR-thermocycler capable of measuring the fluorescence of SYBR Green I. during PCR amplification. qRT-PCR should be run at least in duplicates. To quantify the expression of *dbat* and *dbntbt* gene run the qRT-PCR in 30 μL reaction volume containing: 1× iQ™ SYBR Green Supermix (0.2 mM dNTP; 3 mM $MgCl_2$); 0.5 μM forward and reverse primer (as described in materials) and 50 ng of cDNA. The reaction conditions are as follows: 95°C 3 min; 40 cycles (94°C 30 s; 59°C (*dbat*) or 59.5°C (*dbtnbt*) 25 s; 72°C 25 s); 74°C 4 min; followed by melting curve analysis to confirm amplification of single PCR product. An example of qRT-PCR of *dbat* and *dbtnbt* genes with SYBR Green I detection is shown in **Fig. 2ac**.
2. Verify the length of PCR products (430-bp for *dbat* and 199-bp for *dbtnbt* gene fragment) by gel electrophoresis in EB (0.5 μg/mL) stained 1.6% agarose gel prepared in 1X TAE buffer. An example is shown in **Fig. 2d**.

3.3.3. Evaluation of the Expression Level

The relative amount of both gene transcripts is evaluated according to standard curves obtained by amplification of a serially diluted mixture of cDNA samples, diluted twofold or fourfold, with 5–6 dilutions, each one in two replicates.

3.4 .Time Course Experiment

In order to reveal the time-course of expression of taxane biosynthesis genes (*dbat* and *dbtnbt*) and subsequent intracellular metabolite accumulation during early seedling development and during the growth cycle of callus culture the samples should be analyzed at regular intervals. To assign the reproducibility and homogeneity of the plant material, the samples should be prepared as a mixture of calli or a mixture of seedling's tissues grown

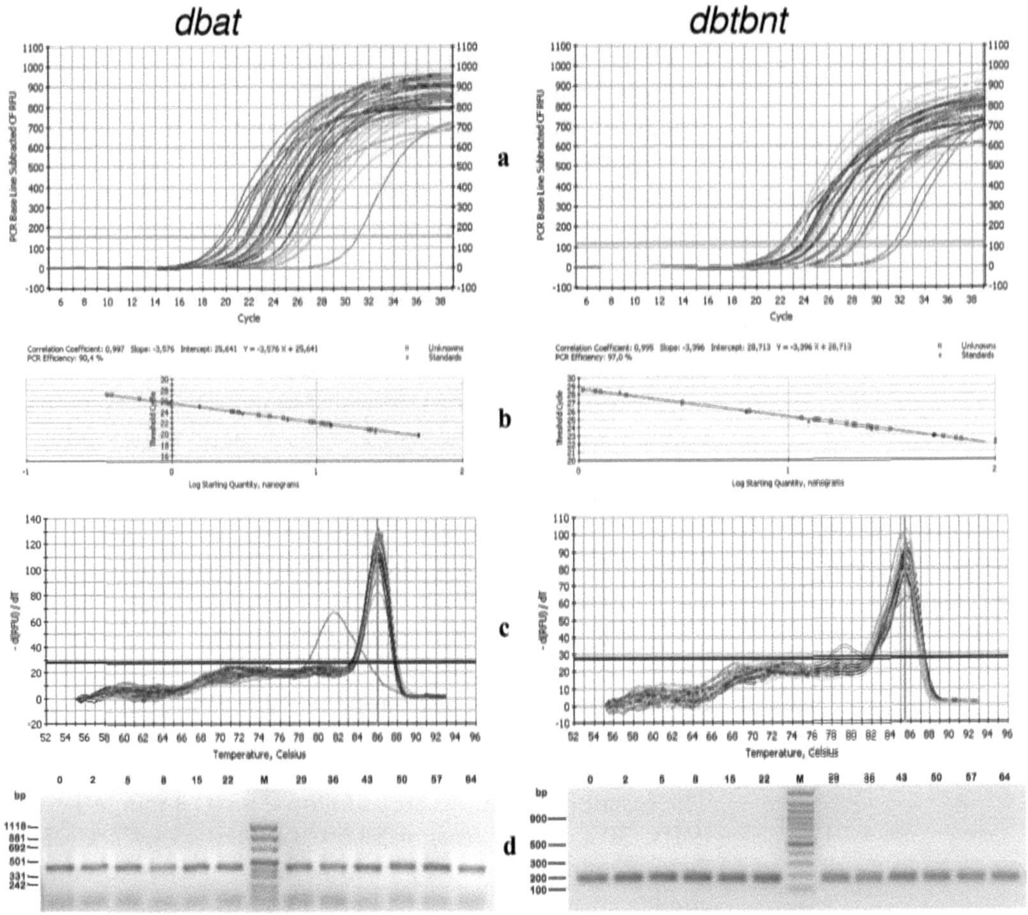

Fig. 2. Quantitative real time RT-PCR of *dbat* and *dbtnbt* genes in *Taxus baccata* L. callus culture during subculture interval (on days: 0, 2, 5, 8, 15, 22, 29, 36, 43, 50, 57, and 64) detected with SYBR Green I.; (**a**) RT-PCR, (**b**) calibration curve, (**c**) melt curve analysis, (**d**) gel electrophoresis of the amplicones.

in several cultivation flasks representing a stock specimen for all analyses in the respective days.

3.4.1. Callus Cultures

The time-course of biomass accumulation, gene expression and intracellular taxane accumulation are determined during the growth cycle of callus culture. The studied kinetic characteristics of culture are recorded during a prolonged growth period depending on growth characteristics; the samples for the analyses should be taken on days 0, 1, 2, 3, 4, 5, 8, 15, 22, 29, 36, 43, 50, 57, and 64 (*see* **Note 12**).

1. For the experiments callus cultures with appropriate growth characteristics (e.g. the doubling time at 10- to –15-d interval) are selected. Sufficient amount of calli must be propagated

before the time-course experiment to assign homogeneity of samples on each experimental day and to have enough material for extractions. Minimum of 150 g calli should be used for inoculation with 1 g inoculum per 100-mL Erlenmeyer flask for a total of 150 flasks for experiment. On each experimental day mix of calli grown in 15 flasks are prepared to determine the growth characteristics and for biochemical and gene expression analyses.

2. The dynamics of callus growth is characterized by growth curves based on fresh and dry weights. On each experimental day the exact fresh and dry weights of ten callus samples, grown in ten flasks, are determined. Lyophilized samples of calli taken on the same day of culture should be mixed to obtain a homogeneous mass and split up to several aliquots and used for biochemical analyses.
3. Remaining five fresh callus samples, grown in five flasks, are mixed and used for gene expression analyses. Approximately 1 g homogeneous callus is grinded in liquid nitrogen using mortar and pestle to fine powder. 100–120 mg frozen material should be weighted immediately into pre-frozen Eppendorf tubes and stored at –80°C until analyses.

3.4.2. Embryo Cultures

1. To obtain sufficient data to determine the relation between steady-state of mRNA transcript abundance, taxane accumulation and developmental stage of germinating zygotic embryo, the regular interval of 4 d is suitable.
2. For initiation of germination at least 500 zygotic embryos should be excised; from this amount approximately 200 fully developed seedlings should be obtained.
3. On each experimental day the plant material is stored for biochemical and gene expression analysis. The appropriate amount of germinating embryos or developing seedlings in a sample is dependent on the developmental stage, 50 or more embryos in the first experimental days or ten fully developed seedlings at the later stages.
4. The samples of embryo's tissues are grinded in the liquid nitrogen using mortar and pestle to fine powder. Immediately weigh 50 mg frozen material in pre-frozen Eppendorf tubes and store at –80°C until analyses.

4. Notes

1. Prepare fresh solutions for each sterilization procedure.

2. Polyvinylpyrrolidone (PVP) is a phenolic-binding compound, which improves the growth of yew cell cultures. The initiation of yew cell cultures is difficult because the explants produce and release phenolics; the oxidation of these compounds leads to darkening, inhibition of growth and loss of viability of the tissue *(19)*.
3. For long-term callus culture the content of 2,4-D is reduced to 1.5 mg/L.
4. For cell suspension cultures liquid B5 medium of the same composition as in callus cultures is used. The content of PVP can be enhanced to 3% *(w/v)*. Preparation, pH adjustment and sterilization procedure are the same as for callus cultures. For viability checks a 0.5% solution of trypane blue is used.
5. During the sterilization period, the stems must be fully immersed and agitated. Using this sterilization procedure the typical contamination rates of outdoor-grown donor plants are less then 2% of the total number of explants used. No negative influences on the callus induction were shown. In the case of high infestation of a plant with endophytes, the contamination can be very high and a change of donor plant is recommended.
6. Cut yew tissues are prone to darkening due to the production of phenolics after injury; the cuttings of explants should be made smoothly with a sharp blade and the instruments used in plant preparation must be cooled to room temperature after sterilization to avoid the heat-damage of the tissue.
7. Our observations indicate that all parts of young seedlings such as root segments, hypocotyls, and cotyledones are more prone to cell proliferation on callus-induction media as compared to the stem segments of adult trees. Stem explants display different responses to in vitro culture depending on the season; the explants derived from the youngest shoots, 2- to 4-wk old, collected in May and June show very low or no callogenesis *(20)*.
8. The total volume in the separator funnel should not be greater than 3/4 of the funnel volume. After a few seconds of shaking do not remember to release the air pressure. About 1 min total vigorous shaking is usually sufficient to allow solutes to come to equilibrium between the two solvents. This procedure allows removing most of the lipophilic compounds and most of the chlorophyll. The hexane step is recommended for callus or green plant tissue samples and can be avoided for cleaner extracts (e.g., for cell suspension or if the precolumn for liquid chromatography is used for prevention of irreversible binding impurities to the column) *(19)*.

9. Dichloromethane is a common extraction solvent that is preferred because of its lower toxicity (e.g., in comparison with chloroform). However, all chlorinated solvents tend to form emulsions. In this case the separation of the layers becomes difficult; first try to swirl mildly the solution in the separator funnel; if the emulsion does not disperse, try add a small amount of methanol to the mixture, saturated aqueous sodium chloride or water soluble detergent solution are also effective.
10. All instruments used for extraction and purification should be RNase-free. Store the frozen ground plant material in LN or at –80°C until RNA extraction. Frozen plant material should not thaw during handling. Purified RNA should be stored at –80°C.
11. Because the expression of the most frequently used house-keeping genes varies during ontogenesis accurate quantification of total RNA by fluorescent dye RiboGreen was used to assign equal amounts of total RNA in all compared samples.
12. It is important to analyze the steady-state of transcripts abundance immediately at the time of inoculation (i.e., on day 0). According to our observations rapid changes of expression occur next to transfer of inoculums on the fresh culture medium. Therefore shorter intervals in the beginning of the growth cycle are recommended.

Acknowledgments

This work was supported by grant VEGA 1/1273/04. Authors would like to thank Mrs Hana Michalčová for her technical assistance at taxane isolation and quantification.

References

1. Schiff, P.B., Fant, J., and Horwitz, S.B. (1979). Promotion of microtubule assembly *in vitro* by taxol. *Nature* **277**, 665–667.
2. Rowinsky, E.K., Donehower, R.C., Spivak, J.L., Burke, P.J., Griffin, C.A., and Jones, R.J. (1990). Effects of the differentiating agent hexamethylene bisacetamide on normal and myelodysplastic hematopoietic progenitors. *J. Natl. Cancer Inst.* **82**, 1926–1931.
3. Brown, D.T. (2003). Preclinical and clinical studies of the taxanes. In: *Taxus: The Genus Taxus* (Itokawa, H. and Lee, H.K. eds.), Taylor & Francis, London, pp. 387–435.
4. Sledge, G.W. (2003). Gemcitabine combined with paclitaxel or paclitaxel/trastuzumab in metastatic breast cancer. *Semin. Oncol.* **30**, 19–21.
5. Patel, R.N. (1998). Tour de paclitaxel: Biocatalysis for semisynthesis. *Annu. Rev. Microbiol.* **98**, 361–395.
6. Holton, R.A., Kim, H.B., Somoza, C., Liang, F., Biediger, R.J., Boatman, P.D., Shindo, M.S., Mith, C.C., and Kim, S.C. (1994). First total synthesis of taxol. II. Completion of the C and D rings. *J. Am. Chem. Soc.* **116**, 1599–1600.

7. Nicolaou, K.C., Yang, Z., Liu, J.J., Ueno, H., Nantermet, P.G., Guy, R.K., Clalborne, C.F., Renaud, J., Couladouros, E.A., Paulvannan, K., and Sorensen, E.J. (1994). Total synthesis of taxol. *Nature* **367**, 630–634.
8. Christen, A.A., Gibson, D.M., and Bland, J. (1991). Production of taxol or taxol-like compounds in cell culture. US Patent N° 5019504.
9. Zhong, J.J. (2002). Plant cell culture for production of paclitaxel and other taxanes. Review. *J. Biosci. Bioeng.* **94**, 591–599.
10. Croteau, R., Ketchum, R.E.B., Long, R.M., Kaspera, R., and Wildung, M.R. (2006). Taxol biosynthesis and molecular genetics. *Phytochem. Rev.* **5**, 75–97.
11. Guo, B.H., Kai, G.Y., Jin, H.B., and Tang, K.X. (2006). Taxol synthesis. *Afr. J. Biotechnol.* **5**, 15–20.
12. Hezari, M. and Croteau, R. (1997). Taxol biosynthesis: an update. *Planta Med.* **63**, 291–295.
13. Nims, E., Dubois, C.P., Roberts, S.C. and Walker, E.L. (2006). Expression profiling of genes involved in paclitaxel biosynthesis for targeted metabolic engineering. *Metab. Eng.* **8**, 385–394.
14. Katkovčinová, Z., Lázárová, M., Bruňáková, K., Košuth, J., and Čellárová, E. (2008). Expression of *dbat* and *dbtnbt* genes involved in paclitaxel biosynthesis during the growth cycle of *Taxus baccata* L. callus cultures. *Z. Naturforsch.* **63c**, 721–730.
15. Bruňáková, K., Košuth, J., Katkovčinová, Z., Lázárová, M., and Čellárová, E. Gene expression during germination of *T. baccata* L. zygotic embryos. *Biol. Plantarum* (accepted 10.2008).
16. Linsmaier, E.M. and Skoog, F. (1965) Organic factor requirements of tobacco tissue cultures. *Physiol. Plantarum* 18, 100–127.
17. Gamborg, O.L., Miller, R.A., and Ojima, K. (1968). Nutrient requirements of suspension cultures of soybean root cell. *Exp. Cell Res.* **50**, 151–158.
18. Zhiri, A., Jaziri, M., Homes, J., Vanhaelen, M., and Shimomura, K. (1994). Factors affecting the *in vitro* rapid germination of *Taxus* embryos and the evaluation of taxol content in the plantlets. *Plant Cell Tiss. Org. Cult.* **39**, 261–263.
19. Fett-Neto, A.G. and DiCosmo, F. (1997) Taxol and taxane production by cell culture. In: *Encyclopedia of Molecular Biology and Molecular Medicine 6* (Meyers, R.A. ed.), VCH Verlagsgesellschaft mbH, Germany, pp. 10–17.
20. Bruňáková, K., Babincová, Z., and Čellárová, E. (2000). Nutritional and seasonal requirements for callus growth in *Taxus baccata* L. *Int. J. Hort. Sci.* **6**, 111–114.

Chapter 21

Catapol Production in Chinese Foxglove (*Rehmannia glutinosa* Libos.) Hairy Roots Transformed with *Agrobacterium rhizogenes* ATCC15834

Sung Jin Hwang

Summary

Hairy root clones of *Rehmannia glutinosa* were established via transformation with *Agrobacterium rhizogenes* ATCC15834. To optimize the culturing conditions for both root growth and catalpol production, the effects of various combinations of seven basal media, pH, and carbon sources were examined in the dark. The fastest root growth was obtained in an SH medium containing 4% sucrose, pH 5.8. The highest catalpol content (0.54% dry weight) was achieved in a WPM medium supplemented with 4% sucrose, pH 5.8. Effects of plant growth regulators and chitosan were also investigated. Auxin 2 mg/L IAA significantly increased both root length and the frequency of lateral roots. Both 50 mg/L chitosan and 0.5 mg/L GA_3 induced catalpol production, with contents calculated at 0.7% dry weight and 0.65% dry weight, respectively.

Key words: Catalpol, Chitosan, PGRs, Hairy root cultures, *Rehmannia glutinosa*

1. Introduction

Plant cell culture for the production of useful secondary metabolites has a number of adventages over conventional procedures (e.g., the assurance of a continuous supply of uniform-quality, high specialized, natural components that cannot be produced in equal quality or specificity through biologic means). Cell cultures also offer reliable quality control and availability independent of environmental changes. However, commercialization of the chemicals obtained depends mainly on identifying appropriate techniques for increasing productivity. For most plant species, the synthesis

S. Mohan Jain and Praveen K. Saxena (eds.), *Methods of Molecular Biology, Protocols for* In Vitro *Cultures and Secondary Metabolite Analysis of Aromatic and Medicinal Plants, vol. 547*
© Humana Press, a part of Springer Science+Business Media, LLC 2009
DOI: 10.1007/978-1-60327-287-2_21

of secondary products can be enhanced by allowing the plant tissue to differentiate morphologically and form organs *(1–3)*.

Chinese foxglove (*Rehmannia glutinosa* Libos.) is a perennial herb usually distributed in central China. It is the major ingredient in Chinese medicine for treating dehydration, and is also prescribed as an anti-bacterial and anti-inflammatory agent. Chinese apothecaries use either fresh or dried roots. The major metabolites isolated from this plant are sterol, campesterol, catalpol, rehmannia, and some alkaloids *(4, 5)*. Iridoid glycoside exists in plants of many families and processes a wide variety of biologic properties—including purgative, liver-protective, anti-microbial, analgesic, anti-tumor, sedative, and anti-inframmatory properties *(6)*. One example is catalpol, extracted from the foxglove roots. Studies on the calli from *R. glutinosa* have been published *(7)* but an *Agrobacterium rgizogenes*-mediated transformation protocol for *R. glutinosa* has not previously been reported. Hairy root cultures represent an interesting alternative to dedifferentiated cell cultures for the production of secondary metabolites. Because hairy roots originate from a single plant cell via infection with *A. rhizogeens*, they are usually considered genetically more stable than callus lines. Moreover, in contrast to dedifferentiated cells, metabolite production from hairy roots is not repressed during the growth phase of the culture. Therefore, hairy roots usually produce secondary plant products without the loss of concentration that is frequently observed from callus or cell suspension cultures *(8–10)*.

This chapter describes the establishment of transformed root cultures of *R. glutinosa*, an important medicinal plants, and the effects of media type, initial pH, carbon source, plant growth regulators, and chitosan on the development of hairy roots and the production of iridoid glycoside.

2. Materials

2.1. Plant Materials

1. *R. glutinosa* seeds as a source of in vitro propagation obtained from the Rural Development Administration (Suwan, S. Korea).

2.2. Bacterial Strains

1. *Agrobacterium rhizogenes* strain ATCC15834 is used for gene transfer.
2. Prior to inoculation, culture bacteria on potato extracts medium (PEM) (*see* **Subheading 2.3**) in the dark at 27 ± 1°C.

2.3. Media and Other Solutions

1. PEM medium; 200 g/L potato extracts; 2% sucrose; 1.5% Bacto-agar.
2. Half-strength Murashige and Skoog (½MS) basal salts and vitamins without plant growth regulators.
3. Kanamycin (Sigma,; St. Louis, MO) and cefotaxime (Sigma).

2.4. Special Laboratory Tools and Materials

1. A tissue culture room, maintained at 25 ± 1°C under a light intensity of 54 μmol/m²/s from cool white fluorescent lamps (16 h photoperiod).
2. Horizontal orbital shaker (Jeiotek; Korea).
3. Laminar airflow hood (Shin-il; Korea).
4. High-performance liquid chromatography (HPLC) (Waters Co., USA).
5. General tissue culture and plant tissue culture materials.

3. Methods

3.1. Medium Preparation

1. Prepare MS culture medium containing ½ strength basal salts and vitamins without plant growth regulators
2. Adjust pH of media with 0.1N NaOH before autoclaving at 121°C for 15 min.
3. Plant growth regulators are added after filter sterilization of the medium.
4. Dissolve kanamycin (Sigma) and cefotaxime (Sigma) in water, filter sterile (0.2 μm pore size), and add after autoclaved medium is cooled around 40–50°C.

3.2. Seed Germination

1. Surface-sterilize seeds with 70% *(v/v)* ethanol for 5 min and 1% *(v/v)* sodium hypochlorite for 10 min, and rinse four times in sterile distilled water.
2. Germinate surface sterilized seeds on half-strength MS medium *(11)* (1/2 MS) in 110 × 20-mm Petri dishes in darkness at 25 ± 1°C.
3. Transfer plantlets to solid MS medium containing 2% sucrose at 25 ± 1°C, 16-h photoperiod light intensity 54 μmol/m²/s from cool white fluorescent lamps.

3.3. Genetic Transformation and Selection of Root Clones

1. Cut leaves from in vitro- grown seedlings into small pieces and co-culture with *A. rhizogeens* strain ATCC15834 for 24 h in the dark.
2. Excise induced adventitious roots and culture on a 1/2 MS medium containing 300 mg/L cefotaxime to eliminate bacteria (*see* **Note 1**).
3. Transformed root clones obtained after the excision of single roots are maintained and propagated in the dark on a hormone-free MS medium supplemented with 3% sucrose (*see* **Note 2**).

3.4. Culture of Hairy Root Clone

1. Seven different media are tested for their effects on biomass yield and catalpol production: ½ full-strength MS, B_5 *(12)*, LS *(13)*, SH *(14)*, WPM *(15)*, and NN *(16)* (*see* **Note 3, 4, 5**).

2. Before autoclaving, adjust the pH of the medium with 1N NaOH to 4.8, 5.8, or 6.8 (*see* **Note 8**).
3. All experiments are carried out in 100-mL Erlenmyer flasks containing 40 mL liquid medium (sucrose; 2, 4, 6, 8%), inoculated with ca. 0.05 g fresh weight tissue, and subculture at 3-wk intervals (*see* **Notes 6, 7**).
4. The flasks are incubated on a rotary shaker at 100 rpm at 25 ± 1°C under darkness.
5. After 12 wk of culture, biomass dry weights and catalpol contents are determined.

3.5. Treatment with Plant Growth Regulators and Chitosan

1. Add plant growth regulators after the media were sterilized with a 0.45 μm membrane filter (*see* **Notes 9, 10, 11**).
2. Dissolve chitosan (Sigma) in 6% *(w/v)* acetic acid, and remove insoluble materials by centrifugation.
3. Precipitate dissolved chitosan in 5N NaOH, recover by centrifugation. Repeat this procedure twice.
4. Wash chitosan residue three times with distilled water and freeze-dry. The final chitosan concentration in the hairy root cultures is in the range of 5–500 mg/L (*see* **Notes 12, 13, 14**).

3.6. Sample Preparation and Analytical Methods

1. Harvest hairy roots and lycophilize to measure metabolite contents.
2. Grind lycophilized samples (~ 1 g dry weight) and extract with 80% methanol.
3. Following filtration through a 0.45 μm Millipore filter, the root extracts are subjected to HPLC analysis.
4. The chromatographic separation is carried out on a Capcell pack C_{18} reverse-phase column: UV levels are detected at 218 nm. The mobile phase contains 1:100 *(v/v)* acetonitrile: water, flow rate of 1.2 mL/min.
5. Catalpol concentrations are quantified with an external standard of catalpol
6. For fresh weight determinations, the roots are gently pressed on a filter paper to remove excess water, weigh., and freeze dry for a minimum 48 h at 10^{-1} mbar (–42°C) before recording their dry weights.

4. Notes

1. Transformed roots are easy to identify by their rapid, highly branching, plagiotropic pattern of growth and usually cover the entire surface of the solid medium within 3–4 wk (*see* **Fig. 1a, b**).

Fig. 1. Transformed roots *(arrow)* induced from leaf explants by co-cultuvating with *A. rhizogenes*. (**b**) Hairy root clone RH105 cultured on solid 1/2 MS medium containing 2% sucrose. **C**. Hairy root cultivated in liquid media containing 4% sucrose, pH 5.8, after 12 wk.

2. Fifteen adventitious root clones based on the phenotypic characteristics of their roots on the solid medium containing 2% sucrose are selected. The most actively growing strain RH 105 is isolated and used for determining the optimal culturing conditions.
3. Manipulation of media components is an important strategy for improving the yield of useful secondary metabolites through cell and tissue culture. The optimal growth of *R. glutinosa* hairy roots, expressed as root dry weight per flask, is attained in the hormone-free SH medium. In contrast, neither LS nor B5 medium is suitable for promoting the growth of hairy roots.
4. Catalpol production, determined as percent of dry weight, is also affected by medium choice, with a high content being measured from tissues culture in the hormone-free WPM medium (*see* **Figs. 1c** and **2a, b**).

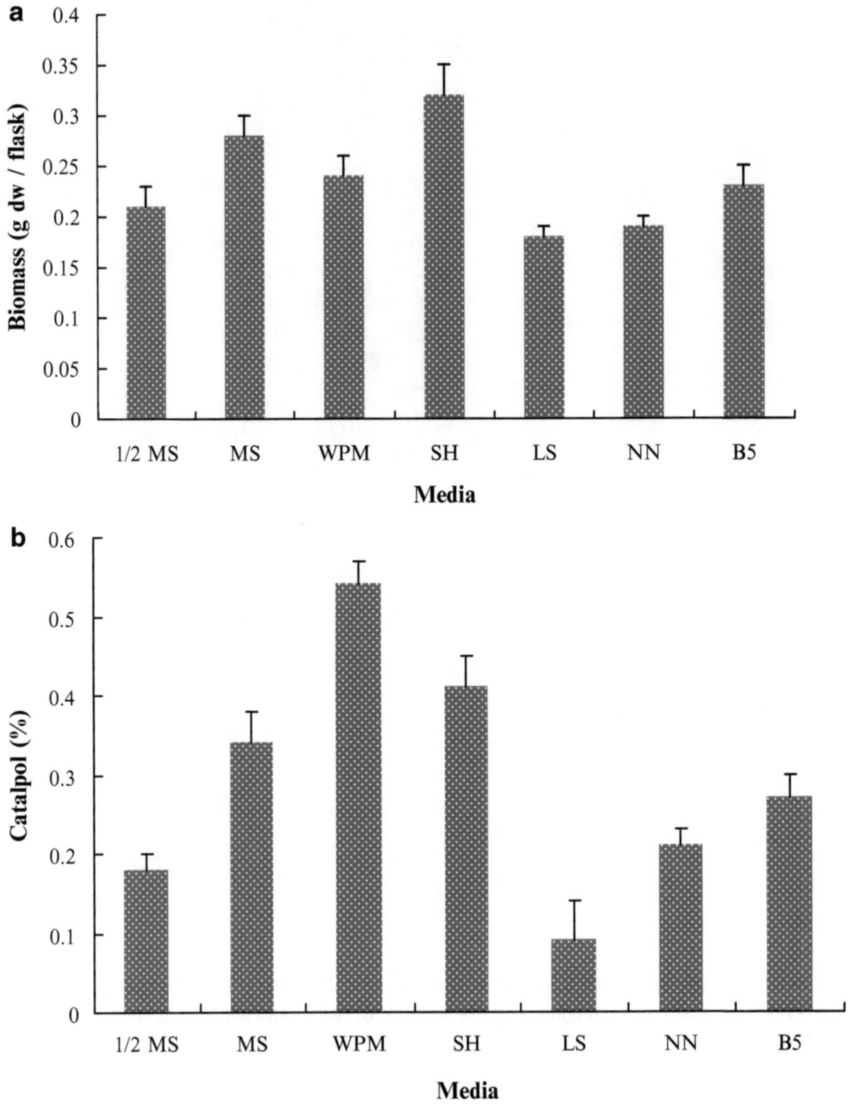

Fig. 2. (**a**) Effects of basal media on growth in hairy root cultures of *R. glutinosa*. (**b**) Effects of basal media on catalpol content in hairy root cultures of *R. glutinosa*.

5. Transformed roots are sensitive to the combination of medium components, especially mineral ions and carbon source, with regard to both growth and productivity. Both growth and catalpol content are highest in the medium containing 4% sucrose (*see* **Fig. 3a, b**).
6. Sucrose concentrations greater than 10% retard early development and cause prolonged root growth.
7. The hairy roots growth and catalpol production are highest in the presence of sucrose as the sole carbon source. All other

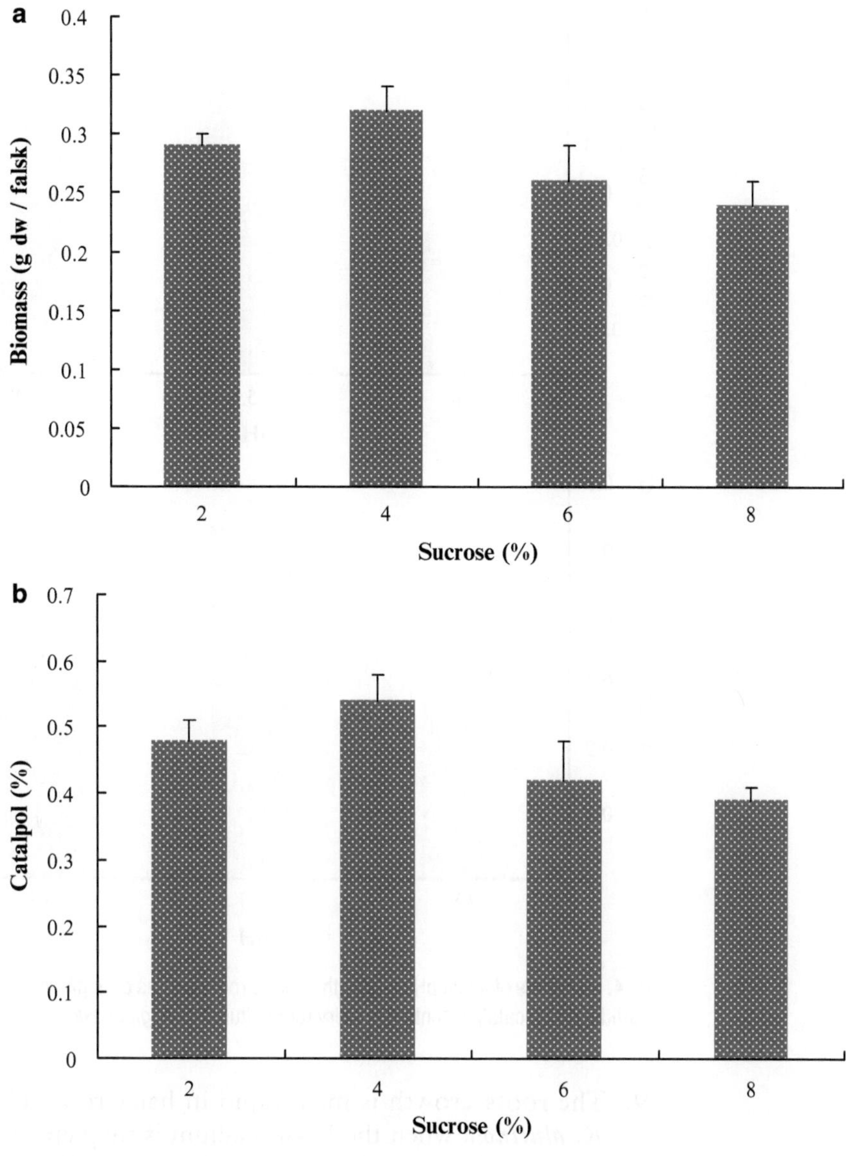

Fig. 3. (**a**) Effects of sucrose on growth in hairy root cultures of *R. glutinosa*. (**b**) Effects of sucrose on catalpol content in hairy root cultures of *R. glutinosa*.

sources (i.e., fructose, glucose, and lactose, have no positive effect on catalpol content).

8. Although hairy root development is similar at pH 5.7 or 5.8, catalpol production is higher (0.54% of dry weight) at pH 5.8 (*see* **Fig. 4a, b**). Root growth is slightly inhibited below pH 5.2 and above pH 6.2. In addition, it is noteworthy that in all cases, the media pH remains stable from the beginning to the end of the culture period.

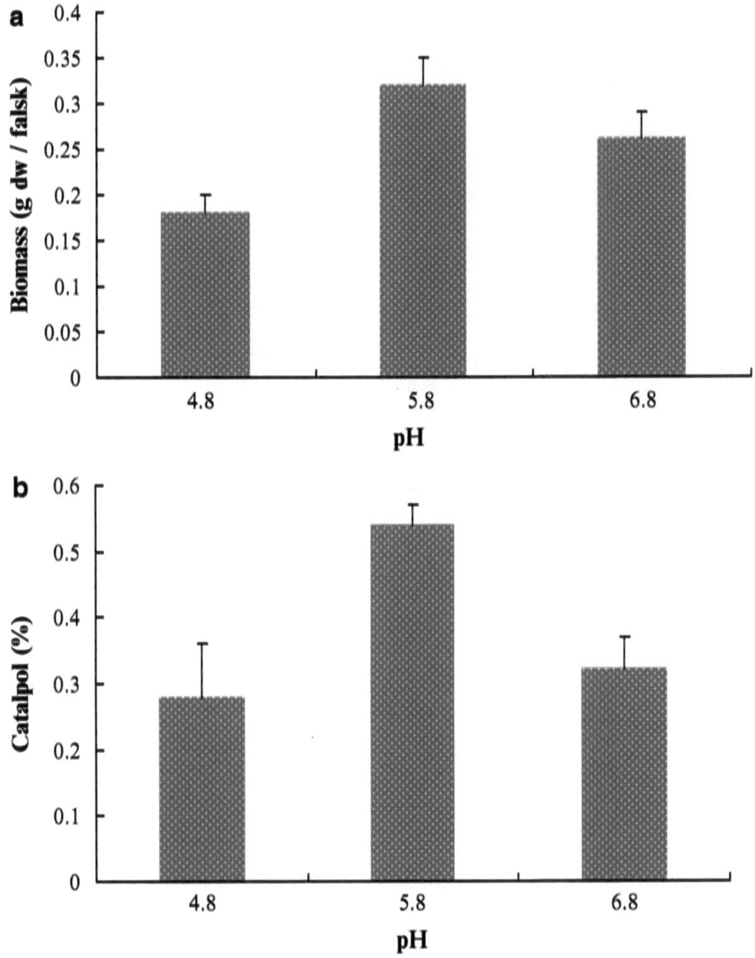

Fig. 4. (**a**) Effects of initial pH on growth in hairy root cultures of *R. glutinosa*. (**b**) Effects of intitial pH on catalpol content in hairy root cultures of *R. glutinosa*.

9. The roots growth is most rapid in hairy root cultures of *R. glutinosa* when the basal medium is supplemented with 2 mg/L IAA (*see* **Table 1**). The presence of IBA and NAA is less effective.
10. The addition of cytokinin decreases hairy root growth and stimulate callus formation: 2,4-D and ABA also reduces biomass and catalpol production. The cytokinin enhances root growth and results in high branching and root elongation rates.
11. In this study, hairy roots of *R. glutinosa* accumulate higher levels of catalpol in WPM medium containing 0.5 mg/L GA_3 (*see* **Table 2**). The effects of plant growth regulators on growth and secondary metabolite production are probably related to the genotype and physical-chemical characteristics of the explants *(17, 18)*.

Table 1
Effects of Plant Growth Regulators on Growth of *R. glutinosa* Hairy Root Cultures

Auxins (mg/L)		Fresh weight (g/flask)[a]	Dry weight (g/flask)[a]
IAA	0.5	14.3 ± 0.2	0.33 ± 0.02
	1.0	14.7 ± 0.2	0.38 ± 0.03
	2.0	15.4 ± 0.3	0.43 ± 0.02
IBA	0.5	14.2 ± 0.3	0.31 ± 0.02
	1.0	14.1 ± 0.4	0.26 ± 0.03
	2.0	13.7 ± 0.4	0.27 ± 0.03
NAA	0.5	14.1 ± 0.1	0.26 ± 0.02
	1.0	14.3 ± 0.3	0.31 ± 0.03
	2.0	14.0 ± 0.3	0.21 ± 0.02
2,4-D	0.5	13.1 ± 0.3	0.15 ± 0.03
	1.0	12.6 ± 0.4	0.13 ± 0.04
	2.0	12.1 ± 0.4	0.17 ± 0.03

[a]Experiment was done in duplicate. Mean ± S.D

Table 2
Effects of Plant Growth Regulators on Catalpol Production in Hairy Root Cultures of *R. glutinosa*

Plant growth regulators (mg/L)		Catalpol contents (% dry weight)[a]
IAA	0.5	0.43 ± 0.02
	1.0	0.48 ± 0.03
	2.0	0.53 ± 0.03
IBA	0.5	0.41 ± 0.02
	1.0	0.36 ± 0.02
	2.0	0.37 ± 0.02
GA_3	0.5	0.65 ± 0.02
	1.0	0.41 ± 0.03
	2.0	0.32 ± 0.03
BA	0.5	0.25 ± 0.04
	1.0	0.21 ± 0.04
	2.0	0.17 ± 0.04

[a]Experiment was done in duplicate. Mean ± S.D

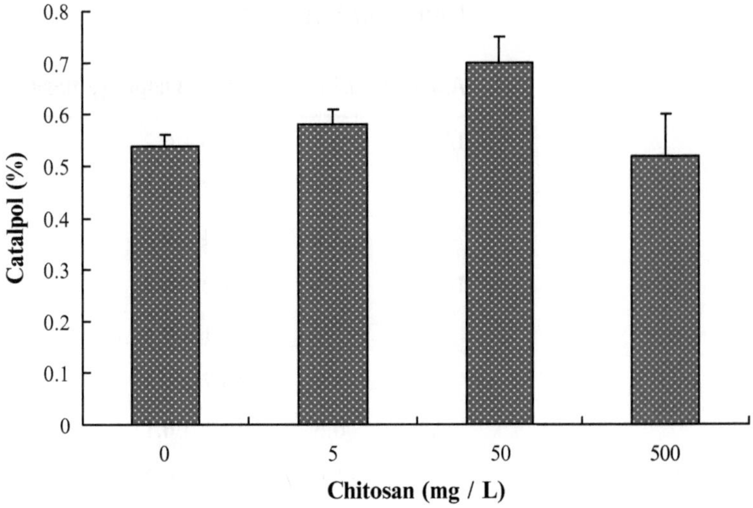

Fig. 5. Effects of chitosan on catalpol content in hairy root cultures of *R. glutinosa*.

12. Chitosan, a polymer of beta-1, 4-D-glucosamine, is obtained by alkaline hydrolysis of shellfish chitin. It is well known as an inducer of plant secondary metabolites *(19)*. After 12 wk of cultures, the additional chitosan has a little effect on growth, but increases catalpol production (*see* **Fig. 5**).
13. The medium supplemented with 50 mg/L chitosan shows the highest catalpol content, 0.70% of dry weight, 1.3-fold greater than the control. However, high chitosan concentration also inhibits hairy root growth and, subsequently leads to decrease in catalpol production.
14. When the effect of various chitosan concentrations on the release of compounds is assayed, catalpol is not detected in any of the medium used.

References

1. Balandrin, M.J. and Klock, J. A. (1998). Medicinal, aromatic and industrial materials from plants. In IY Bajaj, ed, Biotechnology in Agriculture and Forestry, Vol. 4. Medicinal and Aromatic Plants. Springer, Berlin, pp 1–36.
2. Enders, R. (1994). Plant Cell Biotechnology. Springer, Berlin, pp 149–163.
3. Lindsey, K. and Yeomann, M. M. (1983). The relation between growth rate, differentiation and alkaloid accumulation in cell cultures. *J Exp Bot* **34**, 1055–1065.
4. Takaetsu, K., Paul, P. H. B., Guo, J. X., Chung, K. S., and Han, B. H. (1996). International Collection of Traditional and Folk Medicine, Vol. 1. World Scientific, Singapore, pp 148–150.
5. Zhu, U. P. (1998). Tonifying herbs, In Chinese Materia Medica. Harwood Acad. Pub, Groningen, pp 583–584.
6. Ismailoglu, U. B., Saracoglu, I., Harput, U. S., and Sahin-Erdemli, I. (2002). Effects of phenylpropanoid and iridoid glycosides on free radical induced impairment of endothelium dependent relaxation in rat aortic rings. *J Ethnopharmacol* **79**, 193–197.
7. Park, C. H., Seong, N. S., Kim, S. K., and Paek, K. Y. (1999). Characteristics in tissue cultured plants of Chinese foxglove (*Rehmannia glutinosa* Libosch). *Kor J Plant Tiss Cult* **26**, 205–210.
8. Kittipongpatana, N., Hock, R. S., and Porter, J. R. (1998). Production of solasodine by hairy

root, callus, and cell suspension cultures of *Solanum aviculare* Forst. *Plant Cell Tiss Org Cult* **52**, 133–143.

9. Rhodes, M. J., Robin, R. J., Hamill, J. D., Parr, A. J., Hilton, M. J., and Walton, M. J. (1990). Properties of transformed root cultures. In BV Chalwood, MJ Rhodes, eds, Secondary products from Plant Tissue Culture. Clarendon Press, Oxford, pp 201–225
10. Toivonen, L. (1993). Utilization of hairy root cultures for production of secondary metabolites. *Biotechnol Prog* **9**, 12–20.
11. Murashige, T. and Skoog, F. (1962). A revised medium for rapid growth and bioassay with tobacco tissue culture. *Physiol Plant* **15**, 473–479.
12. Gamborg, O. L., Miller, R. A., and Ojima, K. (1968). Nutrient requirements of suspension cultures of soybean root cells. *Exp Cell Res* **50**, 151–158.
13. Linsmaier, E. M. and Skoog, F. (1965). Organic growth factor requirements of tobacco tissue cultures. *Physiol Plant* **18**, 100–127.
14. Schenk, R. U. and Hildebrandt, A. C. (1972). Medium and techniques for production and growth of monocotyledonous and dicotyledonous plant cell cultures. *Can J Bot* **50**, 199–204.
15. Lloyd, G. and McCown, B. (1980). Commercially feasible micropropagation of Mountain laurel, *Kalmia latifolia*, by use of shoot tip culture. *Comb Proc Int Plant Prop Soc* **30**, 421–427.
16. Nitsch, J. P. and Nitsch, C. (1969). Harploid plants from pollen grains. *Science* **164**, 85–87.
17. Liu, C. Z., Wang, Y. C., Outang, F., Ye, H. C., and Li, G. F. (1997). Production of artemisin by hairy root cultures of *Artemisia annua* L. *Biotechnol Lett* **19**,927–929.
18. Robins, R. J., Bent, E. S., and Rhodes, M. J. C. (1991). Studies on the biosynthesis of tropane alkaloids by *Datura stramonium* L. transformed root cultures: 3. The relationship between morphological integrity and alkaloid biosynthesis. *Planta* **185**, 385–390.
19. Funk, C. and Brodelius, P. (1990). Influence of growth regulators and an elicitor on phenylpropanoid metabolism in suspension cultures of *Vanilla plantifolia*. *Phytochemistry* **29**, 845–848.

Part III

Molecular Markers and Microsatellites

Chapter 22

Identification of Medicinal Plants and Plant Sequences: A Multiplexed MLPA Assay

Roger A. Barthelson

Summary

Recognition of medicinal plant species or plant characters can be accomplished through the use of genomic DNA sequences unique to a species, a group of species, or a species variant. An assay well-suited to this application is the *M*ultiplexed *L*igase-dependent *P*robe *A*mplification (MLPA) assay. It uses the sensitivity of the polymerase chain reaction, but increases the specificity by including a key ligation step for those MLPA probes that hybridize to a DNA sequence. The MLPA can be used to perform multiple tests in one tube, but the number of tests is limited when the amplified products are separated by chromatography. The use of hybridization to a microarray as part of the MLPA allows for a potentially greater number of tests to be performed on one sample. We describe the method for the MLPA procedure in detail, including the microarray hybridization protocol.

Key words: MLPA, Ligase, Microarray, CGH, Plants, Genome

1. Introduction

The accurate identification of plants used in herbal medicines, and the identification of the contents of herbal preparations from unknown or poorly controlled sources is critical to the safe use of alternative medicines and to their study. A large majority (85–90% according to the WHO) of the world's population employ herbal medicines, which can lead to poisoning or mistreatment of illnesses because of inaccurately characterized products *(1)*. Polymerase chain reaction (PCR)-based characterization of genomic DNA sequences can provide a highly sensitive and accurate means for identification of herbal materials.

S. Mohan Jain and Praveen K. Saxena (eds.), *Methods in Molecular Biology, Protocols for* In Vitro *Cultures and Secondary Metabolite Analysis of Aromatic and Medicinal Plants, vol. 547*

© Humana Press, a part of Springer Science+Business Media, LLC 2009

DOI: 10.1007/978-1-60327-287-2_22

*M*ultiplexed *L*igase-dependent *P*robe *A*mplification (MLPA) or multiplexed ligase-dependent probe amplification is a molecular method that has improved reliability over PCR for identification of sequences, because it increases the selectivity of the hybridization probe. The MLPA probes (or primers) must match their target sequence closely at their adjacent ends (**Fig. 1**) to allow the proper alignment for ligation *(2)*. For the MLPA, probes are added to a sample for hybridization, followed by ligation of the two probes, which then allows PCR amplification. A limitation of the MLPA is the need to produce different-sized final PCR products for multiplexing multiple reactions together in a tube. Subtle differences can be incorporated into 40 or more synthetic oligo pairs used as MLPA probes when capillary electrophoresis *(3, 4)* or high-performance liquid chromatography (HPLC) *(5)* is used for separating the PCR products. The specificity of the MLPA, coupled with quantitative analysis on chromatograms provides an effective platform for analyzing chromosome copy number *(6)*, chromosome rearrangements, deletions or duplications *(3–5)*, and chromosome methylation *(7)*.

The MLPA can also be coupled with DNA microarray technology to monitor for potentially a greater number of targets *(8)*. In previous studies, we demonstrated that plant species can be differentiated by MLPA, and that a combined mixture of MLPA probes can identify plant genomic DNA correctly when a microarray is used to recognize the resulting amplified genomic sequence. For these studies we used sequences from cytochrome P450 enzymes obtained by the CODEHOP procedure *(9)*. Cytochrome P450 enzyme sequences were exploited because of their common use in plants for the synthesis of relatively unique, secondary metabolites. Thus, in comparison with enzymes important to core metabolism of an organism, many of these enzymes are not as likely to be consistently conserved in sequence homology between related medicinal plant species. The use of cytochrome P450 sequences is not essential to the assay, and indeed, the protocol

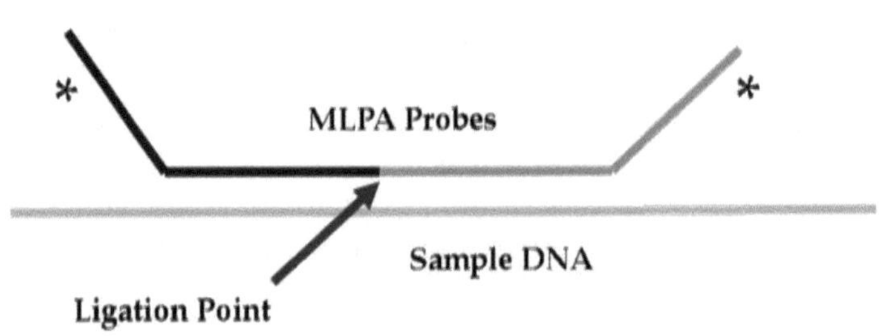

Fig. 1. MLPA Ligation Reaction. For a MLPA product to be formed, both probes must be brought together by hybridization to the target DNA, which allows ligation to occur. The MLPA probes have PCR primer sequences (*asterisk*) added to the outer ends, so that the ligation product can be amplified by PCR.

is suited to any application where identification of the presence or absence of multiple genetic markers is desired.

This chapter details the procedures necessary for use of microarrays with the MLPA in order to achieve high degrees of multiplexing of assays in one tube. This method offers the potential for identifying plant species, single feature polymorphisms, or other markers on a large scale. Recognition of the correct plant species, the focus of our work, is important to studies of medicinal plants, where related species have different properties, and can vary from therapeutic to purely toxic. In general, the identification of key genetic markers could be critical to finding the appropriate species or strains to be used as resources in the development of more reliable medicinal plant strains, and in the development of other improved crops. Further, the accurate identification of genetic markers is critical to the development of plant-based pharmaceuticals, biofuels, or other desirable chemical products.

2. Materials

2.1. Genomic DNA

1. It is assumed for this method that genomic DNA samples from diverse plants of interest are available for testing. These samples can be prepared by whatever method provides clean, genomic DNA for plant samples. We have used the Qiagen DNeasy Plant kits (Qiagen; Valencia, CA) for this purpose with good success.

2.2. MLPA Reaction

1. The MLPA is designed around a specific, Bacillus-derived, moderately thermostable DNA ligase. This ligase, Ligase-65, is available from MRC (Holland, Amsterdam). We used the reagents in the Lig-5b kit, which contains Ligase-65, Buffer A (NAD buffer), and SALSA PCR Buffer. These reagents are proprietary, but the contents likely include approximately 2 mM nicotinamide adenine dinucleotide (NAD) in the NAD buffer, and typical PCR buffer contents such as 20 mM $MgCl_2$; 40 mM Tris–HCl, pH 8.5; and 0.1% Triton X-100. The buffers from MRC Holland are stable at −20°C for long periods (e.g., 1 yr).
2. Surestart *Taq* DNA polymerase a hot start *Taq* polymerase (Stratagene, La Jolla, CA).
3. MLPA Buffer: 1.5M KCl; 0.25M Tris–HCl, pH 8.5; and 1mM ethylenediamine tetraacetic acid (EDTA). This buffer is stable at room temp for long periods, e.g. up to 1 yr.
4. Molecular grade water (Fisher Scientific).
5. MLPA Probes: 250 nM stock solution of both ligation primers. These are oligonucleotides specific for a plant species or other plant sequence (*see* **Notes 1** and **2**). The design of the MLPA

Table 1
Example Sequences for Oligonucleotides

Ligation Probe 1 for *D. metel*:
GGGTTCCCTAAGGGTTGGAACAATCTAAAATCATGACCTTTCATGTCAACATCTT
Ligation Probe 2 for *D. metel*:
CTTCAATGAATCTTTCAGGCCTGAATTGTAAAGTCTAGATTGGATCTTGCTGGCAC
Forward PCR primer:
GGGTTCCCTAAGGGTTGGA
Reverse PCR primer:
GTGCCAGCAAGATCCAATCTAGA
Microarray Probe for *D. metel*:
ACAATCTAAAATCATGACCTTTCATGTCAACATCTTCTTCAATGAATCTTTCAG-GCCTGAATTGTAAAG

oligos requires that the forward and reverse PCR primer sequences be integrated in the MLPA ligation probes so that PCR amplification can be achieved (**Fig. 1**). See **Table 1** for an example.

6. Oligonucleotide PCR primers. For use with microarrays, the PCR primers should be modified with a 5′ fluorophore such as cyanine 5.
7. dNTPs: 2 mM deoxyribonucleic acid triphosphates (dGTP, dATP, dTTP, and dCTP; 2 mM each) (Fermentas, Inc; Glen Burnie, MD).
8. $MgCl_2$ solution: 50 mM solution of $MgCl_2$ (Teknova; Hollister, CA).

2.3. Agarose Electrophoresis

1. NuSieve agarose (Cambrex; East Rutherford, NJ).
2. Seakem GTG agarose, (Cambrex).
3. 100-bp DNA ladder (Novagen; San Diego, CA).
4. DNA gel loading buffer, 6X (Novagen).
5. Tris–Acetate–EDTA (TAE) Buffer: 1M Tris–Acetate and 0.025M 25X EDTA, pH 8.0 (Fisher Scientific).
6. Sybr Gold DNA Nucleic Acid Stain (10,000X in dimethyl sulfoxide [DMSO]) (Invitrogen; Carlsbad, CA).

2.4. Microarray Printing

1. 20X SSC: 3.0M NaCl; 0.3M NaCitrate, pH 7.0; sterile, molecular grade (nuclease-free); (Fisher Scientific).
2. 70-mer oligonucleotides.

3. Aminosilane-coated glass microarray blank slides (Telechem International; Sunnyvale, CA).

2.5. Microarray Hybridization

1. 10% sodium dodecyl sulfate (SDS) Solution (Teknova).
2. 20X SSC (Fisher Scientific).
3. Blocking reagent: Western blocking reagent (Roche Applied Sciences; Indianapolis, IN) (cat. no. 11921673001).
4. Molecular grade water (Fisher Scientific).
5. Lifterslips (Erie Scientific Co; cat. no 24X601–2–4733).
6. Hybridization Cassette (Telechem International; cat. no. AHCXD).

3. Methods

The MLPA is an established method for identification of DNA sequences by a combination of hybridization and ligation. The ligation step requires a perfect match between oligos and the target DNA sequence near the ligation site for ligation to proceed (*see* **Note 3**), which enhances the specificity of the reaction. Recognition of a specific sequence by hybridization and ligation allows PCR amplification of the ligated oligos to occur. In this way, PCR amplification of the ligation product is an indicator for the presence of a specific DNA sequence. Multiple plant sequences (*see* **Note 1**) can be tested for in a single tube if the MLPA probes include spacers that create a different size PCR product for each species, so they may can be separated by electrophoresis (*see* **Note 4**). Creation of the MLPA probes in assorted lengths can be time-consuming. Alternatively, a single plant species can be determined in a single tube, which is what is described below. For testing for multiple species, we modified the protocol to allow for a microarray hybridization step, which is also described.

3.1. MLPA Assay: Ligation Reaction

1. Start with a 3.4 μL genomic DNA sample containing 50–200 ng DNA. The following steps are performed with a 500-μL PCR tube in a thermocycler, or with suitable heating blocks set for the appropriate temperatures.
2. Heat the genomic DNA to 98°C for 5 min, then cool and hold at 25°C.
3. Combine 50 μL MLPA probes/50 μL MLPA Buffer in a separate tube and add 4.6 μL of this mixture to each sample.

4. Incubate the sample mixture to 95°C 1 min, and then incubate it for 16h at 60°C.
5. After approximately 16h, combine in a separate tube the following reagents to make the Ligase-65 mix (one reaction): 3 μL buffer A; 3 μL SALSA PCR buffer; 25 μL molecular grade water; and 1 μL Ligase-65.
6. Reduce the temperature of the incubating sample to 54°C, and add 32 μL Ligase-65 mix. Incubate the sample 15 min and then heat it to 98°C 5 min, and hold it at 4°C.

3.2. MLPA Assay: PCR Reaction

1. Prepare the PCR mix by combining the following reagents (1 reaction): 25 μL molecular grade water; 5 μL 10X SureStart Taq Polymerase Buffer; 2 μL 50 mM $MgCl_2$; 5 μL 2 mM dNTPs; 1.25 μL 10 mM PCR forward primer; 1.25 μL 10 mM PCR reverse primer; and 0.5μL SureStart Taq Polymerase.
2. Combine in a 500-μL PCR tube 10 μL of the ligation reaction mixture (**step 6**; MLPA Ligation Reaction) and 40 μL PCR mix.
3. Incubate the PCR reaction according to the following protocol in a thermocycler: 9 min 95°C; 35 total cycles × (45s 94°C, 1 min 60°C, 1 min 72°C); 7 min 72°C; and hold at 4°C.

3.3. Agarose Electrophoresis

1. Prepare enough agarose solution to fill the gel tray of a horizontal gel electrophoresis apparatus (e.g., 7 × 8cm): 1% agarose, 3% NuSieve agarose (*see* **Note 4**), in 1X TAE buffer prepared in distilled or deionized water.
2. Heat this mixture until the agarose is dissolved in a microwave oven, or boiling water bath, cool for 5 min at room temp, and pour the gel into the tray fitted with a comb designed for small wells (10–15 μL).
3. When the gel is hardened (~1h), remove the comb and fill the apparatus with 1X TAE buffer.
4. Combine 10 μL of the PCR samples with 2 μL 6X DNA gel loading buffer.
5. Apply the PCR samples and 100-bp ladder to the wells of the agarose gel, and electrophorese at approximately 5V/cm until the dark blue, bromophenol blue band has migrated approximately two-thirds of the length of the agarose gel.
6. Remove the agarose gel from the apparatus and incubate in 1X TAE buffer containing a 1/10,000 dilution of the Sybr Gold stock solution.
7. Place the gel on a ultraviolet (UV) transilluminator and photograph the fluorescent bands through an ethidium bromide filter or through a Sybr Gold filter (*see* **Fig. 2** for a typical result).

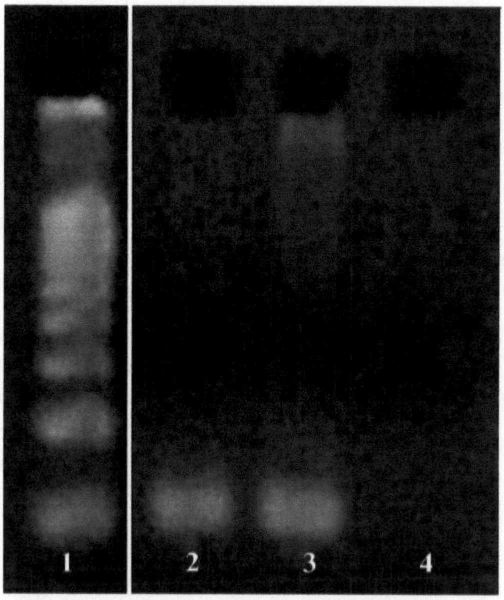

Fig. 2. A typical MLPA Reaction. In testing MLPA probes, we test with both positive and negative controls. The MLPA amplification products were separated on a 3% NuSieve, 1% agarose electrophoresis gel and stained with Sybr Gold. The first lane contained a 100-bp ladder. The MLPA was conducted with probes designed for *A. absinthium* and tested with *A. absinthium* (*lane 2*), *A. vulgaris* (*lane 3*), and *C. thalictroides* (*lane 4*) genomic DNA. The product in the lane for *A. vulgaris* genomic DNA indicates that these MLPA probes were not specific to *A. absinthium*.

3.4. Microarray Printing

1. The use of a microarray to determine which plant sample is detected by the MLPA requires that a custom microarray must be printed (*see* **Note 5**). Printing a microarray requires a microarray printer, and could be performed by a core facility or commercial lab. A simple description of printing a microarray is provided here, but the details of printing depend on what type of printer is used and what type of chemistry is on the slide surface.
2. Design oligonucleotides to match the gene-specific portions of each amplicon that will be produced by the MLPA (*see* **Table 1** for an example). If only one PCR primer is labeled, then care must be taken that the oligos printed on the microarray are complementary to the strand of the PCR product that is labeled with a fluorophore.
3. Dilute the oligonucleotides to 20 μM final concentration in 3X SSC and distribute them into microplate wells for printing.
4. Print the microarrays with a microarray printer (*see* **Note 6**).

3.5. Microarray Hybridization

1. For preparing a sample for microarray hybridization, the PCR reaction described above should use one PCR primer that has a fluorophore attached to the 5'-end, and be scaled up to 100 μL total, with a template volume of 25 μL. The fluorophore should be one that can be detected with the scanner to be used for imaging the hybridized microarray slide.
2. After amplification, purify the PCR sample from the MLPA using a typical PCR cleanup kit, such as the QIAquick kit.
3. If the microarray slides were printed in aqueous media, the array can be rehydrated for a more consistent and often stronger signal, but this step may be omitted (*see* **Note 6**). Rehydration requires that the printed side of the slide be held in the vapor from a boiling beaker of water for approximately one second, just enough time for the salts from the buffer printed on the array to imbibe moisture and form small individual droplets for each spot. The slide is quickly dried by warming (e.g., on a 55°C heating block).
4. All slides should be cross-linked to attach the DNA to the microarray surface. This is accomplished by exposing the printed surface to 120 mJ UV in a cross-linking machine.
5. Wash the microarray slides by incubating with a rocking motion in 0.1% SDS solution in molecular grade water for 10 min. Rinse with molecular grade water and dry under a stream of nitrogen or with a slide drier.
6. Prepare the hybridization solution by combining the clean, fluorescently-labeled PCR product with 9 μL 20X SSC, 4.5 μL blocking reagent, 0.9 μL 10% SDS, and molecular grade water to a final volume of 90 μL.
7. To initiate the hybridization, place a Lifterslip on the microarray so that it covers evenly the printed area, heat the hybridization solution to 95°C for 2 min, cool to room temperature, then apply the solution to one end of the Lifterslip so that the solution is drawn under it by capillary action **(Fig. 3)**.
8. Place the microarray in a hybridization cassette with 2X SSC added in the wells provided to keep the array humidified during incubation. Close the cassette and incubate in the dark at 58°C for 5 h or overnight.
9. After incubation, remove the microarray from the cassette and immerse (with the Lifterslip in place) in 2X SSC, 0.5% SDS at 58°C and remove the Lifterslip. Incubate the microarray with agitation for 5 min.
10. Wash twice for 5 min with 0.5X SSC at room temp, and dry the slide (as above).
11. Scan the slide (*see* **Note 7**) immediately at the appropriate wavelength for the fluorophore used (**Fig. 4** uses 635 nm for cyanine 5).

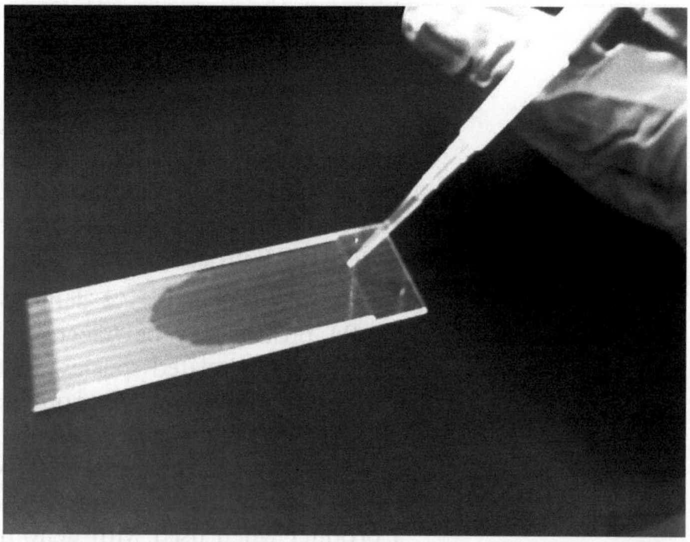

Fig. 3. Hybridization of the microarray. The microarray is covered with the Lifterslip, which has raised rails on the underside to provide extra space underneath for the hybridization solution. The hybridization solution is added to the edge of the Lifterslip at an angle to allow the liquid to be pulled underneath by capillary action. In this way, the hybridization solution can be added without forming bubbles.

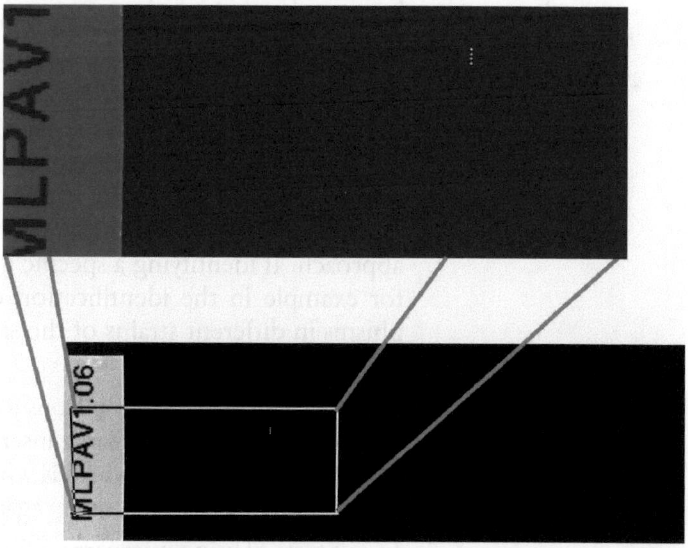

Fig. 4. Hybridization of a custom microarray with a MLPA sample. A sample of *Datura metel* genomic DNA was incubated with a combined mixture of 15 pairs of MLPA probes for diverse plant sequences including ones for *D. metel* (see Table 1) and *Lobelia inflata*. The microarray had oligonucleotide probes matching *D. metel* and *L. inflata* sequences, and eight oligos matching *A. thaliana*, all printed in triplicate. The amplified MLPA product was labeled with Cy5 (see Subheading 3). The only probes that became fluorescent after hybridization were the ones for *D. metel*.

4. Notes

1. To use the MLPA with species that have little or no sequence information available, then sequencing of cloned DNA samples is required. To accomplish this, PCR amplification of sequences for cloning can be effectively targeted by using the CODEHOP procedure *(8)*.
2. The MLPA test in **Fig. 2** demonstrates how a typical set of MLPA probes may be selective, but still not adequate for distinguishing related species. In this case, the probes for a cytochrome P450 enzyme from *Artemisia absinthium* detect the cognate sequence in a sample of *A. absinthium* genomic DNA, but not in the genomic DNA of *Caulophyllum thalictroides*. This same set of probes forms an equal amount of a product when used with genomic DNA from *A. vulgaris*. It is possible that if the hybridization temperature prior to ligation were increased, the probes would form product only with genomic DNA from *A. absinthium*. This change would require testing and adjusting the protocol, which may not be a viable approach when many probes need to be used under the same condition to achieve a highly multiplexed assay. Often multiple sets of probes need to be tested for each target to find probe sequences that provide adequate selectivity. In all cases, every probe pair needs to be tested.
3. Before any other technical consideration is made to prepare for the MLPA, the most important question is, is the MLPA the appropriate tool? The strength of the MLPA lies in its selectivity for a specific sequence. If the overall goal is to find related sequences in a new species to be targeted, then PCR or cross-species hybridization on microarrays may be a better approach. If identifying a specific gene sequence is important, for example in the identification of single feature polymorphisms in different strains of the same species, the MLPA is a good choice.
4. As shown in **Fig. 2**, the amplicons for the MLPA are small, 110–120 bp, in the absence of large inserted sequences in the probes to adjust their sizes *(2)*, which is why a 3% NuSieve, 1% agarose gel is required for separation by agarose electrophoresis.
5. In our tests of using microarrays for detecting amplified MLPA products, we found that microarrays can be sufficiently specific and allow for potentially greater numbers of tests combined in one tube than has been achieved with capillary electrophoresis (**Fig. 4**). This particular version of the MLPA would be useful for increasing the selectivity of comparative genomic hybridization (CGH) studies. The single feature polymorphisms

represented on CGH arrays could be hybridized with amplified MLPA ligation products.

6. In most cases, the use of microarrays in conjunction with the MLPA implies that a custom microarray has been printed to match the genome-specific portion of the ligation products. Details of the microarray printing process were not addressed here because this is generally performed by a facility equipped to do this on a routine basis. The buffer used for suspension of the oligonucleotides can be 3X SSC as described here, or dimethyl sulfoxide, depending on what type of slide surface chemistry is used in preparing the slides. For slides printed with DMSO, the spot size tends to be slightly larger, and rehydration of the spots is neither necessary nor possible. In using commercially prepared blank microarray slides for printing, the manufacturer's instructions should be used to guide how the slides are printed and hybridized. In some cases, the slide manufacturer provides solutions for printing, too.
7. Scanning of the microarrays after hybridization requires the use of a laser scanner and will yield an image file that documents the fluorescence of the slide surface in detail. In the small example given here, interpretation requires only a quick manual comparison to a chart of how the slide was printed. In the case that a much larger number of probes are used, analysis of the image file requires software for correlating the spots on the image to a grid consisting of the projected, relative locations of the probes on the slide, which is typically recorded by the microarray printer in a **. gal** file. "Spot finding" requires that the grid created from the spot printing information be superimposed over the actual image. The software then extracts the fluorescent intensity for each spot in the image and provides it in a spreadsheet along side the identifier for the corresponding probe. We used the software Genepix 6.0 (Axon/Molecular Devices, Sunnyvale, CA) for scanning the slide, because it accompanied our Axon 4000 AL scanner. The same software works well for extracting fluorescent intensity information for larger microarray slides.

Acknowledgments

We appreciate the guidance and resources provided by David Galbraith, and Raymond Woosley for this research. This research was supported by the Centers for Disease Control (Grant no. H75/CCH923339–01).

References

1. Kumar, C. (2007). Pharmacognosy can help minimize accidental misuse of herbal medicine. *Curr Sci* **93**, 1356–1358
2. Schouten, J. P., McElgunn, C. J., Waaijer, R., Zwijnenburg, D., Diepvens, F., and Pals, G. (2002). Relative quantification of 40 nucleic acid sequences by multiplex ligation-dependent probe amplification. *Nucleic Acids Res* **30**, e57
3. Lai, K. K. S., Lo, I. F. M., Tong, T. M. F., Cheng, L. Y. L., and Lam, S. T. S. (2006). Detecting exon deletions and duplications of the DMD gene using Multiplex Ligation-dependent Probe Amplification (MLPA). *Clin Biochem* **39**, 367–372
4. Slater, H. R., Bruno, D. L., Ren, H., Pertile, M., Schouten, J. P., and Choo, K. H. (2003). Rapid, high throughput prenatal detection of aneuploidy using a novel quantitative method (MLPA). *J Med Genet* **40**, 907–912
5. De Lellis, L., Curia, M. C., Catalano, T., De Toffol, S., Bassi, C., Mareni, C., Bertario, L., Battista, P., Mariani-Costantini, R., Radice, P., and Cama, A. (2006). Combined use of MLPA and nonfluorescent multiplex PCR analysis by high performance liquid chromatography for the detection of genomic rearrangements. *Hum Mutat* **27**, 1047–1056
6. Buffart, T. E., Coffa, J., Hermsen, M. A., Carvalho, B., van der Sijp, J. R., Ylstra, B., Pals, G., Schouten, J. P., and Meijer, G. A. (2005). DNA copy number changes at 8q11–24 in metastasized colorectal cancer. *Cell Oncol* **27**, 57–65
7. Dikow, N., Nygren, A. O. H., Schouten, J. P., Hartmann, C., Kramer, N., Janssen, B., and Zschocke, J. (2007). Quantification of the methylation status of the PWS/AS imprinted region: Comparison of two approaches based on bisulfite sequencing and methylation-sensitive MLPA. *Mol Cell Probes* **21**, 208–215
8. Barthelson, R. A., Sundareshan, P., Galbraith, D. W., and Woosley, R. L. (2006). Development of a comprehensive detection method for medicinal and toxic plant species. *Am J Bot* **93**, 566–574
9. Morant, M., Hehn, A., and Werck-Reichhart, D. (2002). Conservation and diversity of gene families explored using the CODEHOP strategy in higher plants. *BMC Plant Biol* **2**, 7

Chapter 23

Isolation of Microsatellites from *Catharanthus roseus* (L.) G. Don Using Enriched Libraries

Sabhyata Bhatia and Bhumika Shokeen

Summary

Catharanthus roseus is an indispensable source of the anticancerous alkaloids-vincristine and vinblastine, even though they are produced in trace amounts in vivo. In order to increase the yield of alkaloids, in vitro tissue culture studies are carried out which result in a large number of lines/cultures. For identification and characterization of the in vitro cultures, microsatellites in the form of STMS (Sequenced Tagged Microsatellite Sites) markers are used for identification of genetic polymorphism. STMS markers are also used for assessment of genetic diversity within natural populations as well as for construction of genetic linkage maps. Isolation of microsatellites and development of STMS markers typically involves library construction and screening, DNA sequencing, polymerase chain reaction (PCR) primer design, and PCR optimization. This chapter details two approaches for the isolation of microsatellite loci. The first approach is based on PCR using microsatellite containing primers which also have degenerate bases at the 5′-end that act as anchors preventing the primers from slippage to the 3′-end and the subsequent loss of polymorphism. The multi-locus PCR amplified product is cloned and sequenced. Though this method generates a large number of microsatellites, the major drawback is the high redundancy observed in this method. The second approach described in this chapter is based on the construction of a microsatellite enriched library which involves preferential cloning of the microsatellite enriched fraction of genomic DNA. This method therefore necessitates the isolation of microsatellites through hybridization with biotin labeled oligoprobe followed by their capture with streptavidin-coated magnetic beads. In comparison to the first approach, this approach yields less redundant clones with high microsatellite enrichment. Moreover enriched libraries are 40–60 times more efficient than the conventional small insert genomic libraries.

Key words: STMS, Microsatellites, Biotinylated, Streptavidin, *C. roseus*, Enriched library, Degenerate primers

1. Introduction

The advent of techniques based on detection and exploitation of naturally occurring DNA sequence polymorphism has led to the inception of a new field in molecular biology— the molecular

S. Mohan Jain and Praveen K. Saxena (eds.), *Methods in Molecular Biology, Protocols for* In Vitro *Cultures and Secondary Metabolite Analysis of Aromatic and Medicinal Plants, vol. 547*

© Humana Press, a part of Springer Science+Business Media, LLC 2009

DOI: 10.1007/978-1-60327-287-2_23

markers – which are considered ideal tools in the assessment of genetic polymorphisms within and across species particularly because of the absence of the confounding effects of the environment *(1)*. Of the innumerable markers available, microsatellites (also called Simple Sequence Repeats or SSRs), which are iterations of 1- to 6-bp nucleotide motifs, dispersed randomly and ubiquitously throughout the genome, are extensively utilized as a source of various marker systems. Most polymorphic and useful amongst these are the STMS (Sequence Tagged Microsatellite Sites) markers which are a set of primers designed from the conserved sequence regions flanking the microsatellite motifs. For DNA polymorphism analysis, STMS markers are used to amplify microsatellite loci across genotypes that typically generate size polymorphic fragments because of the difference in the number of repeat motifs at the locus. Hence, STMS markers have emerged as ideal tools for deciphering the innate genetic polymorphisms and are increasingly being used in a wide range of molecular genetic studies such as genetic mapping, QTL detection, forensic studies, conservation/management of biologic resources, and phylogenetic and pedigree analyses.

Despite their wide applicability, the major drawback of microsatellites is that they need to be isolated *de novo* from most species being examined for the first time. Traditionally, microsatellites have been isolated by hybridization screening of partial genomic libraries *(2–6)*. However, low yields of microsatellites in traditional small insert genomic libraries led to the advent of microsatellite enriched libraries, which are based on the principle of selectively isolating and preferentially cloning only the microsatellite containing genomic sequences. Construction of a microsatellite enriched library involves restriction digestion and adapter-ligation of genomic DNA followed by its hybridization with the repeat containing probes. The probes can be labeled both by radioactive (^{32}P, ^{33}P) or non-radioactive (e.g., digoxigenin, biotin) methods *(7, 8)*. In one such example of non-radioactive labeling, the microsatellite oligonucleotide probe labelled with biotin is allowed to hybridize with digested genomic DNA which preferentially hybridizes with the microsatellite sequences. This biotinylated, microsatellite rich fraction is immobilized using streptavidin coated magnetic beads and is subsequently eluted after several washings to remove the nonspecifically bound DNA. The eluted DNA is then amplified by PCR and cloned into an appropriate vector. In the enrichment procedures, the ability to efficiently recover the microsatellite- containing DNA after the selective hybridization step is most important. Hence other methods are also available in which the probe is bound to nylon membranes and hybridized with genomic DNA fragments *(9–11)*. Nevertheless, the biotinylated probe method is preferred as in this case liquid medium is used, making the probe fully available for hybridization *(7, 8)*. However, over the years both the techniques have found immense application in plant genomes for *denovo* isolation of microsatellites *(12–18)*.

Apart from the hybridization-based protocols, several fast and simple polymerase chain reaction (PCR) based methods of enriched library construction are also currently being used. One such approach, described by Fisher et al. *(19)*, is a novel technique aimed at generating a microsatellite rich PCR profile from genomic DNA which is cloned to yield a genomic library enriched for microsatellites. In this technique, the primers used consist of microsatellite repeats alongwith 7 degenerate bases at the 5′ end that are used to anchor the primer to the 5′ end of a microsatellite. For this purpose two nucleotides at the extreme 5′ end of the anchor are designed so that they could anneal to any nucleotide. The next five nucleotides, designated as blocking bases, are designed in such a manner that they are unable to pair up with the motifs that are to be amplified. This approach prevents slippage of the primer to the 3′ end of the microsatellite during PCR and hence results in amplification of exact number of repeats. All amplified products in this method contain terminal but complementary microsatellites at the 5′ and 3′ ends. In addition to these, some sequences may have internal microsatellites also, implying the presence of "SSR hotspots" or SSR clustering in some regions of the genome *(19, 20)*. The major advantage of this method is that it yields at least two microsatellites per amplicon reducing the cost per locus for microsatellite development as only one locus specific primer is designed in these cases. However, the major drawback of this method is the high redundancy thereby resulting in a fewer number of STMS markers. Nonetheless, this method has been used in several studies to isolate microsatellites *(21–23)*.

In the present chapter two different approaches for construction of an enriched library are described in detail. Of these, one is a fast and simple method, which involves the use of a degenerate primer *(19)* for amplification of microsatellite rich fragments which are then cloned, while the second approach is the one based on the capturing of microsatellite rich genomic fragments with the help of biotin-streptavidin interactions prior to cloning *(8)*.

2. Materials

2.1. Isolation of Microsatellite Loci Using 5′ Anchored Degenerate Primers

1. Genomic DNA of *C. roseus*.
2. Primers:
   D1: KKVRVRV$(AG)_{10}$ where K = G/T, V = G/C/A and R = G/A.

M13 F: GTA AAA CGA CGG CCA GT.

M13 R: AAC AGC TAT GAC CAT G.

- *Taq* DNA polymerase with 10× buffer + $MgSO_4$ (New England Biolabs).
- 2.5 mM dNTPs: 2.5 mM each of dATP, dCTP, dGTP, and dTTP.
- *Loading dye.* 50% glycerol, 1% SDS, 0.25% bromophenol blue, 0.25% Xylene Cyanol.
- Agarose (Sigma chemicals).
- PCR purification kit (Qiagen).
- pGEM- T cloning kit (Promega).
- Chemically competent cells (Invitrogen).
- Luria Broth (LB) agar + IXA plates: Dissolve 10 g of Tryptone, 5 g of yeast extract, 5 g of NaCl and 16 g of agar in dH_2O and make up the volume to 1 L. Sterilize by autoclaving. Add ampicillin (stock of 50 mg/ml in ddH_2O), IPTG (20% *[w/v]*), X-gal (2% in dimethyl formamide) to a final concentration of 50 µg/mL, 20 µg/mL, 20 µg/mL, respectively, once the medium has cooled to 50°C and then pour the medium into 150-mm petriplates.

2.2. Construction of a Microsatellite Enriched Library Using the Biotinylated Oligonucleotide

1. Nuclear DNA of *C. roseus* (good quality HMW DNA). The nuclear DNA of *C. roseus* can be isolated using the protocol of Malmberg et al. *(24)*.
2. *Restriction enzymes. Rsa* I with 10× buffer (New England Biolabs).
3. T_4 DNA ligase with 10× buffer (New England Biolabs).
4. Taq DNA polymerase with 10× buffer containing $MgSO_4$ (New England Biolabs).
5. 2.5 mM dNTPs (as above).
6. Agarose (as above).
7. Loading dye (as above).
8. LB agar + IXA plates (as above).
9. PCR purification kit (as above).
10. Adaptors and primers:

Biotin-labeled microsatellite oligoprobe (BO2) (50 µM): ATAGAATAT$(GA)_{20}$

Each probe is biotinylated on the 3′ end to prevent polymerase extension from the probe in subsequent PCRs and to allow later capture of target DNA:

Blunt End Primer (BEP A) (10 µM): CTCTTGCTTACGCGTGGACTA

Blunt End Primer (BEP B) (10 μM): pTAGTCCACGCG-TAAGCAAGAGCACA

(BEP B is phosphorylated)

M13 F: GTA AAA CGA CGG CCA GT

M13 R: AAC AGC TAT GAC CAT G

11. Wash buffers. 20X SSC, 0.5X SSC, 0.1X SSC.
12. Elution buffer. Autoclave MQ water.
13. TOPO TA cloning kit (Invitrogen).
14. Streptavidin magnesphere (Promega; Streptavidin Magne-Sphere paramagnetic particles).
15. Streptavidin magnesphere magnetic stand.

3. Methods

3.1. Isolation of Microsatellite Loci Using 5′ Anchored Degenerate Primers

To clone the microsatellite sequences, a library enrichment protocol based on the PCR amplification of genomic DNA using 5′ anchored degenerate microsatellite primers was used *(19)*. The anchor at 5′-end is used to avoid slippage of primer towards the 3′ end of the microsatellite causing a loss of repeat length polymorphism. Thus this method consistently anchors PCR primers at the 5′-ends of microsatellite, amplifying two close and inverted SSRs and the region between them.

3.1.1. PCR Amplification of Genomic DNA with Degenerate Primer

1. Amplify 50 ng *C. roseus* genomic DNA in a PCR reaction mix containing 0.125U Taq DNA Polymerase (NEB), 1 μM of D1 primer, 2.5 mM $MgSO_4$, 0.125 mM each of dNTPs and 1X Taq DNA polymerase buffer (NEB).
2. The amplification profile consists of an initial denaturation of 2 min at 94°C followed by 40 cycles (*see* **Note 1**) of denaturation for 10 s at 94°C, annealing for 10 s at 57°C (*see* **Note 2**) and extension for 10 s at 72°C with a final extension for 20 min at 72°C (*see* **Note 3**).

3.1.2. Purification and Cloning of PCR Amplified Products

1. Analyze an aliquot of PCR products on a 1% agarose gel (**Fig. 1**).
2. Purify the remaining product using the PCR purification kit (Qiagen) and finally elute in 20 μl of autoclaved MQ water (*see* **Note 4**). To check the concentrations of the purified product analyze an aliquot on a 1% agarose gel.
3. Ligate 150 ng of the purified PCR product with 50 ng of pGEM-T vector (Promega) in a reaction volume of 10 μl along with 1X T_4 DNA ligase buffer (Promega) and 20U of T_4 DNA ligase (Promega). Incubate the reaction mixture at 16°C for 12–16 h.

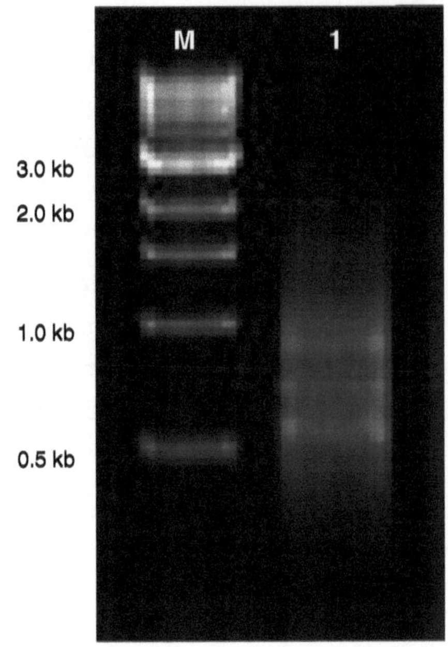

Fig. 1. The PCR amplification profile of *C. roseus* genomic DNA using the degenerate primer D1 on a 1% agarose gel. *Lane 1:* the PCR amplified product; *M:* 1-kb ladder.

4. Transform 100 µl of chemically competent cells (Invitrogen) with the total ligation mix. Briefly, add the ligated mix to the competent cells after thawing the cells on ice. Incubate the mixture on ice for half an hour. Give a heat shock at 42°C for 90 s. Snap freeze on ice for 2 min. Add 900 µL of LB and incubate at 37°C for 1 h to allow growth of cells. Spread 100 µL of this transformed product on an IXA plate (*see* **Note 5**). Incubate the plate at 37°C for 12–16 h. Look for the blue and white colonies (*see* **Note 6**). Pick and grid the white colonies on an IXA plate for further analysis.
5. Identify the putative recombinants via colony PCR. Briefly, with a toothpick, lift a few cells from the colony in 10 µL of autoclaved MQ water and lyse at 94°C for 10 min. After the lysis of colonies add 10 µL of PCR reaction mix containing 1X Taq DNA polymerase buffer, NEB: with Mg^{2+}, 0.125 mM of each dNTPs, 0.5 µM each of M13 Forward and Reverse primers and 0.1U of Taq DNA polymerase. The PCR reaction profile consists of an initial denaturation at 94°C for 2 min followed by 35 cycles of denaturation at 94°C for 30 s, 55°C for 1 min, 72°C for 1 min. This is then followed by a final extension of 7 min at 72°C. Check for the insert size by electrophoresis of the colony PCR products on a 1.2% agarose gel.
6. Isolate the plasmid DNA using the alkaline lysis method *(25)* from the recombinants identified to contain the desired inserts.

7. Obtain the nucleotide sequence of the insert by sequencing using ABI3700 Prism automated DNA sequencer, based on Sanger's sequencing method, with the help of Big Dye Terminator reaction kit.

3.1.3. Sequence Analysis and Primer Design

1. Analyze the sequences and remove the vector sequences and the adapter sequences using the VecScreen program (http://www.ncbi.nlm.nih.gov/VecScreen/VecScreen.html).
2. Select unique sequences by discarding the redundant clones using the CAP3 *(26)* (http://www.pbil.univ-lyon1.fr/cap3.php) program and identify the microsatellite containing clones (*see* **Note 7**) either manually or by using various microsatellite identifying programs like: Tandem Repeat finder *(27)* and TROLL *(28)*.
3. After selecting the unique clones containing the microsatellites, design primers complementary to the sequence flanking the microsatellites, using the PRIMER 3 (http://www.frodo.wi.mit.edu/) software *(29)*.

3.2. Construction of a Microsatellite Enriched Library Using a Biotinylated Oligonucleotide

In this approach of constructing an enriched library, SSRs are isolated with the help of biotinylated microsatellite oligonucleotides. This protocol is based on the hybridization based capture methodology described by Kijas et al. *(8)*. In this method, the nuclear DNA of *C. roseus* is used as the starting material for the construction of the enriched library. The nuclear DNA is restricted and the microsatellite rich genomic fragments are captured using the biotin-labeled microsatellite oligoprobe [Here BO2: ATAGAATAT $(GA)_{20}$] which is then allowed to bind to the streptavidin-coated magnetic beads. Once the binding is achieved, the nonspecifically bound DNA is removed via washings whereas the bound microsatellite DNA is eluted and used for the construction of the library. This approach results in an enrichment of about 40–60% reducing the cost of screening.

3.2.1. Digestion and Adapter Ligation of Nuclear DNA

1. Restrict 8 μg of *C. roseus* nuclear DNA with 45 units of *Rsa*I in a 100 μL reaction mix at 37°C for 12–16 h along with the appropriate buffer supplied with the enzyme (*see* **Notes 8–11**). Stop the reaction by heat inactivating the enzyme at 70°C for 10 min.
2. Check an aliquot of the digestion mixture (10 μl) for complete digestion of DNA on a 1.2% agarose gel **(Fig. 2)** (*see* **Note 12**).
3. For adapter ligation take 1 μg of the restricted DNA along with 0.1 μM each of Blunt End Primers (BEPA& B) (*see* **Note 13**), in a 200 μLreaction mix with 5 units of T4 DNA ligase. Keep the reaction mixture at 20°C for 2 h to allow simultaneous annealing of the adapter and its ligation with digested DNA.

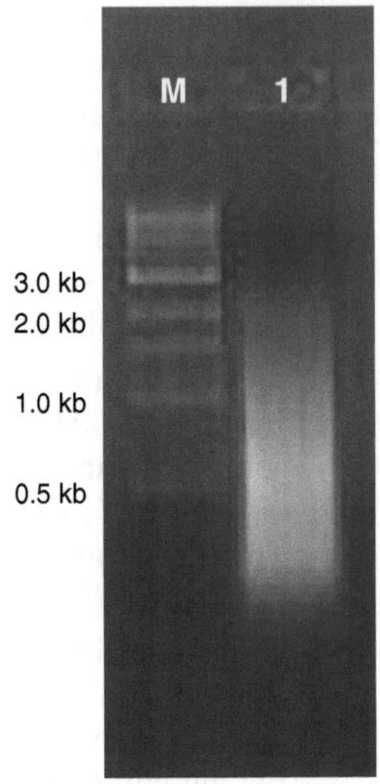

Fig. 2. Restriction digestion of *C. roseus* nuclear DNA with the *Rsa* I enzyme. *Lane 1:* restricted DNA; *M:* 1-kb ladder.

3.2.2. Preamplification of the Digested and Adaptor Ligated DNA

1. Preamplify an aliquot (3 μL) out of the 200 μL of the adapter-ligated DNA, using 0.4 μM of BEP A primer along with 0.2 units of Taq DNA polymerase in a reaction volume of 50 μl having 0.2 mM of dNTP's mix and 1X of Taq Buffer (*see* **Note 14**).
2. The PCR programme consists of an initial denaturation for 4 min at 95°C followed by 20 cycles of denaturation at 94°C for 30 s, annealing at 60°C for 1 min. and extension at 72°C for 2 min. A final extension for 8 min. is carried out at 72°C.
3. Check an aliquot of the pre-amplified product on a 1.2% agarose gel and purify the remaining product (~40 μl) with PCR purification kit according to manufacturer's instructions (*see* **Note 4**).

3.2.3. Binding of Biotin Labeled Microsatellite to DNA

1. Dilute the pre-amplified purified PCR products obtained from the above step to 500 μL with sterile MQ water. Further, denature this DNA at 95°C for 15 min.
2. After denaturation, add 50 μM of the biotinylated microsatellite oligo [BO2: ATAGAATAT $(GA)_{20}$: *see* **Notes 15–17** to the DNA along with 13 μL of 20X SSC.
3. Incubate the mix at ambient temperature, 50°C: depending on the Tm of the oligo, for 20 min while slowly mixing after every 2 min.

3.2.4. Washing of Streptavidin Magnesphere Particles

1. The biotinylated DNA (containing microsatellite sequences) is captured using 600 μl of Streptavidin-coated magnesphere, Promega, Streptavidin MagneSphere paramagnetic particles, which are to be washed prior to their use.
2. Wash the beads thrice with 0.5X SSC using a MagneSphere Magnetic Separation Stand. First magnetize for 30 s and then resuspend again in 600 μL of 0.5X SSC after discarding the supernatant. Repeat this for two more times (*see* **Note 18**).
3. Finally resuspend the streptavidin-coated magnesphere in 100 μL of 0.5X SSC.

3.2.5. Binding of Streptavidin Magnesphere Particles to DNA Hybridized with Biotin Labeled Microsatellite

1. Mix the 100 μL prewashed magnesphere obtained in **Subheading 3.2.4** with 514 μL hybridization mix, biotinylated probe bound to microsatellite rich DNA, obtained in **Subheading 3.2.3**. Incubate the mix at ambient temperature, 50°C: depending on the Tm of the oligo, for 10 min intermittent shaking.
2. Magnetize with the help of a MagneSphere Magnetic Separation Stand for 30 s and discard the supernatant. Resuspend the beads in 300 μL of 0.1X SSC and wash thrice with 0.1XSSC as mentioned in **Subheading 3.2.4** (*see* **Note 19**).

3.2.6. Elution of Microsatellite Bound DNA

1. After the three washings, as mentioned in **Subheading 3.2.5**, finally resuspend the beads in 50 μl of sterile MQ water.
2. This eluate of 50 μl contains microsatellite enriched DNA and is further used for PCR amplification and library construction.
3. Collect a second fraction of 150 μL in a similar manner and keep it at –20°C for future use.

3.2.7. Amplification

1. Amplify a 10 μl aliquot of the above eluate using 0.4 mM of BEP A primer in a 50 μl reaction mix containing 1X buffer, 0.2 mM dNTP's and 0.2 units of Taq DNA polymerase.
2. The PCR cycles consist of an initial denaturation at 95°C for 1 min followed by 20 cycles of denaturation at 94°C for 40 s, annealing at 60°C for 1 min and extension at 72°C for 2 min. A final extension at 72°C was carried out for 20 min (*see* **Note 3**).
3. Check for amplification by electrophoresis of an aliquot of the PCR product on a 1.2% agarose gel.

3.2.8. Ligation and Transformation

1. Ligate an aliquot of the PCR amplified product (3 μl) in pCR 2.1-TOPO vector using the TOPO T/A (Invitrogen) cloning kit according to manufacturer's instructions.
2. Briefly, 3.0 μL of the PCR product is mixed with 1.0 μL of salt solution and 1.0 μL of TOPO T/A Vector and the final volume

is made to 6.0 μL with sterile MQ. Then incubate the reaction mix at room temperature for 30 min and transfer it to ice.

3. Transform 2 μL of the above mix with the help of chemically competent cells supplied along with the Kit. Incubate the mix on ice for 5 min followed by a heat shock at 42°C for 30 s.
4. Again chill on ice and add 250 μL of the SOC medium. The cells are then grown at 37°C for 1 h, after which 50 μL of the transformed product is plated onto one IXA (IPTG, X-gal and Ampicillin) selective plates with the help of a spreader. Similarly spread the transformed product on other plates.
5. Incubate the plates at 37°C for 12–16 h.

3.2.9. Sequencing of Positive Recombinants and Primer Design

1. From the blue and white colonies, pick and grid the white colonies with the help of a tooth pick and grid on a fresh IXA plate.
2. To select the recombinants (*see* **Note 20**) carry out colony PCR **(Fig. 3)** or restriction digestion of the plasmid DNA isolated using the method described by Sambrook et al. *(25)*.
3. Isolate the plasmid DNA of the positive recombinants (with insert) using the alkaline lysis method of Sambrook et al. *(25)*.
4. Obtain the nucleotide sequence of the insert by sequencing in the ABI3700 Prism automated DNA sequencer using the Big Dye Terminator reaction kit (*see* **Note 21**).
5. Analyze the sequences and remove the vector sequences and the adapter sequences using the VecScreen program (http://www.ncbi.nlm.nih.gov/VecScreen/VecScreen.html).
6. Select unique sequences by discarding the redundant clones using the CAP3 (http://pbil.univ-lyon1.fr/cap3.php) *(26)* program and identify the microsatellite containing clones either manually or by using various microsatellite identifying programs like: Tandem Repeats Finder (TRF) *(27)* and TROLL (Tandem Repeat Occurrence Locator) *(28)*.
7. After selecting the unique clones containing the microsatellites, design primers complementary to the sequence flanking the microsatellites, using the PRIMER 3 software (http://frodo.wi.mit.edu/) *(29)*.

3.0 kb
2.0 kb
1.0 kb

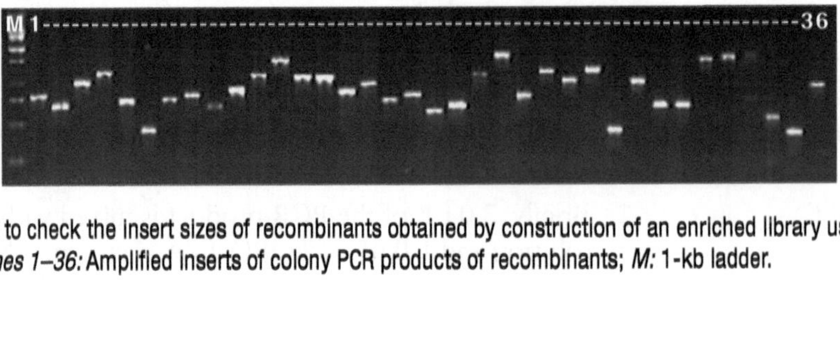

Fig. 3. Colony PCR to check the insert sizes of recombinants obtained by construction of an enriched library using biotinylated probe. *Lanes 1–36:* Amplified inserts of colony PCR products of recombinants; *M:* 1-kb ladder.

4. Notes

1. The number of PCR cycles may be reduced in order to avoid high redundancy of clones. Hence optimization of the number of cycles may have to be done.
2. Annealing temperature needs to be optimized. High temperature (high stringency) will hinder primer binding and amplification whereas low temperature (low stringency) will permit slippage of primer to the 3′-end. Therefore the annealing stringency has to be high enough to allow amplification to occur only when one or more of the five blocking bases pairs with the template.
3. An extension of 20–30 min at 72°C is necessary to ensure 'A' tailing of the PCR amplified fragments. This 'A' tailing helps in cloning of these fragments in a 'T' tailed vector.
4. To avoid dilution of the purified products, always elute in smaller volumes and preferably twice (in separate collection tubes) to prevent any loss.
5. Spread the total transformed cells by pelleting down at 12,000 rpm for 30 s. Decant the extra liquid and resuspend the cells in a small amount (100 μL) of LB. This can then be spread on the IXA plates to obtain blue and white colonies.
6. The IXA plates can be incubated at 4°C for 1–1.30 h to develop the blue color.
7. The presence of terminal microsatellites containing more than six repeats confirms that anchoring has been successful.
8. The choice of restriction enzyme sensitive to methylation should be preferred in order to increase the abundance of microsatellite sequences in the starting material.
9. As the average size of fragments obtained by restriction enzyme depends upon the genome composition, several authors have proposed to overcome this limitation by using multiple restriction enzymes to digest genomic DNA.
10. 4 bp cutters are preferred for digestion in comparison to the 6-bp cutters as they generate maximum number of fragments in the preferred size range to facilitate cloning in plasmid vector.
11. Choice of restriction enzyme will depend on the GC content of the nuclear DNA (i.e., for *Mse* I is used for targeting AT rich regions, *Taq* I, *Rsa* I and *Sau3A* I as intermediates and *Hae* III for GC rich regions).
12. A size selection step can be included to control the size range of digested fragments. This can be carried out after the

digestion *(8)* or after the ligation *(7)*. In the present protocol we omitted the size selection step after digestion or ligation because of the fact that small fragments are preferentially lost during selective hybridization.

13. The adapters should be phosphorylated to prevent formation of nicks in the DNA strands.
14. PCR amplification is carried out to obtain a sufficient amount of DNA for selective hybridization, however care must be taken to avoid over amplification of digested genomic DNA, which may lead to overabundance of some fragments.
15. The additional sequence attached with the biotin labeled primer is to separate the repeat sequence from the biotinylated part of the oligonucleotide. This favors efficient hybridization of the DNA fragments with the microsatellites and that of biotin with the streptavidin by decreasing the steric hindrance.
16. The size of the synthetic microsatellite oligonucleotides used as probes should be long enough to capture the genomic microsatellites.
17. Choose a microsatellite probe that is well distributed throughout the genome. Some of the microsatellite probes are also chosen because of the technical ease of their hybridization. GA/GT probes are easier to hybridize compared to the AT or GC probes, which are difficult to hybridize owing to self-complementarity and low annealing temperatures. Further optimization of the method includes usage of multiple probes in the hybridization step which seems to increase the overall enrichment efficiency.
18. Each wash entailed resuspension of the beads in 0.5X SSC followed by the use of a magnetic stand (Promega) to draw the beads aside allowing for removal and disposal of the wash buffer. Remove the supernatant slowly using a pipet without moving the tube or touching the beads. Further, remove the tube from the stand and add buffer as required to resuspend the beads. Slowly resuspend the beads with a pipet if required.
19. After the hybridization, washings should be done carefully, as the conditions applied at this step affect the level of enrichment for microsatellite sequences in each library.
20. PCR screening of the recombinant clones could also be carried out using two primers-one specific to the vector and the second a repeat containing oligonucleotide.
21. Alternatively the PCR product can also be sequenced directly after the PCR purification. It is also advisable to add DMSO (5% final concentration) to the sequencing reaction mix in order to prevent poor sequencing results due to presence of repeats.

Acknowledgments

The authors thankfully acknowledge NIPGR (National Institute of Plant Genome Research), an autonomous institute fully funded by the Department of Biotechnology, Government of India, for providing financial support for this work. The fellowship provided to BS by Council for Scientific and Industrial Research (CSIR), India is duly acknowledged.

References

1. Waugh, R., and Powell, W. (1992). Using RAPD markers for crop improvement. *Trends in Biotechnol.* **10**:186–191.
2. Panaud, O., Chen, X., and McCouch, S.R. (1996). Development of microsatellite markers and characterization of simple sequence length polymorphism (SSLP) in rice (Oryza sativa L.). *Mol. Gen. Genet.* **252**:597–607.
3. Chen, X., Temnykh, S., Xu, Y., Cho, Y.G., and McCouch, S.R. (1997). Development of a microsatellite framework providing genome-wide coverage in rice (Oryza sativa L.). *Theor. Appl. Genet.* **95**:553–567.
4. Paniego, N., Echaide, M., Marianne, M., Ferrandez, L., Tolales, S., Faccio, P., Fuxan, I., Carrera, M., Zandomeni, R., Suarez, E., and Hopp, H.E. (2002). Microsatellite isolation and characterization in sunflower (*Helianthus annuus* L). *Genome* **45**:34–43.
5. Suwabe, K., Iketani, H., Nunome, T., Kage, T., and Hirai, M. (2002). Isolation and characterization of microsatellites in *Brassica rapa* L. *Theor. Appl. Genet.* **104**:1092–1098.
6. Lee, J.M., Nahm, S.H., Kim, Y.M., and Kim, B.D. (2004). Characterization and molecular genetic mapping of microsatellite loci in pepper. *Theor. Appl. Genet.* **108**:619–627.
7. Kandpal, R.P., Kandpal, G., and Weissman, S.M. (1994). Construction of libraries enriched for sequence repeats and jumping clones, and hybridization selection for region-specific markers. *Proc Natl Acad Sci USA* **91**:88–92.
8. Kijas, J.M.H., Fowler, J.C.S., Garbett, C.A., and Thomas, M.R. (1994). Enrichment of microsatellites from the citrus genome using biotinylated oligonucleotide sequences bound to streptavidin-coated magnetic particles. *BioTechniques* **16**:657–662.
9. Karagyozov, L., Kalcheva, I.D., and Chapman, V.M. (1993). Construction of random small-insert genomic libraries highly enriched for simple sequence repeats. *Nucleic Acids Res.* **21**:3911–3912.
10. Armour, J.A., Neumann, R., Gobert, S., and Jeffrey, A.J. (1994). Isolation of human simple repeat loci by hybridization selection. *Hum. Mol. Genet.* **3**:599–565.
11. Edwards, K.J., Barker, J.H.A., Daly, A., Jones, C., and Karp, A. (1996). Microsatellite libraries enriched for several microsatellite sequences in plants. *Biotechniques* **20**:758.
12. Rallo, P., Dorado, G., and Martin, A. (2000). Development of simple sequence repeats (SSRs) in the olive tree (*Olea europea* L.). *Theor. Appl. Genet.* **101**:984–989.
13. Alghanim, H.J., and Almirall, J.R. (2003). Development of microsatellite markers in Cannabis sativa for DNA typing and genetic relatedness analyses. *Anal. Bioanal. Chem.* **376**:1225–33.
14. Lowe, A.J., Moule, C., Trick, M., and Edwards, K.J. (2004). Efficient large-scale development of microsatellites for marker and mapping applications in *Brassica* crop species. *Theor. Appl. Genet.* **108**:1103–12.
15. Stajner, N., Jakse, J., Kozjak, P., and Javornik, B. (2005). The isolation and characterization of microsatellites in hop (*Humulus lupulus* L.). *Plant Sci.* **168**:213–221.
16. Sethy, N.K., Shokeen, B., Edwards, K.J., and Bhatia, S. (2006). Development of microsatellite markers and analysis of intra-specific genetic variability in chickpea (Cicer arietinum L.). *Theor. Appl. Genet.* **112**:1416–1428.
17. Metais, I., Hamon, B., Jalouzot, R., and Peltier, D. (2002). Structure and level of genetic diversity in various bean types evidenced with microsatellite markers isolated from a genomic enriched library. *Theor. Appl. Genet.* **104**:1346–1352.
18. Viruel, M.A., and Hormaza, J.I. (2004). Development, characterization and variability analysis of microsatellites in lychee (Litchi chinensis Sonn., Sapindaceae). *Theor. Appl. Genet.* **108**:896–902.

19. Fisher, P.J., Gardner, R.C., and Richardson, T.E. (1996). Single locus microsatellite isolated using 5′ anchored PCR. *Nucl. Acids Res.* **24**:4369–4371.

20. Bornet, B., Muller, C., Paulus, F., and Branchard, M. (2002). Highly informative nature of inter simple sequence repeat (ISSR) sequences amplified using tri- and tetra-nucleotide primers from DNA of cauliflower (*Brassica oleracea* var. botrytis L.). *Genome* **45**:890–896.

21. Kumar, S.V., Tan, S.G., Quah, S.C., and Yusoff, K. (2002). Isolation and characterization of seven tetranucleotide microsatellite loci in mungbean (*Vigna radiata*). *Mol. Ecol. Notes* **2**:293–295.

22. Choudhary, S., Sethy, N.K., Shokeen, B., and Bhatia, S. (2006). Development of sequence-tagged microsatellites site markers for chickpea (*Cicer arietinum* L.). *Mol. Ecol. Notes* **6**:93–95

23. Sethy, N.K., Choudhary, S., Shokeen, B., and Bhatia, S. (2006). Identification of microsatellite markers from Cicer reticulatum: Molecular variation and phylogenetic analysis. *Theor. Appl. Genet.* **112**:347–357.

24. Malmberg, R., Messing, J., and Sussex, I. (1985). Molecular Biology of Plants: A laboratory course manual, Cold Spring Harbor Laboratory Press, Cold Spring Harbor, New York, NY.

25. Sambrook, J., Fritsch, E.F., and Maniatis, T. (1989). Molecular cloning: A laboratory manual, 2nd Edn. Cold Spring Harbor Laboratory Press, New York, NY.

26. Huang, X., and Madan, A. (1999). CAP3: A DNA sequence assembly program. *Genome Res.* **9**:868–877.

27. Benson, G. (1999). Tandem repeats finder: a program to analyze DNA sequences. *Nucl. Acids Res.* **27**:573–580.

28. Castelo, A.T., Martins, W., and Gao, G.R. (2002). TROLL – Tandem repeat occurrence locator. *Bioinformatics* **18**:634–636.

29. Rozen, S., and Skaletsky, H.J. (1997). *Primer 3.* Code available at http://www.genome.wi.mit.edu/genome_software/other/primer3.html.

Part IV

Biotransformation, Bioreactors, and Metabolomics

Chapter 24

Production of Cinnamyl Glycosides in Compact Callus Aggregate Cultures of *Rhodiola rosea* Through Biotransformation of Cinnamyl Alcohol

Zsuzsanna György and Anja Hohtola

Summary

Rhodiola rosea is a multipurpose medicinal plant with adaptogenic properties: it increases the body's nonspecific resistance and normalizes body functions. The commercial interest for roseroot-based products has quickly increased worldwide. Nearly all raw-materials originate from natural populations. As a result of the intensive collection the species has become endangered. Production of the pharmaceutical compounds from the medicinal plants in cell cultures is an alternative to field cultivation. The present protocol describes the establishment of compact callus aggregate culture of *Rhodiola rosea* and the production of cinnamyl alcohol glycosides through biotransformation.

Key words: *Rhodiola rosea*, Cinnamyl alcohol, Compact callus aggregate, Biotransformation, Cinnamyl glycoside

1. Introduction

Rhodiola rosea (roseroot) is a medicinal plant of which the most important secondary metabolites – salidroside and the cinnamyl alcohol glycosides rosin, rosavin, rosarin – are accumulated in the rhizome of the plant. There are more than 200 *Rhodiola* species but cinnamyl alcohol glycosides are found only in *R. rosea*. Other *Rhodiola* species contain only salidroside *(1)*. This plant has been described as an adaptogenic herb because it offers generalized resistance to physical, chemical, and biological stressors *(2)*.

S. Mohan Jain and Praveen K. Saxena (eds.), *Methods in Molecular Biology, Protocols for* In Vitro *Cultures and Secondary Metabolite Analysis of Aromatic and Medicinal Plants, vol. 547*
© Humana Press, a part of Springer Science+Business Media, LLC 2009
DOI: 10.1007/978-1-60327-287-2_24

Furthermore, animal and human studies revealed anticancer, antioxidant, immune enhancing, and sexual stimulating effects *(3)*.

Field cultivation of roseroot takes several years to obtain a satisfactory content of the pharmaceutical compounds *(4)*. The alternative source of producing these compounds is cell cultures. However, in callus and in vitro grown plants no salidroside or rosavin was found *(5)*. The phenylpropanoids are probably transformed to *p*-hydroxyrosine (triandrin) by phenylhydroxylases *(4)*. Furmanova et al. added 2.5 mM trans-cinnamyl alcohol to the media and found that 90% of it was transformed into several products, but only rosavin was identified *(6)*. In case of *Rhodiola sachalinensis* compact callus aggregates (CCA) are used instead of cell suspensions for producing salidroside *(7)*.

This chapter describes the establishment of compact callus aggregate cultures of roseroot and the examination of the possibility of stimulating the production of cinnamyl glycosides through the biotransformation of cinnamyl alcohol.

2. Materials

2.1. Callus Induction from Seeds

2.1.1. Stock Preparation

1. The composition of Murashige and Skoog (MS) *(8)* medium is provided in **Table 1**.
2. Culture vessels: Glass jars and polypropylene Petri dishes.
3. For the preparation of stock solutions, analytical grade chemicals and distilled water are used. Prepare separate stock solutions for macronutrients, micronutrients, iron, vitamins, and *myo*-inositol by dissolving the weighed constituent chemicals individually in 500-mL or 1-L bottles with screw-on caps.
4. For half strength MS media, take half of the prescribed quantities from each chemical.
5. Growth regulator stocks: 6-benzylaminopurine (BA) and 1-naphthaleneacetic acid (NAA).

2.1.2. Inoculation

1. Laminar airflow cabinet.
2. Forceps, scalpels, sterile blades, burner, autoclaved plates, 4% sodium hypochlorite and 70% rectified spirit.

2.1.3. Culture Conditions

1. Culture room conditions are: 26 ± 2°C; 16 h photoperiod; light intensity of approximately 45 μM/m^2 s provided by cool, white fluorescent tubes.

Table 1
Composition of the Murashige and Skoog Medium

Constituents	Amount (mg/L medium)
Major salts	
NH_4NO_3 (Merck, Germany)	1650
KNO_3 (Merck, Germany)	1900
$CaCl_2 \cdot 2H_2O$ (Merck, Germany)	440
$MgSO_4 \cdot 7H_2O$ (Merck, Germany)	370
KH_2PO_4 (Merck, Germany)	170
Minor salts	
H_3BO_3 (Merck, Germany)	6.2
$MnSO_4$ (Sigma, U.S.A.)	22.3
$ZnSO_4 \cdot 7H_2O$ (Sigma, U.S.A.)	8.6
KI (Merck, Germany)	0.83
$Na_2MoO_4 \cdot 2H_2O$ (Merck, Germany)	0.25
$CuSO_4 \cdot 5H_2O$ (Merck, Germany)	0.025
$CoCl_2 \cdot 6H_2O$ (Merck, Germany)	0.025
Iron	
Te Na EDTA (Fluka, Switzerland)	36,7
Organics	
Glycine (Merck, Germany)	2.0
Pyridoxine (Merck, Germany)	0.5
Nicotinic acid (Sigma, U.S.A.)	0.5
Thiamine HCl (Sigma, U.S.A.)	0.1
Myoinositol (Sigma, U.S.A.)	100
Sucrose (Fluka, Switzerland)	30,000
Concentration of gelling agent	
Agar (Oxaid, England)	8000

2.2. Establishing and Culturing CCA Cultures, Experiments with CCAs

2.2.1. Stock Preparation

See **Subheading 2.1.1**, in addition cinnamyl alcohol stock is needed.

2.2.2. Inoculation

See **Subheading 2.1.2**.

2.2.3. Culture Conditions

See **Subheading 2.1.3**. Cultures are incubated on a shaker set at 135 rpm.

3. Methods

3.1. Stock Solutions for Media Preparation

1. To avoid precipitation, store all stock solutions at 4°C and for the same reason, stock solutions of plant growth regulators should be stored at –20°C in aliquots.
2. To prepare BA stock, weigh the hormone and dissolve it in a few drops of 1 M NaOH . Make-up the desired volume with distilled water. NAA stock is prepared by dissolving the weighed quantity in minimum amount of ethanol followed by making up the volume with distilled water.

3.2. Media Preparation

1. Measure the stock solutions according to their dilution factors and in case of solid media add agar (0.8%). Make-up the desired volume with distilled water.
2. Adjust the pH of the media to 5.8 with the addition of 1 M NaOH and 1 M HCl.
3. Autoclave the media at 121°C and 1.05 kg/cm^2 pressure for 20 min.
4. In case of solid media pour 20 mL culture medium into sterile Petri dishes or 50 mL into 300-mL ordinary glass jars. In case of liquid media pour 100 mL into 250-mL Erlenmeyer flasks or 20 mL into 50-mL Erlenmeyer flasks. Use tinfoil to cap the culture vessels and seal with Parafilm.

3.3. Inoculation

1. Sterilize forceps and scalpels by controlled flaming with rectified spirit, 95% ethanol, inside the laminar airflow cabinet.

3.4. Initiation of Aseptic Cultures of *Rhodiola rosea*

1. Sterilize seeds in 70% ethanol in a sterile flask by shaking for 30 s. Discard alcohol and add 4% sodium hypochlorite and leave it for 10 min shaking once in a while. Discard sodium hypochlorite and wash seeds four times in sterile distilled water.

2. Germinate seeds on half strength MS media (*see* **Notes 1** and **2**).
3. Callus is induced from the leaves of the in vitro grown plants (*see* **Note 3**).

3.5. Initiation and Maintenance of Callus Cultures of Rhodiola rosea

1. Excise leaves from the in vitro grown shoots. Cut off the leaf edges on a sterile Petri dish and place the wounded leaves onto MS media containing 1.5 mg/L BA and 0.5 mg/L NAA.
2. Callus develops on the wounded edges in 3–4 wk. Transfer the callus with sterile forceps onto fresh media of the same composition. After 4 wk the callus should be at least twice as large as in the beginning.

3.6. Initiation and Maintenance of CCA Cultures

1. Separate the callus from media pieces on a sterile Petri dish with sterile forceps and knives and gently brake with the forceps (*see* **Note 4**).
2. About 10 g fresh weight of these small pieces are transferred into 250-mL Erlenmeyer flasks filled with 100 mL liquid MS medium containing 0.5 mg/L BA and 1 mg/L NAA, heat-sterilize the opening of the flask, close with sterile tinfoil and seal with Parafilm. Incubate the flasks on a shaker set at 135 rpm.
3. Subcultures are carried out in every 8–10 d by decanting all medium from the flask and adding fresh medium to the cultures.
4. Compact callus aggregates start to form in the liquid culture 3–5 wk later. The established CCA culture is composed of green or light green, spherical, smooth surfaced callus aggregates (**Fig. 1**). Cavities are formed in the centre of large aggregates, diameter about 1 cm.

Fig. 1. Compact callus aggregates of *Rhodiola rosea*. Bar = 1 cm.

3.7. Growth Kinetics Experiment

1. Inoculate 1 g fresh weight CCA into 30 Erlenmeyer flasks of 50 mL containing 20 mL MS medium supplemented with 0.5 mg/L BA and 1 mg/L NAA (*see* **Note 5**).
2. Three flasks are harvested every 3 d for 30 d to determinate fresh and dry weights and to carry out their chemical analysis (**Fig. 2**).

3.8. Fresh and Dry Weight Measurements

1. The fresh weight is measured after vacuum filtration. Cut filter paper pieces fitting in the vacuum filtrate. Wet a filter paper piece and vacuum filtrate it, then place it into the oven at 40°C for 24 h. Place the paper in desiccator until the dry weight measurement. Weigh it on the scale. Wet it again, vacuum filtrate it, weigh it on the scale. Put the paper back to the vacuum filtrate, pour over the content of a flask, vacuum filtrate it, and weigh it on the scale. Determine the fresh weight of the callus after subtracting the weight of the wet filter paper. After fresh weight determination, put the filter paper with the callus into the oven at 40°C for 24 h.
2. When the samples are dried, place them into desiccator until the dry weight measurement. While measuring the dry weight try to work fast. Determine the dry weight of the callus after subtracting the weight of the dry filter paper.
3. The samples are stored in brown paper bags in a desiccator until the chemical analysis is performed.
4. Media samples are taken after the vacuum filtration and are kept at -20°C until the chemical analysis.
5. Viability is determined on the basis of the callus color (green, light green, white or brown) and the results of the fresh weight measurements.

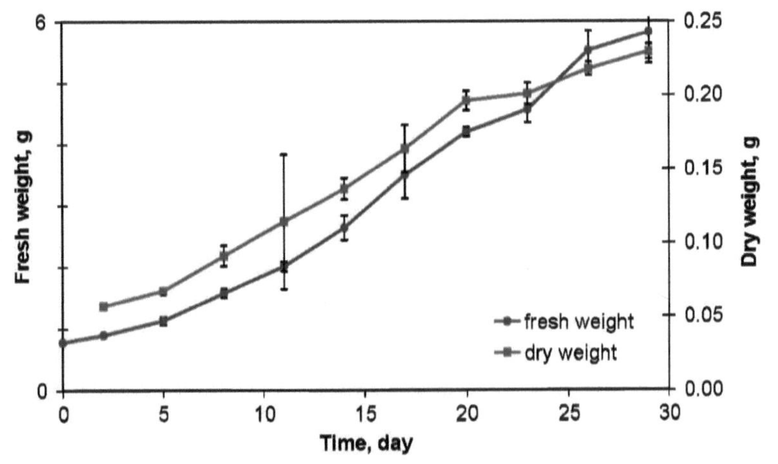

Fig. 2. Growth curve of roseroot compact callus aggregates.

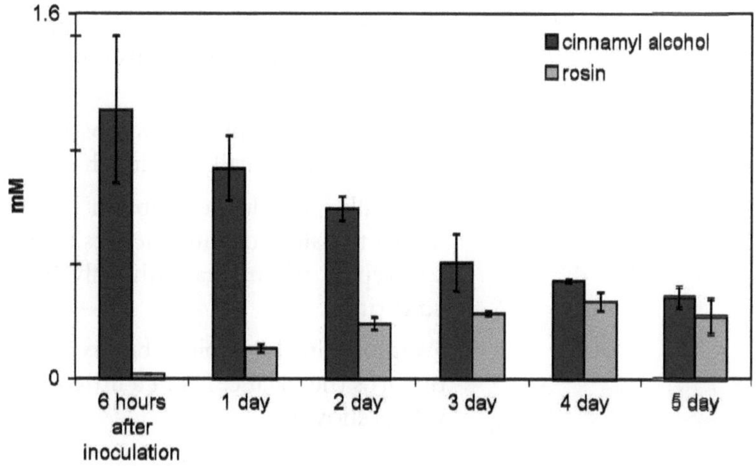

Fig. 3. The concentration of cinnamyl alcohol and biotransformation product rosin in the culture media 6 h and 1, 2, 3, 4, 5 d after the inoculation.

3.9. Determination of the Optimal Concentration of Cinnamyl Alcohol

1. Inoculate 1 g fresh weight CCA into 50-mL Erlenmeyer flasks containing 20 mL MS medium amended with 0.5 mg/L BA and 1 mg/L NAA (*see* **Note 5**).
2. Cinnamyl alcohol is added to the media in final concentrations of 0, 0.05, 0.1, 0.5, 1.0, 1.5, 2, 3, 4, and 5 mM. The cinnamyl alcohol is dissolved in ethanol, 0.1 g in 1 mL EtOH, and water and sterilized through a 0.2 nm filter.
3. The CCAs are cultured in this media for 1 wk. The experiment is made in triplicate: three flasks of each composition are harvested for the determination of fresh and dry weights, viability and for the chemical analysis.

3.10. Cinnamyl Alcohol Consumption

1. Inoculate 2 g fresh weight CCAs in 100-mL Erlenmeyer flasks containing 33 mL MS medium supplemented with 0.5 mg/L BA, 1 mg/L NAA (*see* **Note 5**).
2. Add cinnamyl alcohol to the media at the final concentration of 2 mM in the previously described way.
3. Three flasks are harvested every day during 7 d to determine fresh and dry weights, viability and to carry out their chemical analysis (**Fig. 3**).

4. Notes

1. It is wise to place relatively few seeds per dish (10–20), because a single contaminated seed may contaminate all the rest in the dish.

2. Roseroot seeds germinate quite poorly and slowly and may take several weeks. It is advantageous to start germination in several batches.
3. Wait until a little plant develops at least with 6–8 leaves, as the leaves are very small, and it is hard to work with them.
4. The callus should be friable if it is regularly (in about every 4 wk) transferred onto the fresh media. If it is very hard, try to prepare the media with only 0.6% agar for the next subculturing.
5. Weighing in the laminar box is never accurate as the air flow disturbs the balance. Wait a little and consider the mean of the values shown.

References

1. Ramazanov, Z. and Mar Bernal Suarez, M. (1999). New secrets of effective natural stress and weight management using *Rhodiola rosea* and *Rhododendron caucasicum*. ATN/Safe Goods Publishing, East Canaan, CT, pp. 9–11
2. Kelly, G. S. (2001). *Rhodiola rosea*: a possible plant adaptogen. *Alternative Medicine Review* **6 (3)**: 293–302
3. Brown, R. P., Gerbarg, P. L. and Ramazanov, Z. (2002). *Rhodiola rosea*: a phytomedical overview. *HerbalGram* **56**: 40–52
4. Furmanova, M., Oledzka, H. and Michalska, M. (1995) XXIII *Rhodiola rosea* L. (roseroot): in vitro regeneration and the biological activity of roots. In: Bajaj, Y.P.S. (ed.) *Biotechnology in Agriculture and Forestry*, Vol. 33, Medicinal and Aromatic Plants VIII. Berlin and Heidelberg, Germany: Springer-Verlag, pp. 412–425
5. Kurkin, V.A., Zapesochnaya, G.G. and Dubichev, A.G. (1991). Phenylpropanoids of callus culture of *Rhodiola rosea*. *Khimiya prirodnykh soedinenii* **4**: 481–490
6. Furmanova, M., Hartwich, M., Alfermann, A.W., Kozminski, W. and Olejnik, M. (1999). Rosavin as a product of glycosylation by *Rhodiola rosea* (roseroot) cell cultures. *Plant Cell, Tissue Organ and Culture* **56**: 105–110
7. Jianfeng, X., Zhiguo, S., and Pusun, F. (1998). Suspension culture of compact callus aggregate of *Rhodiola sachalinensis* for improved salidroside production. *Enzyme and Microbial Technology* **23**: 20–27
8. Murashige, T. and Skoog, F. (1962). A revised medium for rapid growth and bioassays with tobacco tissue cultures. *Physiologia Plantarum* **15**: 473–497

Chapter 25

Spearmint Plantlet Culture System as a Means to Study Secondary Metabolism

Brent Tisserat, Mark Berhow, and Steven F. Vaughn

Summary

Spearmint has one major monoterpene, (−)-carvone, that constitutes up to 90% of all the monoterpenes present. Likewise, the major phenylpropanoid—rosmarinic acid—in spearmint accounts for up to 70% of the phenylpropanoids produced from the plant. These two compounds are each produced by separate distinct biosynthetic pathways which provide an excellent opportunity to study the influence of a wide number of environmental and chemical conditions on secondary metabolism and plant growth. The techniques presented in this chapter employ 1 g of fresh weight material for each secondary metabolite analyses. Analysis of single compounds obtained from the two distinct metabolic pathways simplifies the interpretation of the metabolic results allowing for direct correlations of culture factors on secondary metabolism.

Key words: Carvone, Rosmarinic acid, Plantlet culture, Natural plant products, Phenylpropanoids, Monoterpenoids

1. Introduction

Natural plant products can be generally divided into three major biosynthetic groups: the terpenoids (about 25,000), the alkaloids (about 12,000), and phenylpropanoids and allied phenolic compounds (about 8,000) *(1)*. Spearmint plantlets grown in vitro constitutively accumulate a significant amount of two compounds, carvone (a monoterpene) *(2)*, and rosmarinic acid (a substituted phenolic acid) *(3)*. (-)-Carvone constitutes 90% of the monoterpenes produced in the cultured spearmint plants whereas rosmarinic acid constitutes up to 70% of the phenolics produced in the plant. Cultured spearmint plants also accumulate varying

S. Mohan Jain and Praveen K. Saxena (eds.), *Methods in Molecular Biology, Protocols for* In Vitro *Cultures and Secondary Metabolite Analysis of Aromatic and Medicinal Plants, vol. 547*
© Humana Press, a part of Springer Science+Business Media, LLC 2009
DOI: 10.1007/978-1-60327-287-2_25

levels of several flavones and flavonols *(4, 5)*. The production of these two compounds from two distinctly different secondary pathways makes the spearmint plantlet culture a very useful experimental system. The accumulation of secondary metabolites in mint plantlets is very responsive to cultural factors *(6, 7)*. The occurrence of dominate metabolites simplifies the interpretation of the metabolic results allowing for a direct correlation of the cultural factors on essential oil production.

Mint plantlets are miniature plants as they have the same morphologic features and chemical compositions as soil grown plants. Monoterpenes, being volatile essential oils, are manufactured exclusively in the green shoot foliage (especially in the leaves) and are stored in the oil glands. Monoterpenes are not found in roots or in callus, although some altered terpenes may occur within the callus. Rosmarinic acid, in contrast, is found in all plant organs *(8)*, and has been reported to be particularly high in hairy roots *(9)*, cell suspensions, and undifferentiated callus *(10, 11)*. Within the sterile environment we can study the influence of various environmental and nutritional factors on the growth of plantlets (i.e., primary metabolism) and secondary metabolism. Various forms of chromatography have been used to separate extracts prepared from plants *(12, 13)*. The distinctive light absorption characteristics were used to identify and quantify plant phenolics and terpenoids *(13)*. Current qualitative and quantitative work is done using gas phase-driven chromatography (GC) for volatile compounds, such as the monoterpenoids, and high-performance liquid phase chromatography (HPLC) for nonvolatile compounds. Natural plant products are typically extracted in a suitable solvent or mixture of solvents and separated, identified, and quantified in one operation by coupling the HPLC or GC to a detecting system, such as a light absorbance detector (ultraviolet [UV], visible, or fluorescence), refractive index (RI), evaporative light scattering detector, a flame ionization detector (FID), or a spectrometer capable of detecting different mass ions (MS).

GC-FID is a standard and proven method for the detection and quantification of volatile monoterpenes such as carvone *(13)*. HPLC reverse phase methodology using UV-VIS absorption detectors is the workhorse for phytochemical analysis with compounds that have good chromophores (i.e., conjugated double bond systems and/or phenolic ring structures). Reproducible HPLC methodology has been used for a variety of phenylpropanoid compounds such as the flavonoids *(14, 15)*. This methodology can also be easily adapted for substituted phenolic acids such as rosmarinic acid. Low cost mass-spectrometry (MS) systems and their associated computer software packages now allow for the characterization of mass ion patterns produced by electron-impact ionization or soft ionization sources with quadrople, ion trap, or time of flight MS system *(16)*.

The problem of reproducible results from sample to sample and lab to lab still depends on the key steps of sample preparation and extraction. To maximize analyte extraction from the mark, the sample should be dry and ground to as fine a powder as possible. Depending on the time frame and the number of samples involved, extraction may be made exhaustive by either extracting the sample multiple times or by an increased solvent to solid ratio with time and provide enough heat to maximize the solubilization of the phytochemicals of interest. This is often a fine line between the amount of time and heat needed to maximize the efficiency of extraction or chemical alteration of the native state of the phytochemicals. This situation should be properly examined prior to the implementation of any new analytical procedure. We have found that most phenolic compounds are fairly stable when extracted for 1–2 d in methanol:dimethyl sulfoxide , with an extraction efficiency above 90% using a single step extraction method where the ratio of solvent to solid sample is greater than 10–1.

2. Materials

2.1. Plantlet Culture

1. Murashige and Skoog (Medium (MS) (Sigma; St. Louis, MO) supplemented with: 0.4 mg/L thiamine HCl, 100 mg/L *myo*-inositol and 30,000 mg/L sucrose. Adjust agar MS pH to 5.7 ± 0.1 with 0.1 N HCl or NaOH before adding 8 g/L agar (Difco Laboratories; Detroit, MI). Liquid MS pH was adjusted to 5.0 ± 0.1.
2. Magenta containers: GA-7 polycarbonate box, 77 mm × 77 mm × 97 mm H and GA-7-3 polycarbonate box, 77 mm × 77 mm × 77 mm H and polypropylene couplers to join boxes (Sigma).
3. 100-mL Glass baby food jar, 62 mm diam × 66 mm H (Sigma; St. Louis, MO).
4. Polypropylene translucent Magenta B-cap drilled with a 28-mm diam hole and Magenta 2-way cap (Sigma) drilled in the center top with a 5-mm diam hole.
5. Stainless steel breather/vent filter, 1/8" NPT, 11.11 mm H × 11.11 mm diam; 100 μm filtration (McMaster-Carr; Chicago, IL).
6. Polypropylene washers and miniature barbed fittings, 1/8" in. NPT threaded (McMaster-Carr).

2.2. Carvone Analysis

1. Dichloromethane.

2. Jars wrapped with aluminium foil seals.
3. 0.45-μm nylon 66 filters.
4. Rotoevaporation apparatus.
5. Gas chromatograph equipped with a flame ionization detector (GC-FID).
6. Gas chromatograph equipped with electron impact mass selective detector (GC-MS).
7. Fused silica HP-5MS capillary column (0.25 μm film thickness, 30 mL × 0.25 mm ID).
8. He carrier gas.

2.3. For Rosmarinic Acid Analysis

1. Mortar and pestle.
2. 30 mesh metal sieve.
3. 15 × 45-mm glass vial and locking cap.
4. 2 mL methanol:dimethyl sulfoxide (DMSO) (1:1).
5. Sonic water bath.
6. 0.45-μm nylon 66 filter.
7. A gradient-capable HPLC system with autoinjector, solvent mixer, column oven, and photodiode array detector.
8. 250 × 4.6 ODS-5, 5 micron reverse phase C-18 column).
9. Pure RA sample (Chromadex; Santa Ana, CA).
10. Electrospray ionization source, ion trap mass spectrometer with a gradient capable HPLC system (autoinjector, pump, degasser, and photodiode array or UV/VIS detector).
11. Flowmeter (Cole-Parmer, Chicago, IL)

3. Methods

3.1. Plantlet Culture Methodology

1. Prepare mint plantlet stocks by culturing a single 3-cm long shoot per 25 × 150 mm culture tube on agar medium for 2 wk in order to initiate roots (*see* **Note 1**).
2. Preparation of the plantlet culture system involves the interlocking of two magenta containers via a polypropylene coupler. A separate media container, baby food jar, is positioned within the bottom Magenta box container. Modify a B-cap closure with centrally positioned 28-mm diam drilled hole and fitted with a 25 mm diam translucent 2-way closure predrilled with a 5-mm hole **(Fig. 1)**.

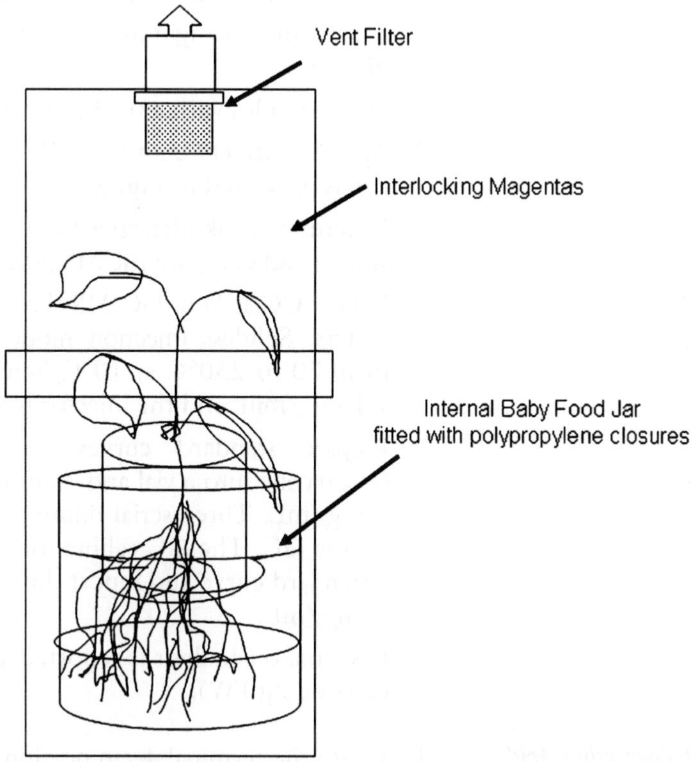

Fig. 1. The plantlet culture chamber. Plantlet is positioned in baby food jar holes drilled in the polypropylene closures within two interlocking Magenta containers. Root portion of plantlet is completely submerged in media and separated by a polypropylene barrier from the shoot portion which is erect in the atmosphere. Atmospheric alteration is achieved by administrations of air treatments through the vent filter.

3. Transfer plantlets into Magenta containers containing an internal baby food jar with the root portion immersed in 47 mL of nutrient medium.
4. Physically separate media within the internal baby food jars from the plantlet foliage entirely. Any uptake of media occurs through the immersed roots. In addition, media is shielded from the atmospheric portion by polypropylene closure.
5. Treat atmospheric portion of the container with air applications administered through a stainless steel breather/filter fitted in the top magenta container (**Fig. 1**).
6. For demonstrative purposes of employing this system, give plantlets 0 and 16 times air applications for 4 wk. Air is generated by an aquarium pump, and monitor and maintain with a flowmeter and apply equally space intervals of 30 min at 100 mL/min (*see* **Note 2**).

3.2. Carvone Analysis

1. For analysis, excise the terminal 4-cm portion of 3–5 mint plantlets and mix together, and then sample by weighing out 1 g and immersing it in 15 mL dichloromethane for 72 h (*see* **Note 3**).
2. Filter samples with a 0.45 μm nylon 66 filter.
3. Analyze extracts on a GC-FID. An example GC-FID trace run is presented in **Fig. 2.**
4. Determine peak identities by running samples on a GC-MS system and compare the MS peaks with standards.
5. Follow GC-FID or GC-MS chromatography operation parameters: Splitless injection mode temperature programmed from 70 to 250°C at 10°C/min; He carrier gas flow rate 1.1 mL/min; and the injector temperature set at 250°C.
6. Prepare standard curves monthly by weighing pure (-)-carvone into a vial and diluting with dichloromethane to 1 mg/mL. Three serial dilutions are prepared and injected on the GC. The GC calibration software is used to prepare a standard curve and report the unknown carvone peak data as mg/mL.
7. Essential oil analysis is reported as carvone/ treatment (mg carvone /g FW).

3.3. Rosmarinic Acid Analysis

1. Excise the terminal 4-cm portion of 3–5 mint plantlets, mix together, and then weigh 1 g sample.

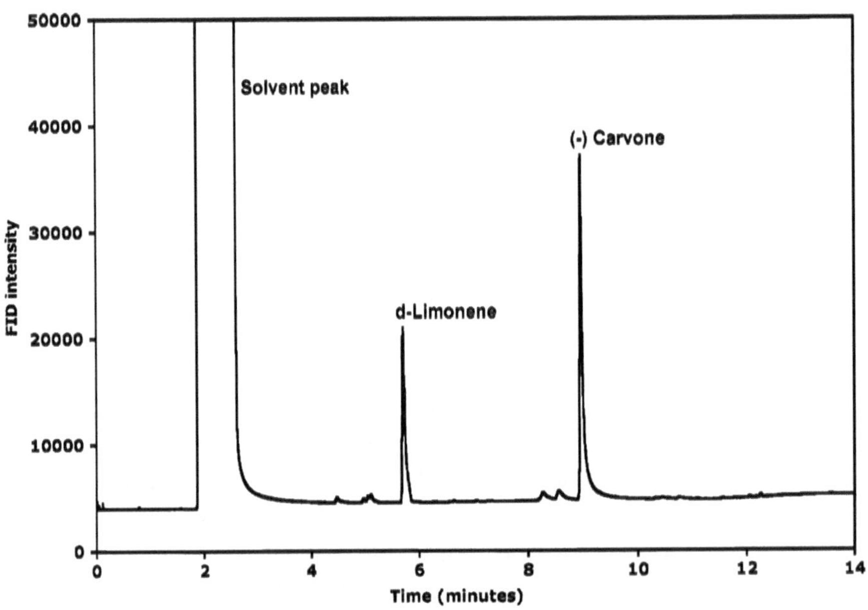

Fig. 2. GC-FID trace of a spearmint extract injection.

2. Plant materials are then freeze-dried for at least 24 h (*see* **Note 4**).
3. Grind samples into a fine powder with a mortar and pestle, pass through a number 30 mesh sieve to remove stem materials.
4. For analysis, samples typically 0.1 g are placed in a glass vial with 2 mL DMSO:methanol (1:1). Cap vials and wrap with a sealing tape.
5. Samples are sonicated at 42 Hz for 15 min at room temperature.
6. Allow samples to stand overnight at room temperature.
7. Remove aliquot samples from the vial and then filter through a 0.45-μm nylon 66 filter.
8. Samples are run a dual pump HPLC system with an autoinjector, and with the column oven set to 40°C.
9. Monitor peaks using a photodiode array detector running under a chromatography software control program.
10. Initial solvent conditions are 20% methanol and 80% 0.01 M phosphoric acid in water.
11. Effluent is monitored at 285 nm; after injections of 15 μL, the column is held at the initial conditions for two min, then developed to 100% methanol in a linear gradient over 55 min.
12. Standard curves are prepared from a series of dilutions of freshly prepared pure RA dissolved in methanol, 1 mg/mL, 2.78 nM/μL, and a series of serial dilutions are made. After running the standards at the same injection volume as the unknowns, the extinction coefficient based on peak absorbance area verses concentration in nM is determined. Alternatively, use the calibration program in the chromatography software only if reliable RA retention times can be obtained over time.
13. The spectra and retention time of the RA in the mint extracts is identical to that of the standard. An example trace HPLC injection is presented in **Fig. 3**.
14. After identification, peaks are integrated to obtain the area under the peak in mAbs units.
15. The presence of RA is confirmed by LC-MS by comparison of retention time and mass spectra.
16. Samples are run on ThermoFinnigan LCQ DECA XP Plus LC-MS system with a Surveyor HPLC system-autoinjector, pump, degasser, and PDA detector and a nitrogen generator all running under the Xcaliber 1.3 software system.

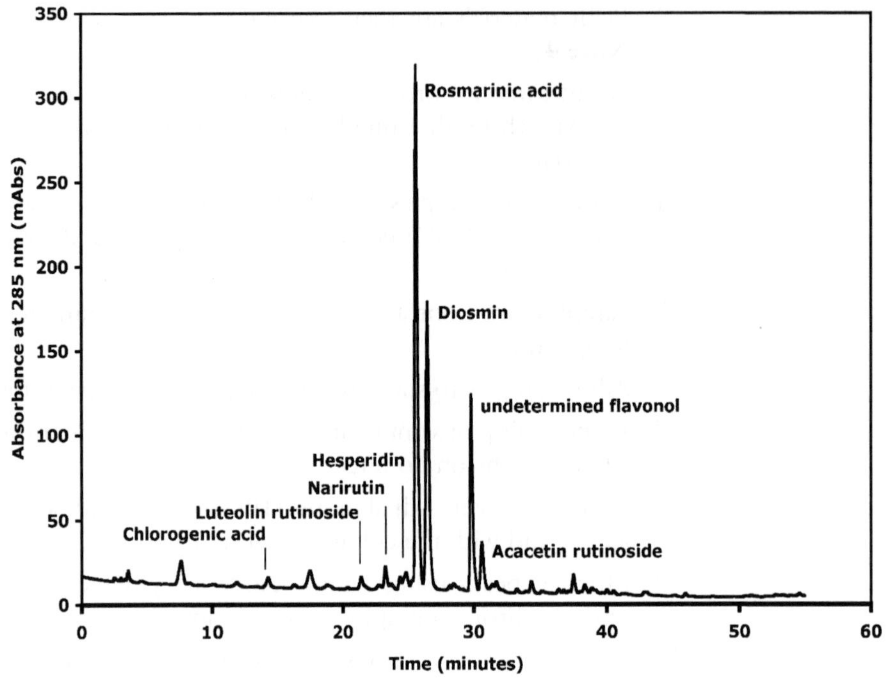

Fig. 3. Trace of HPLC injection of a spearmint extract monitored at 285 nm.

17. The MS is run with the ESI probe in the negative mode, with the source temperature set at 300°C and the needle gas set at 80 arbitrary units and the sheath gas set at 20 arbitrary units.
18. The MS is optimized for the detection of RA by using the autotune feature of the software while infusing a solution of RA standard in with the effluent of the column and tuning on 359 m/z $[M-H]^-$.
19. The initial HPLC conditions are 20% methanol and 0.25% acetic acid in water, at the flow rate 0.3 mL/min. The column is then developed to 100% methanol and 0.25% acetic acid over 50 min.
20. RA in selected samples is confirmed by comparison of retention and MS spectra of the sample RA peak to that of a standard (*see* **Note 5**).

4. Notes

1. These protocols can be adapted for many other plant materials and species.

2. For illustrative purposes, spearmint plant growth and secondary responses employing the plantlet culture systems are shown in **Figs. 4** and **5**. Note that both RA and Carvone quantities can be dramatically altered depending on the treatment given to plantlets, even through the plantlet growth rate remains relatively unaffected **(Fig. 5)**.
3. Carvone levels can be deceivingly unreliable to determine unless the investigator consistently employs the same plant material types and its location of growth. Please note in this investigation we have used terminal 4-cm shoot sections only.
4. RA analysis likewise can be difficult to determine accurately. RA is susceptible to chemical degradation and care must be taken to avoid exposure to excessive heat and acidic conditions. The quicker and efficient removal of water from the samples, HPLC will determine more accurately RA values. In general freeze-drying seems to yield the most reproducible results.
5. Because of the time required for chromatographic analysis, a large number of samples generated during the course of a plant culture study can be overwhelming. We have used near infrared spectroscopic (NIRS) analysis for rapidly determination of relative concentrations of RA in dried mint samples.

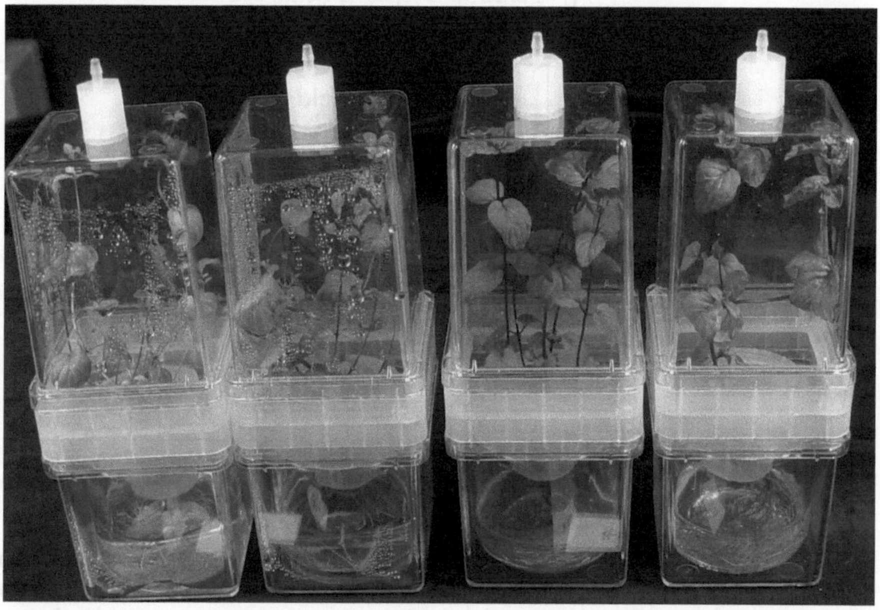

Fig. 4. Comparison of the morphological growth responses from spearmint plantlets grown under non-ventilated (two chambers on far *left*) and ventilated (two chambers on far *right*) conditions. Ventilated chambers were given 16 applications of 100 mL/min air/day. Note that the ventilated chambers contain plantlets with darker, more normal-appearing leaves than the non-ventilated chambers. Also, water vapor common to the non-ventilated chamber was minimal in the ventilated chambers.

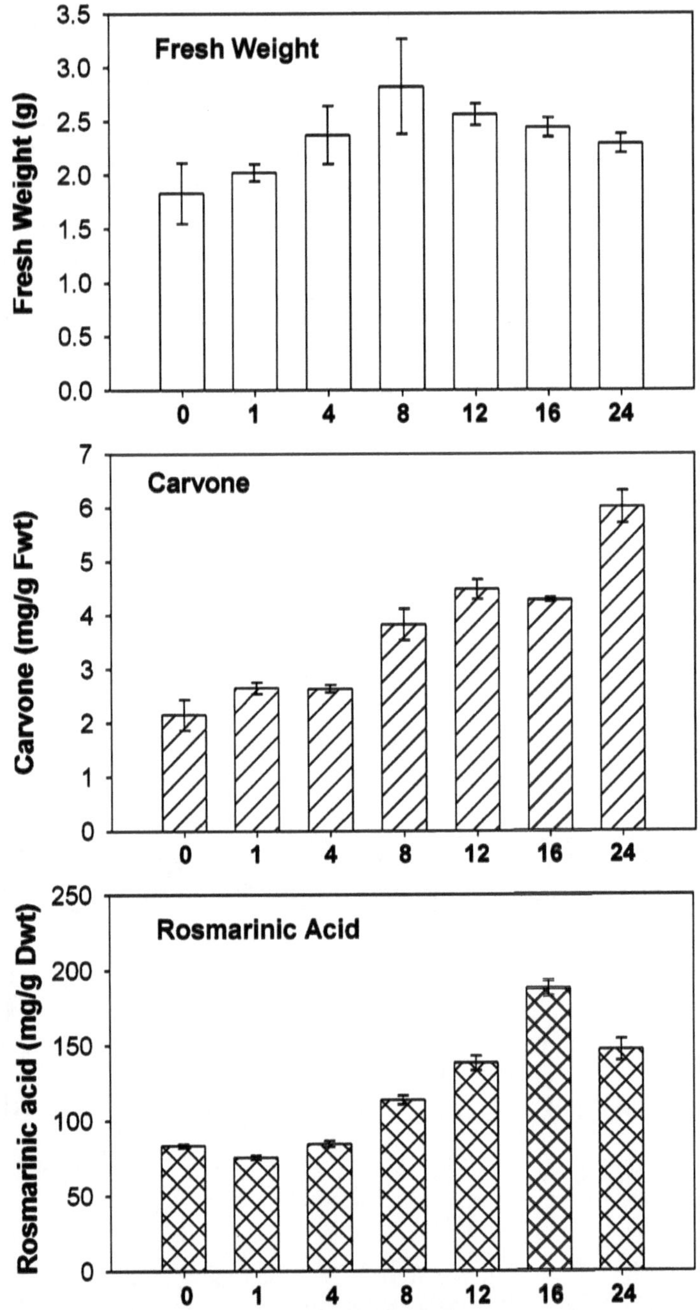

Fig. 5. Influence of ventilation treatments on growth and secondary metabolism of spearmint plantlets. Ventilation treatments consisted of a single 30 min exposure to 100 mL/min air equally spaced apart for a day. Data were averaged for five replications per treatments. Note that higher air treatments promoted more secondary metabolites to be produced while growth was relatively unaffected.

Once a set of samples has been analyzed by HPLC, use them to prepare a calibration curve in the NIRS instrument. Powdered samples can be quickly read on the NIRS instrument for rapid determination of RA levels.

Acknowledgments

The authors would like to thank A. Peterson for plant tissue culture techniques and R. K. Holloway, S. Tisserat and T. Tisserat for chemical analysis. Names are necessary to report factually on available data; however, the USDA neither guarantees nor warrants the standard of the product, and the use of the name by USDA implies no approval of the product to the exclusion of others that may also be suitable.

References

1. Croteau, R. T., Kutchan, M., and Lewis, N. G. (2000). Natural products (secondary metabolites), in Biochemistry & Molecular Biology of Plants (Buchan, B., Gruissem, W. and Jones, R., eds.), American Society of Plant Physiologists, Rockville, MD, pp.1250–1318.
2. Hegnauer, R. (1953). The content of essential oil and carvone in various species of mint. *Ber. Schweiz. Bot. Ges.* **63**, 90–102.
3. Sticher, V. O., and Fluck, H. (1968). Composition of genuine extracted and distilled essential oils of some *Mentha* species. *Pharmaceutica Acta Helvetica* **43**, 411.
4. Bourweig, D., and Pohl, R. (1973). Flavonoids from *Mentha longifolia. Planta Medica* **24**, 304–314.
5. Inoye, T., Sugimoto, Y., Masuda, H., and Kamei, C. (2002). Antiallergic effect of flavonoid glycosides obtained from *Mentha piperita* L. *Biol. Pharm. Bull.* **25**, 256–259.
6. Tisserat, B., and Vaughn, S. F. (2001). Essential oils enhanced by ultra-high carbon dioxide levels from Lamiaceae species grown *in vitro* and *in vivo. Plant Cell Rep.* **20**, 361–368.
7. Tisserat, B., Vaughn, S. F., and Silman, R. (2002). Influence of modified oxygen and carbon dioxide atmospheres on mint and thyme plant growth, morphogenesis and secondary metabolism *in vitro. Plant Cell Rep.* **20**, 912–916.
8. Yang, R., and Shetty, K. (1998). Stimulation of rosmarinic acid in shoot cultures or oregano (*Origanum vulgare*) clonal line in response to proline, proline analog and proline precursors. *J. Agric. Food Chem.* **46**, 2888–2893.
9. Li, W., Koike, K., Academy, Yoshikawa, T., and Nikaido, T. (2005). Rosmarinic acid production of *Coleus forskohlii* hairy root cultures. *Plant Cell Tiss. Org. Cult.* **80**, 151–155.
10. Kim, H. K., Oh, S. -R., Lee, H. -K., and Huh, H. (2001). Benzothiadiazole enhances the elicitation of rosmarinic acid production in a suspension culture of *Agastache rugosa* O. Kuntze. *Biotechnol. Lett.* **23**, 55–60.
11. Karam, N. S., Jawad, F. M., Arikat, N. A., and Shibli, R. A. (2003). Growth and rosmarinic acid accumulation in callus, cell suspension, and root cultures of wild *Salvia fruticosa. Plant Cell Tiss. Org. Cult.* **73**, 117–121.
12. Kirchner, J. G. (1978). Thin Layer Chromatography (2nd Edition), Techniques of Chromatography Series Vol. XIV, Wiley, NY.
13. Harborne, J. B. (1984). Phytochemical Methods. Chapman and Hall, NY.
14. Berhow, M. A. (2002). Modern analytical techniques for flavonoid determination, in Flavonoids in the Living Cell (Buslig, B. S. and Manthey, J. A., eds.), Adv. Exp. Med. Biol. Vol. 505, Kluwer/Plenum, NY, pp. 61–76.

15. Marston, A., and Hostettmann, K. (2006). Separation and quantification of flavonoids, in Flavonoids: Chemistry, Biochemistry and Applications, (Anderson, O. M. and Markham, K. R., eds.), CRC, Boca Raton, FL, pp. 1–36.

16. Fossen, T., and Anderson, O. M. (2006). Spectroscopic techniques applied to flavonoids, in Flavonoids: Chemistry, Biochemistry and Applications, (Anderson, O. M. and Markham, K. R., eds.), CRC, Boca Raton, FL, pp. 37–142.

Chapter 26

Bioreactor Production of Secondary Metabolites from Cell Cultures of Periwinkle and Sandalwood

Jagan V. Valluri

Summary

A bench-top bioreactor allowing continuous extraction of secondary metabolites is designed for *Catharanthus roseus* L. (G.) Don (periwinkle) and *Santalum album* L. (sandalwood) plant cell suspensions. Periwinkle cell cultures are exposed to biotic elicitors (*Aspergillus niger*, crude chitin) and abiotic elicitors (mannitol, methyl jasmonate) to induce alkaloid production. Whereas most of the biotic elicitors are effective when added on day 15 of culture, the abiotic elicitors are effective when added on day 20. The use of *trans*-cinnamic acid, an inhibitor of phenylalanine ammonia lyase (PAL) activity, results in significant increase in the alkaloid production of periwinkle cell cultures. Exposure of the cells to mannitol-induced osmotic stress produced marked increment in the total alkaloid production. When biotic and abiotic stress treatments are applied sequentially, an additive effect in alkaloid accumulation is observed. Although no essential oils are detected, secondary metabolites in the form of phenolics are produced by the sandalwood cell cultures in the bioreactor environment. The use of morphologic modification such as organ cultures and transformed cultures is believed to be required for both production and storage of essential oil constituents in sandalwood. The present chapter demonstrates that periwinkle and sandalwood cell suspensions could be developed and successfully cultured in a modified air-lift bioreactor. The exploitation of variant cell strains and biotransformation of added precursors can certainly improve the use of periwinkle and sandalwood cell cultures for the bioproduction of desired compounds.

Key words: Bioreactor, *Catharanthus roseus*, *Santalum album*, alkaloids, phenolics, medicinal plant cells

1. Introduction

The use of plant cell culture for chemical production has long been considered an attractive solution to the problems of extracting secondary metabolites from whole plants *(1)*. Cell suspension

S. Mohan Jain and Praveen K. Saxena (eds.), *Methods in Molecular Biology, Protocols for* In Vitro *Cultures and Secondary Metabolite Analysis of Aromatic and Medicinal Plants, vol. 547*
© Humana Press, a part of Springer Science+Business Media, LLC 2009
DOI: 10.1007/978-1-60327-287-2_26

culture systems could be used for large scale culturing of plant cells from which secondary metabolites could be extracted. The advantage of this method is that it can ultimately provide a continuous, reliable source of natural products. Plant cell bioreactor technology promises further exploitation of the plant kingdom as a rich resource of important and unique specialty chemicals *(2)*. Commercial implementation of large-scale plant cell suspensions for chemicals is still in the development stage with few exceptions *(3)*. Before the full impact of this industry can be realized, fundamental aspects of tissue culture physiology and regulation of secondary metabolism as they relate to bioreactor design must be determined *(4)*.

Plant cell bioreactors provide favorable conditions for improved secondary metabolite formation. There have been significant developments in novel bioreactor configurations in the past ten years. Multistage processes have been developed and various immobilization techniques (e.g., gel entrapment, hollow fibers, and immobilization on membranes) were investigated *(5)*. As research tools and small production units, bench-top bioreactors offer advantages over shake flask cultures *(6)*. Modern bioreactor systems are designed to give simultaneous control over medium composition, pH, dissolved O_2 and CO_2, medium oxidation-reduction potential (Eh), lighting, temperature, impeller mixing, and shear rate and can be operated in either continuous or batch mode *(7)*. Manipulation of these parameters may optimize cell growth and product yield. Despite all these developments, our understanding of plant cell differentiation, intracellular organization, and cell physiology as it is related to secondary metabolite production, is rather limited.

We have been investigating novel bioreactor configurations that provide favorable conditions for the production of secondary metabolites from economically important plant cell cultures. Investigations were undertaken on periwinkle and sandalwood to exploit bioreactor technology for secondary metabolite production. This chapter details standardized protocols for bioreactor culture, use of elicitors, and cultural factors for enhanced production of alkaloids and phenols from periwinkle and sandalwood cell cultures.

2. Materials

2.1. Establishment of Periwinkle and Sandalwood Suspension Cultures

1. Grow Periwinkle (*Catharanthus roseus* L. (G. Don) in greenhouse with an adequate supply of water and light. Use young fully expanded leaves from the apex as the source for explants.

2. Murashige and Skoog (MS) *(8)* supplemented with 1 mg/L 2, 4-dichlorophenoxyacetic acid, 1 mg/mL indole acetic acid, 0.5 mg/L kinetin, 1100 mg/L KNO_3, 2 mg/L thiamine-HCI, 0.1 mg/L riboflavin, 0.1 mg/L biotin, 0.1 mg/L folic acid, 30 g/L and sucrose (Sigma; St. Louis, MO).
3. Photomixotrophic sandalwood calli (*Santalum album* L.) cultures.
4. Use sandalwood calli maintained on MS medium modified with 20 g/L sucrose, 1 mg/L 2, 4-d, and 1.0 mg/L BA (Sigma (*see* **Note 1**).

2.2. Bioreactor Culture

1. New Brunswick Celligen-Plus mechanically agitated bioreactor with a 2.5 L working volume (New Brunswick Sci. Co.; New Brunswick, NJ) (**Fig. 1**).
2. Dissolved oxygen (DO) tension in the medium is measured by galvanic oxygen electrode (Phoenix instruments; Houston, TX).
3. Data is continuously monitored and logged by Loggernaut Data Logging System (Cyborg Corporation; Newton, MA) (*see* **Note 2**).

2.3. Biotic and Abiotic Elicitor Preparation

1. Fungal elicitor *Aspergillus niger* is obtained from the American Type Culture Collection (Rockville, MD; cat. no. ATCC-10787) (*see* **Note 3**).

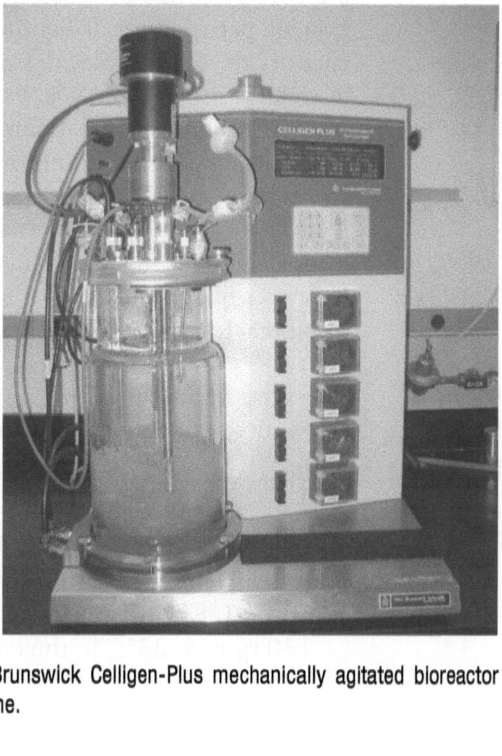

Fig. 1. New Brunswick Celligen-Plus mechanically agitated bioreactor with a 2.5 L working volume.

2. A 10 g/L crude chitin solution is prepared by extraction, crushing crude chitin (Sigma) in deionized water (*see* **Note 4**).
3. Mannitol 2.5% *(w/v)* (Fisher Scientific; PA) is added to periwinkle plant cell suspensions to stimulate osmotic stress (*see* **Note 5**).
4. Jasmonic acid (Sigma) (*see* **Note 6**).

2.4. Determination of Alkaloids

1. Resuspend 1 g freeze-dried cells in 50 mL methanol and homogenize with a polytron at maximum speed for 3 min. Reagent-grade methanol (MeOH) (Fisher Scientific) is used for extractions (*see* **Note 7**).
2. The mixture of alkaloids composed of 2 mg ajmalicine, 2 mg ajmaline, 1 mg vincamine, 1 mg vindoline 2 mg catharanthine and 2 mg yohimbine (Sigma)

2.5. Determination of Phenols

1. 2 N Folin-Cicalteu reagent (Sigma). Store at 4°C.
2. Gallic acid (Sigma). Generally 1000 ppm stock is prepared and stored at 4°C in the dark (*see* **Note 8**).

3. Methods

Elicitations are considered to be an important strategy towards improved in vitro production of secondary metabolites. Elicitation of periwinkle cell cultures with various abiotic and biotic elicitors or signal molecules often results in a dramatic increase in yield of indole alkaloids, likely resulting from the defense role of these secondary metabolites *(9, 10)*. Elicitation of periwinkle cell cultures not only improves indole alkaloid biosynthesis in short time, but also causes excretion of the products into the medium. Combination of two or more elicitors that can synergistically induce metabolic fluxes towards indole alkaloids further improves the productivity of target compounds and performance of bioreactor processing.

3.1. Establishment of Periwinkle and Sandalwood Suspension Cultures

1. Excise approximately 1-cm leaf segments from young leaves to serve as explants (*see* **Note 9**).
2. Place explants on MS medium supplemented with 1 mg/L 2,4-D, 0.5 mg/L kinetin, 250 mg/L casein hydrolysate, and 30 g/L sucrose on 0.8% *(w/v)* agar. Adjust pH to 5.8.
3. Inoculate three to four explants into 100 mL fresh medium in 500-mL Erlenmeyer flasks. Incubate flasks on a rotary shaker, 130 rpm, at 23°C in the dark.

4. Use cell aggregates from swollen leaf explants to developing cell suspension cultures.
5. Initiate photomixotrophic sandalwood suspension cultures (SW-1) by transferring calli to liquid MS medium modified with 20 g/L sucrose, 1 mg/L 2,4-D, and 1.0 mg/L BA and culture on an orbital shaker, 120 rpm, at 25°C and a 16 h photoperiod using cool-white fluorescent lighting (4–6 W/m^2) *(11)*. Initiate a second batch of sandalwood suspensions (SW-2) from calli cultured under a (12–15 W/m^2) light intensity under a continuous cool-white fluorescent lighting *(12)*.

3.2. Bioreactor Culture

1. Carry out all bioreactor work in a positive-pressure clean room with temperature controlled at 26°C. Use 10- to 14-d-old suspension cultures to inoculate the bioreactor. Inoculum volume is 350 mL which is equivalent to 5–8 g/L fresh weight and 17.5% of reactor volume (*see* **Note 10**).
2. Impeller speeds between 0.04 and 0.11 g produce best mixing without excessive foaming for periwinkle and sandalwood cell lines.
3. The lag phase typically observed in plant suspensions after inoculation is not readily apparent (**Fig. 2**). The exponential growth phase for periwinkle and sandalwood cell suspensions appears to begin almost immediately and lasts for 10–12 days. The stationary phase lasts 3–5 d after which a rapid increase in dead cells is observed as well as an increase in medium pH. During exponential growth, fresh weight Td is 5.0 d and the specific growth rate (μ) is 0.10 days/L for periwinkle cell cultures. Fresh weight Td is 5.8 d and the specific growth rate (μ) is 0.12 d/L for sandalwood cell cultures.

3.3. Total Alkaloids

1. Extract cell samples using a standard percolation of MeOH; separate alkaloids into acidic medium and re-extract with ethyl acetate. Resuspend 1–5 g freeze-dried cells in 50–100 mL methanol and homogenize with a Polytron at maximum speed for 3 min. Incubate the homogenates in a water bath with reciprocal shaker, 150 rpm, at 50°C for 2 h (*See* **Note 7**).
2. Determine total alkaloid content by measuring the sample's absorbance at λ = 280 nm and compare this value with a standard curve of a mixture of known alkaloids, 0–75 μg alkaloids/mL.

3.4. Elicitation with Abiotic and Biotic Agents

1. Abiotic stress agents included mannitol and jasmonic acid. Prepare a sucrose plus mannitol solution to yield a final concentration of 2.5% mannitol for inducing osmotic stress. Then add mannitol solution to 20-d-old bioreactor cell suspensions. Add 25 mg/L jasmonic acid to induce alkaloid accumulation.

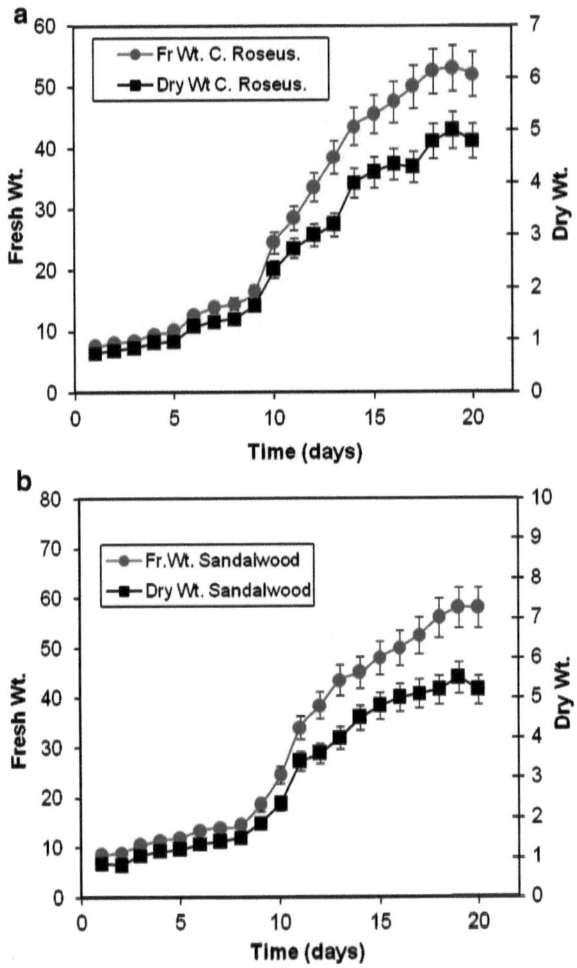

Fig. 2. Growth curves of *C. roseus* (**a**) and sandalwood (**b**) cell lines based on fresh and dry weight. Average of three replications; the maximum standard deviation was 7%.

Sample the cultures after 24, 48, and 72 h incubations. Data shown in the **Fig. 3** represent the average of three replicates.

2. Grow spores of *A. niger* on potato-dextrose- agar, to inoculate into 100 mL MS medium in a 250-mL Erlenmeyer flasks. Keep the cultures on a gyratory shaker, 200 rpm, at 30°C in the dark for 7 d. Collect the mycelium by filtration and homogenize with a Polytron at maximum speed for 10 min with 100 mL distilled water. The homogenates are autoclaved for 20 min, thus yielding the elicitor preparation. Add *A. niger* homogenates to the 15-d-old cultures to render a final concentration of 0.5 g FW/L alone or in combination with the enzyme inhibitor 1 mM *trans*-cinnamic acid (*see* **Note 11**). Adjust the pH of the enzyme inhibitor solutions to 5.8. Data shown in the figures represent the average of three replicates (**Fig. 4**).

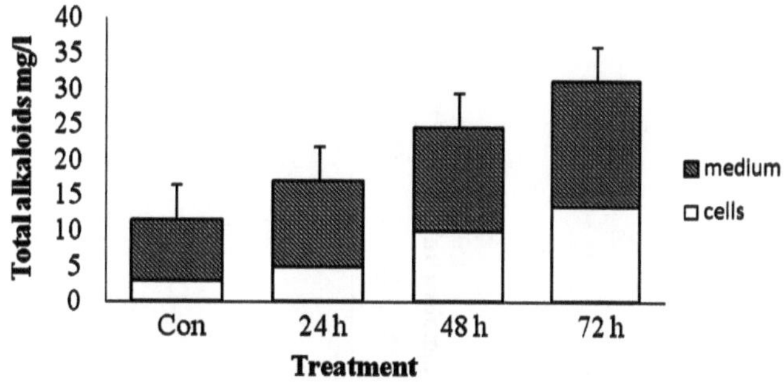

Fig. 3. Total alkaloid production of bioreactor cultured *C. roseus* cell lines exposed to Jasmonic acid. (Con) untreated 20-d-old cells; (24 h) cells after a 24-h treatment; (48 h) cells after a 48-h treatment; (72 h) cells after a 72-h treatment.

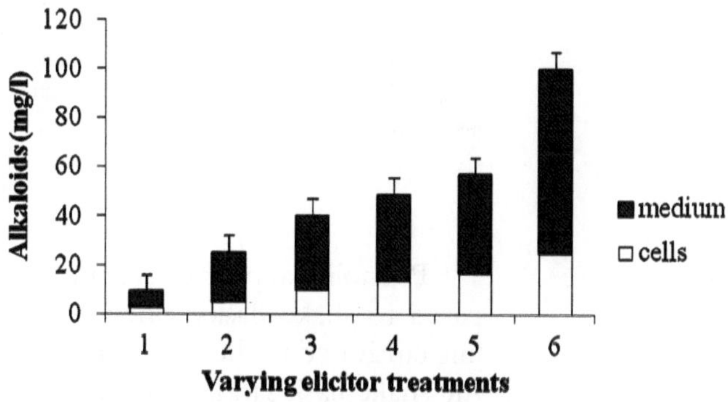

Fig. 4. Total alkaloid production of bioreactor cultured *C. roseus* cell lines exposed to varying elicitor treatments. *(1)* Control at day 15; *(2) A.niger* homogenate; *(3) A.niger* homogenate + trans-cinnamic acid; *(4)* Osmotic stress; *(5)* Osmotic stress + chitin; *(6)* Osmotic stress + trans-cinnamic acid + chitin.

3.5. Essential Oils and Phenols

The chemical composition of the volatile oil derived from the root and heartwood of sandalwood has been subject to numerous investigations. The chief constituent, amounting to more than 90% of the oil is, *santalol*, a mixture of two primary sesquiterpene alcohols $C_{15}H_{24}O$, viz., and β santalol, in which the form predominates *(13)*. The structure of these sesquiterpene alcohols has been elucidated by Brunke and Tumbrink *(14)*. Very few tissue cultures from higher plants are capable of producing volatile oils in considerable quantities with similar components as compared to the whole plant. At present, our analytical investigations by TLC and gas chromatography-mass spectrometry (GC-MS) of undifferentiated suspension cells cultured in the bioreactor did not reveal any essential oil accumulation. However, we have noticed production of phenols which are secondary metabolites in suspension cells of sandalwood over three growth periods.

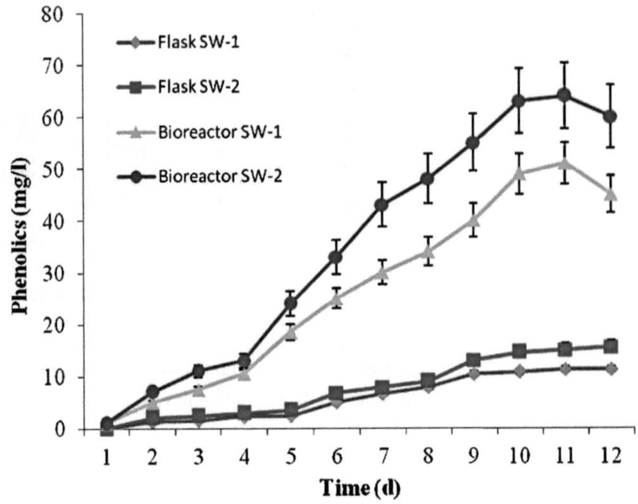

Fig. 5. Phenol production in shake flask cultures vs bioreactor grown cell suspensions of sandalwood. Average of three replications; the maximum standard deviation was less than 8%.

Phenolics appear to accumulate faster in the bioreactor compared to shake flasks. Phenolic content of the sandalwood cell line cultivated in the bioreactor is approximately 2.5 times that of the shake flask sandalwood cell line, 31.5 mg/L vs 12.5 mg/L, respectively. Phenolics begin accumulating during the exponential phase and peaked during the stationary phase **(Fig. 5)**.

4. Notes

1. One-liter packet of pre-mixed medium (MS-salts) is added to approximately 800 mL distilled water in a 2-L flask. When all of the powder is in solution sucrose is added and the solution is stirred to dissolve the sucrose completely. Adjust the pH to 5.7. The plant tissue culture medium is not buffered so add the base or acid in small portions (about 1 mL per dose). Generally plant growth hormones can be added before the medium is sterilized. Usually stock solutions of hormones are made ahead and frozen for storage. Generally 1 mg/mL or 10 mg/mL stocks work best, because most of the hormones are needed in very low concentrations, approximately 1 mg/L. Stock solution composition varies for each hormone. Some have to be dissolved in a very concentrated way using acid or base, then brought to volume with distilled water. After addition of plant growth hormones,

distilled water is added to raise volume in Erlenmeyer flask to 1 L, autoclave for 15 min at 15 psi, and cool to approximately 60°C before pouring into appropriate size flasks.

2. Reactor conditions are 6 h illumination, 180–200 $\mu E/m^2/s$ measured at inside glass surface, 26°C and an initial pH of 5.7. Agitation 0.04–0.11 g of the bioreactor is provided by a "cell-lift" impeller which generates lower shear rates than Rhuston type impellers. The air flow rate is measured by flow-meter and expressed as volume of air per volume of medium per min (vvm). The gas composition is 21% O_2, 2% CO_2 and the bulk N_2. A modified sparger equipped with a fritted glass filter is used to aerate the medium at 0.025–0.5 vvm.
3. Grow spores of *A. niger* on potato-dextrose agar and use to inoculate 250 mL MS medium in a 500-mL Erlenmeyer flasks. Maintain cultures on a gyratory shaker, 150 rpm, at 35°C in the dark for 1 wk before use.
4. Collect the mycelium by filtration and homogenize with a Polytron at maximum speed for 10 min with 100 mL distilled water. The homogenate is autoclaved for 15 min at 121°C at 15 psi to create the fungal elicitor.
5. Autoclave the crude chitin at 121°C for 30 min, stirring during cool down and filter under vacuum on two layers of Miracloth filter membrane.. This solution is ready to use for elicitation.
6. The addition of mannitol to the nutrient medium simulates osmotic stress by acting as a nonpenetrating agent which in turn lowers the water potential of the medium. The cultures are sampled after 1–3 d incubation with mannitol.
7. Jasmonic acid solution is prepared by mixing 250 mg jasmonic acid in 10 mL Milli-Q water. Filter sterilize jasmonic acid solution using a 0.22-μm filter and add to late exponential stage cultures at 25 mg/L.
8. Concentrate the methanolic extracts under vacuum. Resuspend the residue in 20 mL 2.5% sulfuric acid *(v/v)* and wash three times with 20 mL ethyl acetate each time. Adjust the pH of the aqueous phase to 9.5 with concentrated 28% ammonium hydroxide and extract three times with 20 mL ethyl acetate each time. Concentrate the organic phase under vacuum and washed with 6 mL chloroform. Dry the chloroform phase under vacuum, resuspend in 3 mL methanol and store for total alkaloid determination in a Shimadzu 240 ultraviolet (UV) double-beam spectrophotometer.
9. Separate 0.5 mL culture medium from cells by filtration by using Whatman No. 1 filter paper under suction and dilute

the filtrate in 6 mL of distilled water and 0.5 mL of folin-ciocalteu reagent. After 5 min, add 1 mL 5% Na_2CO_3, mix the samples and allow to stand for 1 h in the dark and measure the absorbencies at 725 nm using a Shimadzu 240 UV double-beam spectrophotometer in 1.0 cm quartz cells against 95% ethanol as blank.

10. Sterilize leaf segments for 1 min with 0.3% *(w/v)* $HgCl_2$ and wash at least three times with sterile water.
11. Monitor biomass accumulation in the bioreactor as fresh and dry weight each day by sampling 30-mL aliquots. Estimate cell viability by fluorescein diacetate exclusion method. In culturing photosynthetic plant cells in bioreactors, it is necessary to maintain optimum mixing and aeration without increasing the hydrodynamic and mechanical stress effects beyond tolerance levels. Maintain impeller speed between 0.04 and 0.11 g.
12. Sterilize *A. Niger* homogenates by filtration, prior to their addition to the cultures. Sample the treated cultures after 24, 48, and 72 h of incubation. Measure cell viability with fluorescein diacetate exclusion method and separate cells from the medium by filtration and wash with distilled water under vacuum. Determine the pH of the media. Freeze-dry liquid media and cells separately.

Acknowledgments

This work was supported by funding from NASA West Virginia Space Grant Consortium.

References

1. Zhong, J. J. (2001). Biochemical engineering of the production of plant-specific secondary metabolites by cell suspension cultures, in *Advances in Biochemical Engineering/Biotechnology*, **Vol 72**, (T. Scheper., ed.), Springer, Heidelberg, pp. 1–26.
2. Rao, S. R. and Ravishanakar, G. A. (2002). Plant cell cultures: chemical factories of secondary metabolites. *Biotechnol. Adv.* **20**, 101–153.
3. Scragg, A. H. (1992). Large-scale plant cell culture: methods, applications and products. *Curr. Opin. Biotech.* **3**, 105–109.
4. Sajc, L., Grubiscic, D., and Vunjak-Novakovic, G. (2000). Bioreactors for plant engineering: an outlook for further research. *Biochem. Eng. J.* **4**, 89–99.
5. Paek, K. Y., Hahn, E. J., and Son, S. H. (2001). Application of bioreactors for large scale micropropagation systems of plants. *In Vitro Cell. Dev. Biol. Plant.* **37**, 149–157.
6. Hsiao, T. Y., Bacani, F. T., Carvalho, E. B., and Curtis, W. R. (1999). Development of a low capital investment reactor system: application for plant cell suspension culture. *Biotechnol. Prog.* **15**, 114–122.
7. Chattopadhyay, S., Farkya, S., Srivastava, A. K., and Bisaria, V. S. (2002). Bioprocess considerations for production of secondary

metabolites by plant cell suspension cultures. *Biotechnol. Bioprocess Eng.* 7, 138–149.

8. Murashige, T., and Skoog, F. (1962). A revised medium for rapid growth and bioassays with tobacco tissue cultures. *Physiol. Plant.* **15**, 473–497.
9. Zhao, J., Hu, Q., and Zhu, W. H. (2001). Enhanced catharanthine production in Catharanthus roseus cell cultures by combined elicitor treatment in shake flasks and bioreactors. *Enzyme Microb. Technol.* **28**, 673–681.
10. Namdeo, A., Patil, S., and Fulzele, D. P. (2000). Influence of fungal elicitors on production of ajmalicine by cell cultures of Catharanthus roseus. *Biotechnol. Prog.* **18**, 159–162.
11. Sita, G. L. (1986). Sandalwood (*Santalum album* L.), in *Biotechnology in Agriculture and Forestry*, **Vol 1** (Y.P.S. Bajaj., ed.), Springer, Heidelberg, pp. 363–374.
12. Valluri, J. V., Treat, W. J., and Soltes, E. J., (1991). Bioreactor culture of heterotrophic sandalwood (*Santalum album* L.) cell suspensions utilizing a cell-lift impeller. *Plant Cell Rep.* **10**, 366–370.
13. Sastry, S. G. (1944). Oil of sandalwood East Indian. *J. Sci. Ind. Res.* **3**, 173–187.
14. Brunke, J. E., and Tumbrink, L. (1986). First total synthesis of spirosantalol, a sesquiterpene alcohol from East Indian sandalwood oil. *Prog Essential Oil Res.* Walter de Gruyter, Berlin pp. 321–327.

Chapter 27

Camptothecin Production by In Vitro Cultures and Plant Regeneration in *Ophiorrhiza* Species

Takashi Asano, Hiroshi Sudo, Mami Yamazaki, and Kazuki Saito

Summary

Camptothecin derivatives are clinically used for the treatment of various human cancers. These derivatives are semi-synthesized from camptothecin which is isolated from the extracts of *Camptotheca acuminata* and *Nothapodytes foetida*. For the feasible production of camptothecin, the protocols for the tissue cultures of *Ophiorrhiza* species, *O. pumila*, *O. liukiuensis* and *O. kuroiwai*, have been established. The established aseptic plants and hairy roots produced camptothecin, and *O. pumila* hairy roots accumulated highest amount of camptothecin. Furthermore, we have established methods of plant regeneration from *O. pumila* hairy roots.

Key words: Camptothecin, Indole alkaloids, *Ophiorrhiza pumila*, *Ophiorrhiza liukiuensis*, *Ophiorrhiza kuroiwai*, Rubiaceae, Hairy root, Aseptic plant, *Agrobacterium rhizogenes*, Regeneration

1. Introduction

Camptothecin is a modified monoterpenoid indole alkaloid, originally isolated from *Camptotheca acuminata*, and it was remarked as a candidate for anti-tumor agents by the National Cancer Institute (NCI) in the United States *(1)*. This anti-tumor property of camptothecin is exhibited by specific inhibition of DNA topoisomerase I *(2)*. Currently, semi-synthetic camptothecin derivatives—including topotecan and irinotecan—are clinically used as anti-tumor agents (**Fig. 1**). The worldwide market of these two agents has reached nearly US$ one billion per year *(3)*. In spite of the rapid growth of the market, camptothecin is

S. Mohan Jain and Praveen K. Saxena (eds.), *Methods in Molecular Biology, Protocols for* In Vitro *Cultures and Secondary Metabolite Analysis of Aromatic and Medicinal Plants , vol. 547*
© Humana Press, a part of Springer Science+Business Media, LLC 2009
DOI: 10.1007/978-1-60327-287-2_27

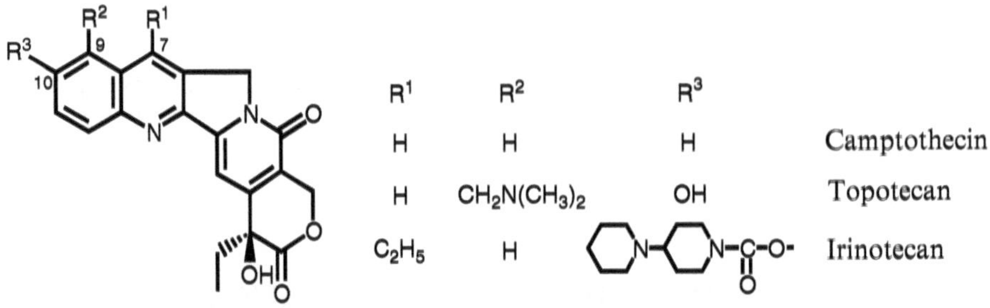

Fig. 1. Clinically used camptothecin derivatives, topotecan and irinotecan.

still supplied exclusively from the intact plants, mainly *C. acuminata* and *Nothapodytes foetida (4)*. However, extraction from intact plants has environmental problems such as the shortage of natural resources. Thus, the production of secondary metabolites by genetically engineered plant cell cultures has become a keen issue *(5)*. However, cell and tissue cultures of *C. acuminate* and *N. foetida* available for feasible camptothecin production system have not been reported.

The genus *Ophiorrhiza* widely distributes about 150 species around the tropical and subtropical Asia *(6)*, and the production of camptothecin-related alkaloids was checked in some *Ophiorrhiza* species distributed in Japan *(7–8)*. These *Ophiorrhiza* species are important resources producing various alkaloids *(9–11)*. This chapter describes methods for tissue cultures of *Ophiorrhiza* species as a feasible system of camptothecin production and further genetic engineering.

2. Materials

2.1. Plant

Ophiorrhiza pumila plants collected in Amamiohshima-Island, Kagoshima, Japan are grown in the greenhouse; seeds of *Ophiorrhiza liukiuensis* collected in Ishigaki-Island, Okinawa, Japan; and *Ophiorrhiza kuroiwai* plants collected in Iriomote-Island, Okinawa, Japan are grown in the greenhouse.

2.2. Culture Medium

2.2.1. Medium for Agrobacterium rhizogenes

1. YEB solid medium: 0.5% *(w/v)* Bacto beef extract (BD Biosciences; Franklin Lakes, NJ).
2. 0.1% *(w/v)* Bacto yeast extract (BD Biosciences).
3. 0.5% *(w/v)* Bacto peptone (BD Biosciences).
4. 0.5% *(w/v)* sucrose (Kanto; Tokyo, Japan).
5. 2 mM $MgSO_4$ (Kanto), pH 7.2; 1.5% agar (Wako; Osaka, Japan).

2.2.2. Medium for Aseptic Plants

1. 100X Half-strength Murashige and Skoog (MS) vitamin solution: 5 g/L myo-inositol (Sigma-Aldrich; St. Louis, MO), 100 mg/L glycine (Wako), 25 mg/L nicotinic acid (Wako), 25 mg/L pyridoxine hydrochloride (Wako), 5 mg/L thiamin hydrochloride (Wako). Store at –20°C.
2. AP solid medium: Half-strength MS plant salt mixture *(12)* (Wako); 1% *(w/v)* sucrose; half-strength MS vitamin solution, pH 5.7; 0.2% gellan gum (Wako).

2.2.3. Medium for Hairy Roots

1. 100X B5 vitamin solution:10 g/L *myo*-inositol, 1 g/L thiamin hydrochloride, 100 mg/L pyridoxine hydrochloride, 100 mg/L nicotinic acid. Store at –20°C.
2. HR liquid medium: Gamborg's B5 medium salt mixture *(13)* (Wako); 2% *(w/v)* sucrose; B5 vitamin solution, pH 5.7.
3. HR solid medium: Liquid medium plus 0.2% gellan gum.

2.3. Reverse-Phase HPLC Analysis

1. Column: TSK gel ODS-80TM (*Φ* 4.6 mm × 150 mm) (TOSOH; Tokyo, Japan).
2. Solvent system: The isocratic elution of 50% MeOH (Wako) in distilled H_2O (Wako) at 1.0 mL/min. All solvents are high-performance liquid chromatography (HPLC) grade.
3. Detection: Camptothecin and 10-methoxycamptothecin exhibiting fluorescence are monitored by their characteristic fluorescence (excitation at 365 nm and emission at 428 nm).
4. Standards for quantification: Camptothecin is purchased from Sigma-Aldrich and 10-methoxycamptothecin is purified from intact plants of *O. liukiuensis (8)*.

3. Methods

In tissue culture, productivities and patterns of secondary metabolites depend on morphology and culture condition. We have established aseptic plants and hairy roots of *Ophiorrhiza* species as one of the effective methods of producing camptothecin.

3.1. Establishment of Aseptic Plants of Ophiorrhiza Species

AP solid medium, half-strength MS plant salt mixture; 1% *(w/v)* sucrose; half-strength MS vitamin solution, pH 5.7; 0.2% gellan gum is divided by 20-mL in a test tube (*Φ* 40 mm × 130 mm) and autoclave it.

Aseptic plants of camptothecin producing *Ophiorrhiza* species, *O. pumila*, *O. liukiuensis* and *O. kuroiwai*, are established according to the following protocols. Established aseptic plants are maintained at 25°C with a photoperiod of 18 h light (22.8

μmol/m²/s)/6 h dark, and sub-culture every 3-month by transferring shoots on the AP solid medium in a test tube **(Fig. 2a)**.

3.1.1. Ophiorrhiza Pumila (14)

1. Sterilize young leaves of *O. pumila* with 1% sodium hypochlorite solution (Wako) for 10 min, and rinse with sterile water.
2. Callus cultures originally derived from the sterilized leaf segments are maintained on MS medium containing 2% sucrose, 0.7% agar (Wako), 1 mg/L indole 3-acetic acid (IAA) (Sigma)

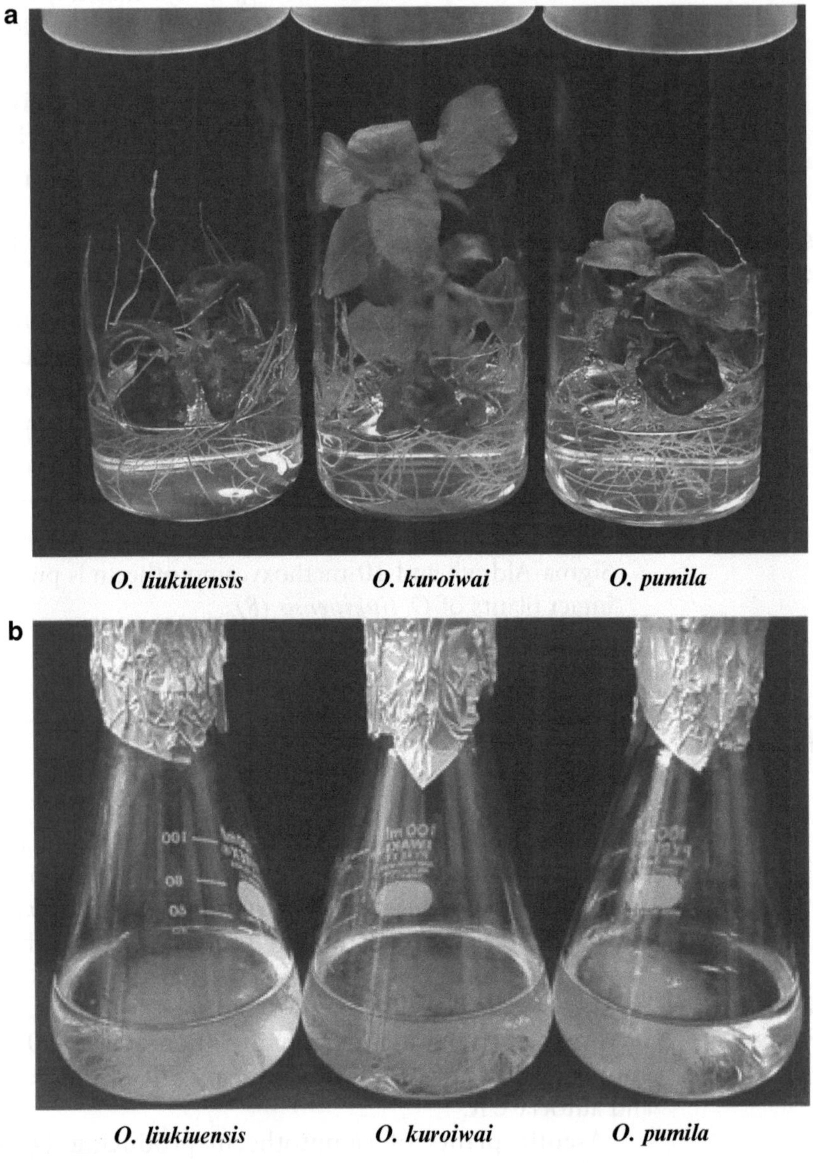

Fig. 2. Established tissue cultures of *Ophiorrhiza liukiuensis O. kuroiwai* and *O. pumila* (**a**) Aseptic plants cultured for 5 wk on 1/2 MS medium containing 1% sucrose and 0.2% gellan gum in test tubes. (**b**) Hairy roots cultured for 4 wk in B5 liquid medium containing 2% sucrose. (Reproduced with permission from **ref.** *15*).

in the light, 8.6 μmol/m^2/s, for 4 wk and then transferred to MS medium supplemented with 1 mg/L kinetin (Wako), 2% sucrose, and 0.7% agar for plant regeneration.

3. After 4 wk, the color of calli will change from light yellow to brown.
4. After 10 wk, the callus differentiates into green leaf buds, and excised regenerated plants are transferred on the AP solid medium in a test tube.

3.1.2. Ophiorrhiza Liukiuensis (15)

1. The seeds are sterilized by 1% sodium hypochlorite solution for 10 min, followed by rinsing with sterilized water.
2. Germinate sterilized seeds on AP solid medium in a test tube.

3.1.3. Ophiorrhiza Kuroiwai (15)

1. By 1% sodium hypochlorite solution treatment for 10 min, young leaves of *O. kuroiwai* are sterilized, and wash out this solution by rinsing with sterile water.
2. Culture sterilized leaf segments on the AP solid medium supplemented with 0.5 μM 1-naphthaleneacetic acid (NAA) (Sigma) and 5 μM kinetin.
3. After 40 d, regenerated shoots are excised and transferred on the AP solid medium in a test tube.

3.2. Induction and Culture of Hairy Roots

1. YEB solid medium: 0.5% *(w/v)* Bacto beef extract; 0.1% *(w/v)* Bacto yeast extract; 0.5% *(w/v)* Bacto peptone; 0.5% *(w/v)* sucrose; 2 mM $MgSO_4$, pH 7.2; 1.5% agar is sterilized by autoclaving and 50 mg/L rifampicin (Wako) is added to the medium. The medium is divided to a plate (Φ 90 mm × 15 mm).
2. Grow *Agrobacterium rhizogenes* 15834 on the YEB solid medium supplemented with 50 mg/L rifampicin at 28°C for 2 d (*see* **Note 1**).
3. The aseptic plants of *Ophiorrhiza* species, *O. pumila*, *O. liukiuensis*, and *O. kuroiwai*, are infected with *A. rhizogenes* 15834 by scratching the stems *(16)*.
4. After several weeks, excise hairy roots which emerged from stem fragments (*see* **Note 2**).
5. HR solid medium: Gamborg's B5 medium salt mixture; 2% *(w/v)* sucrose; B5 vitamin solution, pH 5.7; 0.2% gellan gum is sterilized by autoclaving and 200 mg/L cefotaxime (Claforan®) (Sanofi-Aventis; Paris, France) is added to the medium. The medium is divided to a plate (Φ 90 mm × 20 mm).
6. Culture induced hairy roots on the HR solid medium containing 200 mg/L cefotaxime at 25°C under dark condition for disinfection (*see* **Note 3**).

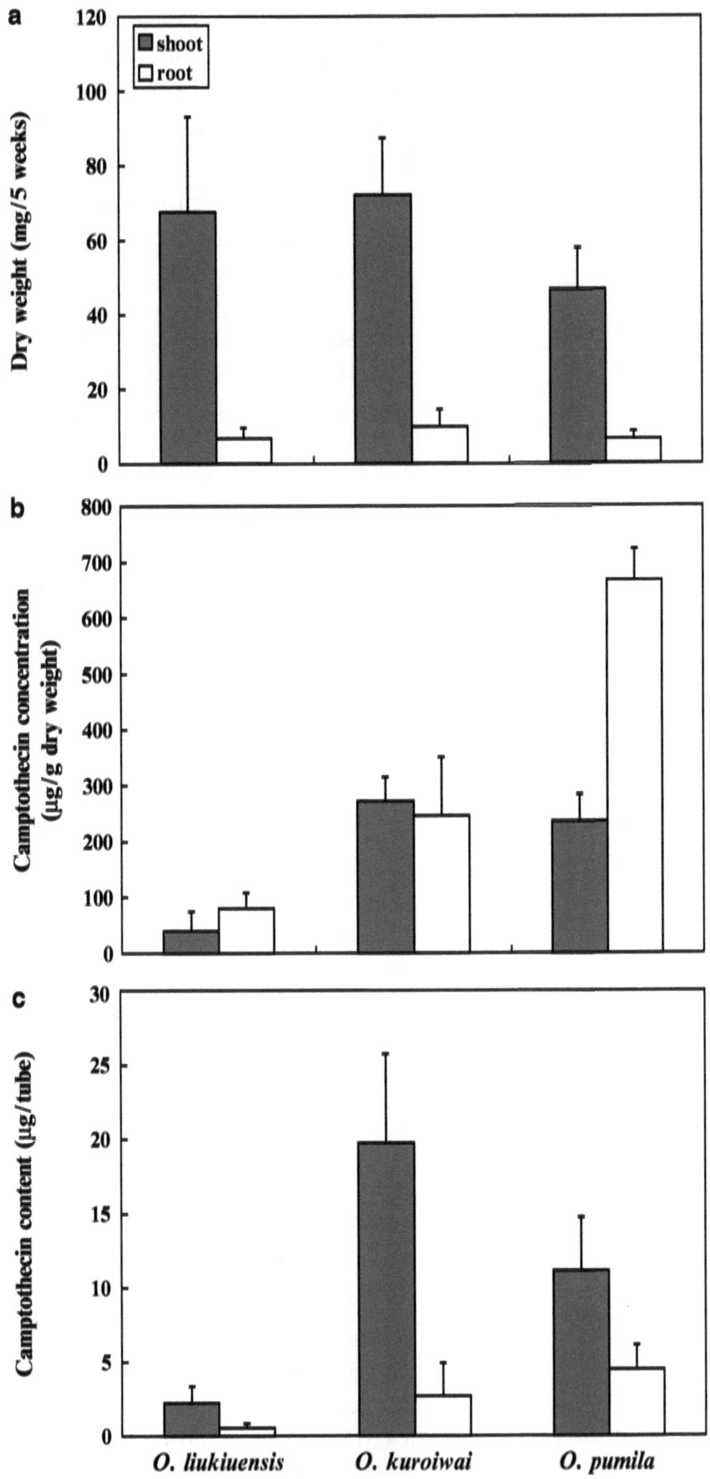

Fig. 3. Growth of aseptic *Ophiorrhiza* plants and camptothecin production for 5 wk. (**a**) Dry weight of shoot and root of a plant. (**b**) Camptothecin concentration per dry weight. (C) Camptothecin content in a whole plant. (Reproduced with permission from **ref.** *15*).

7. HR solid medium is sterilized by autoclaving, and pour in Petri dishes (*Φ* 90 mm × 20 mm).
8. The disinfected hairy roots are cultured on the HR solid medium at 25°C in the darkness.
9. Pour 50 mL HR liquid medium: Gamborg's B5 medium salt mixture; 2% *(w/v)* sucrose; B5 vitamin solution, pH 5.7 in 100-mL Erlenmeyer flasks, and autoclave it.
10. The established hairy roots are sub-cultured every 3–4 wk in the HR liquid medium at 25°C on rotary shaker, 80 rpm, in the darkness *(15, 16)* (**Fig. 2b**).

3.3. HPLC Analysis of Camptothecin-Related Alkaloids

1. For alkaloid extraction, homogenize tissues in a mortar and pestle, add 1 mL MeOH per 100 mg tissue and mix.
2. After sonication for 15 min, homogenates are centrifuged at 10,000*g* for 10 min.
3. Supernatants are diluted to 100 times with MeOH and apply on the reverse-phase HPLC. The results of HPLC analysis for our established cultures are shown in **Fig. 3** and **Table 1** *(15)*.

3.4. Regeneration of Transformed Plants from Hairy Roots (17)

1. Culture clusters, about 1 cm size, of *O. pumila* hairy roots on HR solid medium in the dark for 1 mo. Expose them to light with 16 h photoperiod 22.8 μmol/m^2/s/8 h dark (*see* **Note 4** and **5**).
2. After 5 wk, regenerated shoots buds emerge especially from relatively old parts of the hairy roots.
3. After the regenerated shoots buds are transferred on the AP solid medium under the light for rooting, they grow to greenish shoots.
4. As shown in **Fig. 4**, regenerated plants show various and typical characteristic features of hairy root-derived plants, such as shortened internodes and malformed leaves.

Table 1
Hairy Roots Induced from *Ophiorrhiza* Species

Plant species	Number of established lines	Camptothecin content[a] (μg/g D.W. ± S.D.[b])
O. liukiuensis	11	83.0 ± 27.4
O. kuroiwai	7	219.3 ± 31.4
O. pumila	19[c]	788.5 ± 49.7

[a] Hairy roots were cultured in the HR liquid medium for three weeks
[b] Dry weight ± standard deviation
[c] 19 hairy root lines were randomly selected from over 40 lines

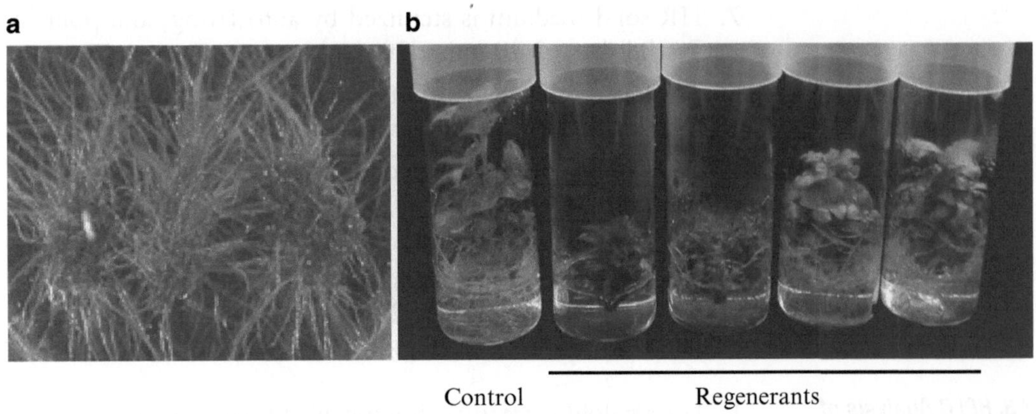

Fig. 4. Regeneration of transgenic *O. pumila* plants from hairy roots. (**a**) Regenerated shoots emerged from hairy roots after 5 wk of culture under light conditions. (**b**) The shapes of regenerated plants cultured on 1/2 MS medium containing 1% sucrose and 0.2% gellan gum in test tubes. (Reproduced with permission from **ref.** *17*).

4. Notes

1. The growth rate of *A. rhizogenes* is different depending on the strain. If another strain is used for the induction of hairy roots, it is necessary to reexamine the culture conditions.
2. The efficiency of hairy roots induced from the infected stems is different according to the growth state of aseptic plants. In our experiments, enough hairy roots have been obtained from the young stems of 2-mo-old aseptic plants.
3. For the disinfection of *A. rhizogenes*, it is necessary to subculture hairy roots once on the same cefotaxime supplemented medium.
4. Both the amount of hairy roots on plates and the light conditions significantly influence the plant regeneration. When young hairy roots are individually cultured under light conditions, they apparently stop growing. In contrast, the cluster of the hairy roots is abundantly regenerated to greenish shoots, under the light within three weeks, much faster than those in the dark.
5. The regenerated shoot buds gradually die during the prolonged cultures in the dark. This indicates that two following conditions are favorable to initiate the shoot regeneration from *O. pumila* hairy roots; sizable mass of the hairy roots and the light exposure after being cultured in the dark.

Acknowledgments

This work was supported in part by Grants-in-Aid for Scientific Research from the Japan Society for the Promotion of Science (JSPS), and by CREST of the Japan Science and Technology Agency (JST).

References

1. Wall, M. E., Wani, M. C., Cook, C. E., Palmer, K. H., McPhail, A. T. and Sim, G. A. (1966). Plant antitumor agents. I. The isolation and structure of camptothecin, a novel alkaloidal leukemia and tumor inhibitor from *Camptotheca acuminata*. *J. Am. Chem. Soc.* **88**, 3888–3890.
2. Hsiang, Y. H., Hertzberg, R., Hecht, S. and Liu, L. F. (1985). Camptothecin induces protein-linked DNA breaks via mammalian DNA topoisomerase I. *J. Biol. Chem.* **260**, 14873–14878.
3. Oberlies, N. H. and Kroll, D. J. (2004). Camptothecin and taxol: historic achievements in natural products research. *J. Nat. Prod.* **67**, 129–135.
4. Govindachari, T. R. and Viswanathan, N. (1972). Alkaloids of *Mappia foetida*. *Phytochemistry* **11**, 3529–3531.
5. Springob, K. and Saito, K. (2002). Metabolic engineering of plant secondary metabolism: a promising approach to the production of pharmaceuticals. *Sci. Cult.* **68**, 76–85.
6. Darwin, S. P. (1976). The pacific species of Ophiorrhiza L. (*Rubiaceae*). *Lyonia* **1**, 47–102.
7. Aimi, N., Nishimura, M., Miwa, A., Hoshino, H., Sakai, S. and Haginiwa, J. (1989). Pumiloside and deoxypumiloside; plausible intermediates of camptothecin biosynthesis. *Tetrahedron Lett.* **30**, 4991–4994.
8. Kitajima, M., Fujii, N., Yoshino, F., Sudo, H., Saito, K., Aimi, N. and Takayama, H. (2005). Camptothecin and two new monoterpene glucosides from *Ophiorrhiza liukiuensis*. *Chem. Pharm. Bull.* **53**, 1355–1358.
9. Aimi, N., Tsuyuki, T., Murakami, H., Sakai, S. and Haginiwa, J. (1985). Structures of ophiorines A and B; novel type gluco indole alkaloids isolated from Ophiorrhiza spp. *Tetrahedron Lett.* **26**, 5299–5302.
10. Aimi, N., Murakami, H., Tsuyuki, T., Nishiyama, T., Sakai, S. and Haginiwa, J. (1986). Hydrolytic degradation of β-carboline-type monoterpenoid glucoindole alkaloids: a possible mechanism for harman formation in *Ophiorrhiza* and related Rubiaceous plants. *Chem. Pharm. Bull.* **34**, 3064–3066.
11. Aimi, N., Hoshino, H., Nishimura, M., Sakai, S. and Haginiwa, J. (1990). Chaboside, first natural glycocamptothecin found from *Ophiorrhiza pumila*. *Tetrahedron Lett.* **31**, 5169–5172.
12. Murashige, T. and Skoog, F. (1962). A revised medium for rapid growth and bioassays with tobacco tissue culture. *Physiol. Plant.* **15**, 473–479.
13. Gamborg, O. L., Miller, R. A. and Ojima, K. (1968). Nutrient requirements of suspension cultures of soybean root cells. *Exp. Cell Res.* **50**, 151–158.
14. Kitajima, M., Nakamura, M., Takayama, H., Saito, K., Stöckigt, J. and Aimi, N. (1997). Constituents of regenerated plants of *Ophiorrhiza pumila*; formation of a new glycocamptothecin and predominant formation of (3R)-deoxypumiloside over (3S)-congener. *Tetrahedron Lett.* **38**, 8997–9000.
15. Asano, T., Watase, I., Sudo, H., Kitajima, M., Takayama, H., Aimi, N., Yamazaki, M. and Saito, K. (2004). Camptothecin production by *in vitro* cultures of *Ophiorrhiza liukiuensis* and *O. kuroiwai*. *Plant Biotech.* **21**, 275–281.
16. Saito, K., Sudo, H., Yamazaki, M., Koseki-Nakamura, M., Kitajima, M., Takayama, H. and Aimi, N. (2001). Feasible production of camptothecin by hairy root culture of *Ophiorrhiza pumila*. *Plant Cell Rep.* **20**, 267–271.
17. Watase, I., Sudo, H., Yamazaki, M. and Saito, K. (2004). Regeneration of transformed *Ophiorrhiza pumila* plants producing camptothecin. *Plant Biotech.* **21**, 337–342.

Chapter 28

Metabolomic Analysis of *Ocotea odorifera* Cell Cultures: A Model Protocol for Acquiring Metabolite Data

Marcelo Maraschin, Paulo Fernando Dias, Ênio Luiz Pedrotti, Hiliana Nunes Ferreira Morais, Ana Maria Viana, and Karl Vernon Wood

Summary

Metabolomics constitutes a quantitative and qualitative survey of the whole metabolites of an organism as well as a tissue, reflecting the genome and proteome of a sample as analyzed. Advanced analytical spectroscopic and chromatographic techniques are used along with uni- or multivariate statistical data analysis, rapidly identifying up- or down-regulated metabolites in complex matrices. In this chapter, protocols for the analysis of *target compounds* (protocol I) and *metabolomics* (protocol II) of *Ocotea odorifera* cell cultures are described. In the first case, the target compound safrole, an aromatic ether used as a flavoring agent and also in the manufacture of insecticides, is analyzed in the organosolvent fraction of stable prototrophic cell lines of *O. odorifera* by gas chromatography-mass spectrometry. For metabolomics studies the protocol is designed to detect and quantify metabolites in the aqueous extract of *O. odorifera* cell lines by using high-resolution 1D- and 2D-nuclear magnetic resonance spectroscopy, followed by chemometric analysis of the 1H NMR spectra dataset. Protocol I has been successfully used, for example, in screening studies of cell lines able of producing safrole. Protocol II is suitable to detect the chemical features of a number of metabolite compounds in aqueous extracts of *O. odorifera* cell lines cultured under certain conditions, leading to new insights into metabolomics of that species.

Key words: Metabolomics, *Ocotea odorifera*, Plant cell cultures, Safrole, Gas chromatography-mass spectrometry, Nuclear magnetic resonance, Chemometrics, Target metabolite, Metabolic profiling

1. Introduction

The advent of genomics and genomic technologies has resulted in a paradigm shift in the emphasis of analytic research of biological systems. The traditional approaches of biochemistry and

S. Mohan Jain and Praveen K. Saxena (eds.), *Methods in Molecular Biology, Protocols for* In Vitro *Cultures and Secondary Metabolite Analysis of Aromatic and Medicinal Plants, vol. 547*
© Humana Press, a part of Springer Science+Business Media, LLC 2009
DOI: 10.1007/978-1-60327-287-2_28

molecular biology, where the cellular processes were investigated individually and often independent of each other, is giving way to a more global concept of analyzing the cellular compositions both in parallel and in its entirety, in order to obtain as complete a picture as possible of the metabolism. In fact, developments in analytic techniques are progressing rapidly to enable simultaneous, nonbiased, high-throughput measurements of several analytes, at the level of transcript (transcriptomics), proteins (proteomics), and metabolites (metabolomics). These developments are being driven primarily by the requirements in the healthcare sector so that the so-called *omics* approaches hold enormous potential for the pharmaceutical, healthcare, and agrochemical industries *(1)*.

In brief, metabolomics constitutes a quantitative and qualitative survey of the whole metabolites of an organism as well as a tissue, thus it reflects the genome and proteome of a sample as analyzed. Generally, such analysis is focused on metabolic differences that evidence responses to a range of extrinsic (ambient; multifactor) and intrinsic (genetic) stimuli. No single analytical method has been found to obtain a complete picture of the metabolome of an organism, thus requiring an adequate association of analytical methods to access that kind of information *(2–4)*. Metabolomics uses advanced analytical techniques gas-chromatography mass spectrometyr (GC-MS), liquid chromatography-mass spectrometry (LC-MS), nuclear magnetic resonance of hydrogen (1H NMR), nuclear magnetic resonance of carbon-13 13C NMR, ultra-high resolution Fourier transform ion cyclotron mass spectrometry (FT-MS) with univariate (analysis of variance [ANOVA], correlation analysis, regression analysis, or multivariate (principal component analysis [PCA], hierarchical analysis [HA], partial least square [PLS]), statistical data analysis to rapidly identify up- or down-regulated endogenous metabolites in complex matrices such as plant extracts.

There are various metabolomics approaches in use, for example, metabolite target analysis[1], metabolite profiling[2] (also referred to as metabolic profiling), metabolomics[3], metabolic fingerprint[4],

---

[1] *Metabolite target analysis*: typically an analysis restricted to metabolites of a particular enzymatic or organelle system (e.g.) that would be directly perturbed by (a)biotic factors *(7)*.

[2] *Metabolite profiling (metabolic profiling)*: analysis focusing on metabolites associated with a specific pathway or on a class of compounds such as alkaloids, terpenoids, amino acids, etc *(7)*.

[3] *Metabolomics*: comprehensive analysis of the whole metabolome under a give set of conditions *(7)*.

[4] *Metabolic fingerprint*. classification of samples on the basis of provenance taking into account their origin or biological relevance *(7)*.

and metabonomics[5] *(5)*. The choice for one or another approach depends on the scientific question to be answered; however, it is possible to design a typical workflow of a metabolomic study, which consists of the following steps *(6)*:

1. Definition of starting conditions, analytical/biological questions, and study design
2. Sample preparation
3. Separation of the analytes, if necessary
4. Analysis and quantification of analytes
5. Validation
6. Documentation
7. Chemometrics (bioinformatics)

By following this workflow, simple protocols for analysis of *target compounds* (protocol I) and *metabolomics* (protocol II) were designed for *Ocotea odorifera* (known in Brazil as "canela sassafras") cell cultures. In the first case, the target compound safrole [5-(2-propenyl)-1,3-benzodioxole – **Fig. 1**], an aromatic ether used as a flavoring agent and also in the manufacture of insecticides and anti-inflammatory drugs, was analyzed in the organosolvent fraction of a stable prototrophic cell line of *O. odorifera* by GC-MS. On the other hand, for metabolomics studies the ultimate goal of the protocol was to detect and quantify every metabolite in the aqueous extract of that cell line by using high-resolution 1D- and 2D-NMR (1H and 13C NMR), followed by chemometric (PCA) analysis of the 1H NMR spectra dataset.

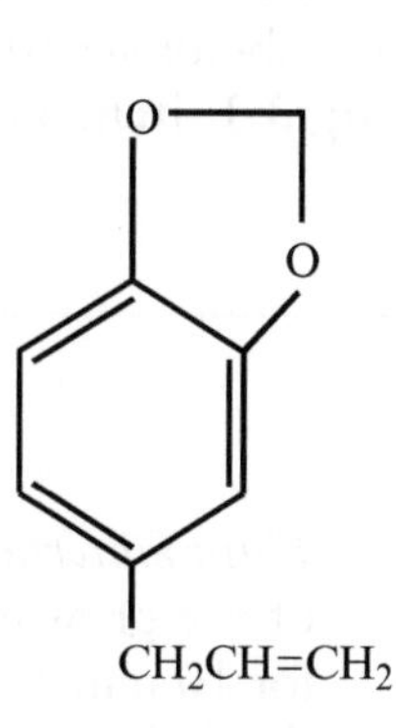

Fig. 1. Safrole [5-(2-propenyl)-1, 3-benzodioxole], the major compound in essential oil fraction of *Ocotea odorifera*.

[5] *Metabonomics*: measure of the fingerprint of biochemical perturbations caused by disease, drugs and toxins *(8)*.

2. Materials

2.1. Protocol I: Analysis of Target Compounds

2.1.1. Zygotic Embryo Culture, Induction of Callus Culture, and Cell Suspension Cultures

Mature fruits, sterile distilled water, 40% *(v/v)* commercial bleach, 0.1% *(v/v)* Tween-20 solution, flow hood, stereomicroscope (100X), orbital shaker, sterilized filter paper, Petri dishes, forceps and scalpel, 25 × 150 mm glass culture tubes, 80 × 80 mm transparent polypropylene (PVC) film.

2.1.2. Safrole Extraction and Gas Chromatography-Mass Spectrometry Analysis

First-grade methanol (MeOH) and chloroform ($CHCl_3$), 97% safrole (Sigma; St. Louis, MO –), rotatory evaporator, gas chromatograph equipped with a DB-1 capillary column (0.18 mm internal Ø, 40 m length) and mass spectrometer equipped with an electron impact ionization interface.

2.2. Protocol II: Metabolomics: Detection and Quantification of Metabolites in the Aqueous Extract by High-Resolution 1D- and 2D-NMR Spectroscopy

2.2.1. Solvents, Chemicals, Equipments, and Softwares

Liquid N_2, lyophilizer, Eppendorf centrifuge, deuterium oxide (D_2O – isotopic purity 99.9 atom % D (Cambridge Isotope Laboratories Inc.; Miami, USA), 3-(trimethylsilyl) propionic-2,2,3,3-δ_4 acid, sodium salt (TSP-δ_4) (Merck; Darmstadt, Germany), sodium deuteroxide (NaOD) (Cortec; Paris, France), potassium dihydrogen phosphate, 5 mm NMR sample tube, NMR spectrometer (Bruker 400 Avance III – 400.13 MHz 1H NMR frequency), workstation (Aspect 2000 data system), XWIN-NMR (v. 3.5, Bruker Biospin), AMIX (v. 3.7, Bruker Biospin), and STATISTICA (v. 5.0) softwares.

2.2.2. Cell Cultures

Prepare callus and cell suspension cultures of *Ocotea odorifera* for metabolomics studies as further described (please, *see* **Subheading 3.1, item 4**).

3. Methods

3.1. Protocol I: Analysis of Target Compounds

1. *Plant material.* Collect mature fruits of *O. odorifera* at the end of the growing season, wash them under running tap water for 10 min and store them in the dark at 4°C until use. Wash the fruits under running tap water for 20 min, transfer them to a 100-mL beaker and add 50 mL 40% *(v/v)* commercial bleach containing 1–2 drops of Tween-20. Agitate them for 25 min. For disinfecting, sequentially immerse into ethanol (EtOH) 70%/5 min, commercial bleach 40%/15 min, and EtOH 70%/1 min. Rinse the fruits five times with sterile distilled water in a flow hood.

2. Culture medium composition: The composition of the culture media used for obtaining cell samples for *target compound* and *metabolomic* analyses are summarized in **Table 1**. In all experiments, growth regulators are added to the MS culture medium *(9)* before autoclaving (121°C, ~110 KPa, 20 min). Adjust media to pH 5.8 before autoclaving.
3. *Zygotic embryo culture:* Remove the fruit and the seed coat with scalpel and forceps and isolate the zygotic embryo, 1- to 2-mm long, from the cotyledon under a stereomicroscope. Sort the excised embryos according to their diameter, discarding immature embryo (Ø < 8 mm). Immerse them in 0.15 g/L ascorbic acid solution and transfer one zygotic embryo/culture tube containing 12 mL Murashige and Skoog (MS) basal medium supplemented with 30 g/L sucrose, 7 g/L agar, 1.5 g/L activated charcoal, and 4.4 μM 6-benzylaminopurine (BAP) *(10)*. Seal the culture tubes with polypropylene film and incubate the cultures at 25 ± 2°C in the absence of light for 4–5 weeks.
4. Cell cultures: Initiate callus cultures from isolating zygotic embryo cultures inoculating *approximately* 250 mg biomass sample fresh weight on MS semi-solid medium (7 g/L agar-agar, 15 mL), supplemented with 30 g/L sucrose, 1.5 g/L activated charcoal, and 144.8 μM 2,4-dichlorophenoxy acetic acid (2,4-D). Seal the culture tubes with polypropylene film and incubate cultures at 25 ± 1°C, in the absence of light, for 30 d. Subsequently, cell suspension cultures are obtained by collecting 1.0 g pale green mass of friable callus and transferring it to a 125-mL Erlenmeyer flask containing 40 mL MS

Table 1
Composition of Culture Media used for Embryo Culture, Callus Culture, and Cell Suspension Cultures in Studies of Target Compound and Metabolomics of Ocotea odorifera

Purpose	Medium	Sucrose (g/l⁻)	Agar-agar (g/l)	Activate charcoal (g/l)	6-BAP (μM)	2,4-D (μM)	Picloram (μM)	Kinetin (μM)
Embryo culture	MS	30	7	1.5	4.4	–	–	–
Callus culture	MS	30	7	1.5	–	144.8	–	–
Cell suspension culture	MS[a]	30	–	–	–	–	10	1
Prototrophic cell culture	MS[a]	30	7	–	–	–	–	–

[a] Also supplemented with vitamins *(11)*

medium supplemented with 30 g/L sucrose, 1.5 g/L activated charcoal, and 144.8 μM 2,4-D. Seal the Erlenmeyer flask with PVC film and incubate cultures at 24 ± 1°C, maintaining under constant agitation, 110 rpm, in the dark. After 30 d, filter (Ø = 2 mm) the suspension cultured cells to remove large cell aggregates and re-inoculate 10 ml filtrate in 40 mL MS medium, supplemented with 30 g/L sucrose, 10 μM picloram, 1 μM kinetin, and vitamins *(11)*. Typical cell suspensions are obtained 21 d after subculturing. Cell suspensions are maintained in 250-mL Erlenmeyer flasks, at 24 ± 1°C, and shaking, 110 rpm, in the dark. Subculture 30 mL cell suspension culture/150 mL culture medium in 3-wk intervals.

5. Prototrophic cell culture: Prototrophic cell lines are obtained by aseptically collecting and filtering *ca.* 100 mL 21-d-old suspension culture cells as above described (**item 4**). The inoculum obtained (~1 g) is resuspended in minimum volume of MS fresh medium and transferred to Petri dishes containing 30 mL MS semi-solid medium (7 g/L agar-agar), supplemented with 30 g/l sucrose and vitamins *(11)*. Seal the Petri dishes with PVC film and incubate the cultures at 24 ± 1°C, 85% relative humidity, under a 16 h-photoperiod and photon flux of 31.5 μmol/m^2/s. Subculture in 3-wk intervals.
6. Safrole extraction and GC-MS analysis: After five generations, collect a cell sample (~2.5 g, dry weight) 30 days after subculturing and extract it with 25 mL $CHCl_3$: MeOH (1:1) solution for 24 h at 4°C, absence of light. The crude extract is filtered through a glass-fiber filter, centrifuged, 6,000 rpm/5 min and concentrated under reduced pressure to a final volume of approximately 0.8 mL. Safrole analysis is performed by directly injecting the organosolvent extract (1 μL) into a ThermoFinnigan GCQ mass spectrometer operated in the electron impact mode (70 eV) using a DB-1 capillary column, 0.18 mm internal diameter, 40 m length for compound separation. The column should be temperature programmed as follows: initial temperature 50°C, held for 0.1 min, and then ramped to 240°C at 10°C/min. The sample mass spectrum (*see* **Fig. 2**) and GC retention time are compared with authentic 97% safrole (Sigma for compound confirmation). Safrole content is evaluated as previously described *(12)* (i.e., calculating the response factor [RF]), the ratio of the peak height of the standard/concentration of the standard (μg), and then measuring the safrole peak height in the sample (SPH). The ratio SPH/RF furnishes the safrole content (μg) in the sample *(13)* (*see* **Note 1**).

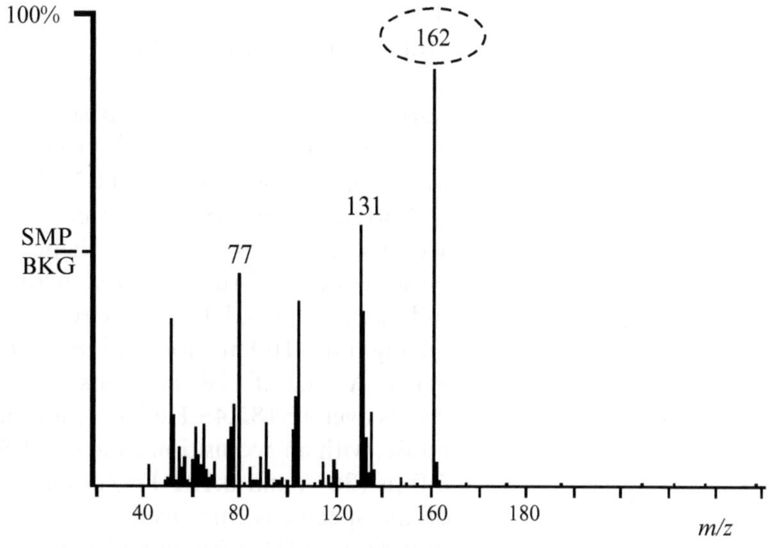

Fig. 2. Mass spectrum (EI mode) of the safrole peak, from the GC-MS analysis of an organosolvent extract of *Ocotea odorifera* cell suspension cultures, showing the base peak being the molecular ion (162 *m/z*), consistent with the mass spectrum of authentic safrole.

3.2. Protocol II: Metabolomics: Detection and Quantification of Metabolites in the Aqueous Extract by High-Resolution 1D- and 2D-NMR Spectroscopy

1. Sample preparation: Aseptically, collect 15 mL cell suspension samples as previously described **(item 4)** at days 5, 10, 15, 20, 25, and 30 after inoculation, following by filtration on a glass-fiber filter. Collect the cell biomass for each sample time and lyophilize it. Grind carefully the lyophilized material, 300 mg/sample, accurately weighted, using liquid N_2 in a mortar and pestle. Add 5 mL distilled-deionized water to the ground material mix it in a Vortex and centrifuge, 5000 rpm/5 min at 8°C. Collect the supernatant and repeat the extraction as before, pooling the aqueous extracts. Lyophilize the aqueous extract followed by addition of 800 μL D_2O buffered with KH_2PO_4. The pH 5.7 is adjusted with 1 N NaOD solution. Mix it in a Vortex and centrifuge, 13,000 rpm/10 min. Collect 650 μL supernatant for further 1D- and 2D-NMR spectroscopy analyses and store the samples at 4°C in a refrigerator with the internal atmosphere of the bottle (headspace) replaced by injection of N_2 gas.
2. High-resolution 1D and 2D nuclear magnetic resonance spectroscopy: Collect 540 μL aqueous extract sample and add it of 60 μl D_2O containing 0.024 g% of sodium 3-(trimethylsilyl)-propionate-2,2,3,3-δ_4 (TSP) as internal chemical shift standard. Transfer the sample to 5 mm NMR tubes and replace the internal atmosphere of the NMR tube by injection of N_2. Insert the sample into the NMR spectrometer, load/set

the acquisition parameters (see below), tune and match the probehead, establish field-frequency lock, and optimize field homogeneity (*shimming*). Start the data collection. High-resolution 1D- and 2D-NMR spectra might be recorded on a Bruker 400 Avance III spectrometer, operating at 400.13 MHz for hydrogen and 100.03 MHz for carbon, at 298 K and nonspinning. For one dimensional 1H NMR spectra are acquired using an modulate shape pulse (NOESYPR1DSP pulse sequence; Bruker library) with triple offset and amplitude scaling applied during the relaxation delay (1.5 s) and the mixing time (100 ms) for saturation of water (4.87 ppm) resonance. A total of 128 transients are collected with 32 K data points over a 5482.45 Hz bandwidth in the absolute intensity mode, with an acquisition time of 2.98 s, pulse width (PW) = 5.0 μs (30°), and 0.121 Hz/point. The signal-to-noise ratio of the spectra is improved by multiplying each free induction decay (FID) with an additional exponential factor corresponding to 0.5 Hz. For quantitative analysis, peak integral is used and the spectra are referenced to TSP resonance (δ–0.00 ppm). The whole peak intensities in every 0.03 ppm of the 1H NMR spectra (spectral window (SW) δ 0.30–11.5 ppm) are used as variables (*see* **Notes 2** and **3**).

3. For 2D 1H/1H NMR spectroscopy, total correlation spectroscopy (TOCSY) is used by acquiring the spectra in phase sensitive mode using time proportional phase incrementation (TPPi), with a MLEV17 pulse sequence for excitation and spin-lock *(14)*. Twenty-four scans are acquired for each of the 512 increments with 1024 data points over a spectral width of 5482.45 Hz in both dimensions, a relaxation delay of 1.5 s and a mixing time of 100 ms. The measurement time is about 6 h. The 2D heteronuclear (1H/13C) single quantum correlation (HSQC) spectra are acquired with inverse detection, 13C decoupling during acquisition, consisting of 2048 data points over a 5482.45–25,153.81 Hz bandwidth in the 1H- and 13C-dimensions, respectively. Forty-eight scans are acquired for each of the 2048 300 increments, with a measurement time of approximately 9 h. HSQC spectra are processed by applying a sine squared function and sine squared constant both F1 and F2 dimension (*see* **Notes 4** and **5**).

4. Chemometrics: The 1H NMR spectra dataset shall be automatically reduced to ASCII files (or JCAMP files) using AMIX software (version 3.7, Bruker Biospin). Scale the spectral intensities to TSP resonance (δ–0.00 ppm) and reduce them to integrated regions of equal width (0.03 ppm) considering the SW of δ 0.30–11.5 ppm. Exclude the region of δ 4.60–5.80 ppm of the data matrix before the analysis because of the effect of the residual signal of water. PCA and

PLS calculations of the 1H NMR spectra dataset might be performed with the AMIX and STATISTICA softwares.

4. Notes

1. Protocol I uses GC-MS analysis to detect the target compound safrole in the organosolvent fraction of *O. odorifera* cell cultures. This technique is both robust and reproducible and utilizes only a minimal amount of sample. Following protocol I, autotrophic and heterotrophic cell lines for growth regulators of *O. odorifera* have been screened regarding safrole content, revealing a wide range of values (19.92–143.3 μg/g dry weight of safrole) according to the sort and concentration of growth regulators supplementing the culture medium. Interestingly, up to now, all the cell lines analyzed (>25, mean content 62.6 μg/g dry weight of safrole) have been able to synthesize the phenylpropanoid safrole, indicating that the adopted cell line selection strategies did not inactivate its biosynthetic pathway. Furthermore, the coupling of GC to MS has also allowed the identification of other compounds in addition to safrole in the crude extract of *O. odorifera* prototrophic cell cultures. Comparing the mass spectra of unknown peaks from the GC-MS analysis with the data available in mass spectral libraries will make it possible to identify secondary metabolites (i.e., metabolic profiling). Finally, when protocol I was applied to *Ocotea catharinensis* (canela preta) cell cultures, no safrole was detected in either MeOH or $CHCl_3$ fractions. This reveals the power of protocol I as a tool for screening potential species for the presence of the phenylpropanoid safrole.
2. It is worth mentioning that the technology platform of choice in metabolomics depends on the type of the sample to be analyzed, but currently the ultimate goal of metabolomics experimentsI (i.e., to quantify all the metabolites in a cellular system) is not realistic given the lack of simple and automated analytic strategies that can make it in a reproducible and robust way *(5)*, especially in combination with statistical evaluation of the data. In fact, NMR spectroscopy has proven to be of enormous potential in metabolomics studies, but one should be aware that care must be taken in setting up the NMR spectrometer prior to acquiring data. Whenever a new sample is placed within the NMR spectrometer the instrument must be optimized for this. The precise nature of the adjustments required and the amount of time spent making these will depend on the sample, the spectrometer and the nature of the experi-

ment. For example, inherent traits of the samples, pH, viscosity, or unexpected static field inhomogeneities might arise during the total sampling period degrading the NMR signals. However, in all cases the aim will be to achieve optimum resolution and sensitivity and to ensure system reproducibility as described in protocol II.

3. For quantitative analysis of compounds by 1H NMR spectroscopy, the longitudinal relaxation time (T_1) should be measured. For spins to relax fully after a 90° magnetization pulse, it is necessary to wait a period of at least $5T_1$. In general, for medium sized organic molecules (e.g., with a mass of a few hundred), proton T_1 s tend to fall in the range 0.5–5 s *(15)*.
4. Metabolite identification in 1D- and 2D-NMR spectra is performed taken into account. For example, the chemical shifts of any given unknown resonance set in comparison to those found for standard compounds analyzed under identical experimental conditions. For that, 1D- and 2D-NMR databanks

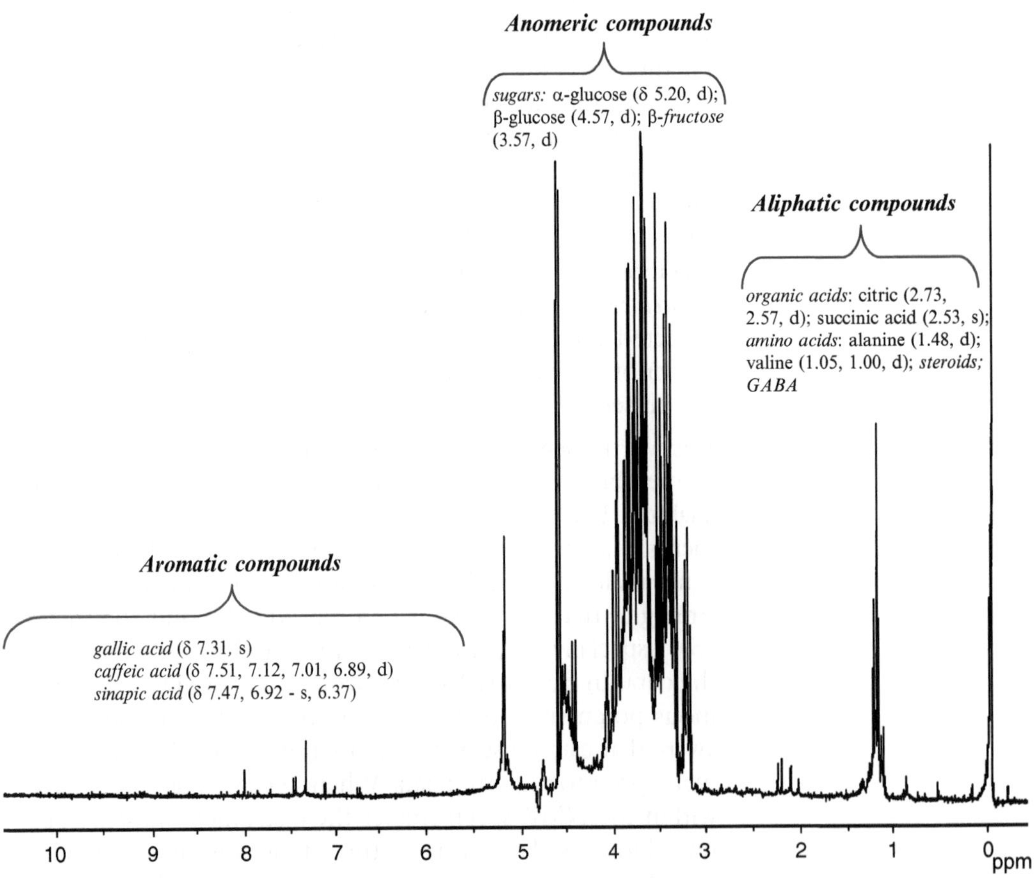

Fig. 3. 1H-nuclear magnetic resonance spectrum of an aqueous extract sample of *Ocotea odorifera* suspension cultured cells. Typical regions of occurrence of resonances of aliphatic, anomeric, and aromatic compounds are shown.

are useful tools for assignments of 1H- and 13C-resonances in respect to the candidate metabolite. However, one should be aware regarding the acquisition conditions (e.g., pH, temperature, solvent) used to collect both spectra (i.e., standard compound and sample spectra for purpose of comparison of the chemical shifts similarities). Homo (1H/1H) and heteronuclear (1H/13C) NMR experiments are indispensables for an unambiguously assignment of resonances and compound identification.

5. Regarding protocol II, in this chapter the metabolomic study under focus was performed in accordance with a standard workflow by using advanced analytic techniques on the extracts of *O. odorifera* cell cultures that included high-resolution 1D- and 2D-NMR (1H and 13C NMR)—a powerful and nonselective spectroscopy. Moreover, after collecting and processing NMR data, PCA was applied to the 1H NMR spectra dataset in order to feasibly detect the chemical features based on a number of metabolite compounds in the complex matrices (i.e., *O. odorifera* aqueous extracts) in a time-course experiment. By following such a protocol approximately 26 primary and secondary metabolites have been unambiguously identified up to now and some of them also quantified (α- and β-glucose, for instance – **Fig. 3**). PCA has shown major discrepancies for *O. odorifera* metabolism in cell samples in initial stages of culturing (0–10-d-old) as compared to those older than 15 d, especially considering the higher contents of sugars as α- and β-glucose/fructose and some phenolic compounds (e.g., gallic and caffeic acids) in early and late stages of the cell cultures, respectively.

References

1. Vaidyanathan, S. and Goodcare, R. (2000). Proteome and metabolome analyses for food authentication, in *Food Authenticity and Traceability* (Lees, M., ed.), CRC, Boca Raton, FL, pp. 69–99.
2. Abhyankar, G., Reddy, V.D., Giri, C.C., Rao, K.V., Lakshmi, V.V.S., Prabhakar, S., Vairamani, M., Thippeswamy, B.S., and Bhattacharya, P.S. (2005). Amplified fragment length polymorphism and metabolomic profiles of hairy roots of *Psoralea corylifolia* L. *Phytochemistry* **66**, 2441–2457.
3. German, J.B., Watkins, S.M., and Fay, L. (2005). Metabolomics in practice: emerging knowledge to guide future dietetic advice toward individualized health. *J. Am. Diet. Assoc.* **105**, 1425–1432.
4. Wind, R.A., Hu, J.Z., and Majors, P.D. (2005). Slow-MAS NMR: a new technology for *in vivo* metabolomic studies. *Drug Discov. Today* **2**, 283–286.
5. Goodcare, R., Seetharaman, V., Dunn, W.B., Harrigan, G.G., and Kell, D.B. (2004). Metabolomics by numbers: acquiring and understanding global metabolite data. *Trends Biotechnol.* **22**, 245–252.
6. Lay, Jr.J.O., Borgmann, S., Liyanage, R., and Wilkins, C.L. (2006). Problems with the "omics". *Trends Anal. Chem.* **25**, 1046–1056.
7. Fiehn, O. (2001). Combining genome, metabolome analysis, and biochemical modeling to understand metabolic networks. *Compr. Funct. Genomics* **2**, 155–168.

8. Nicholson, J.K., Lindon, J.C., and Holmes, E. (1999). 'Metabonomics': understanding the metabolic responses of living systems to pathophysiological stimuli via multivariate statistical analysis of biological NMR spectroscopic data. *Xenobiotica* **29**, 1181–1189.
9. Murashige, T. and Skoog, F. (1962). A revised medium for rapid growth and bioassays with tobacco tissue cultures. *Physiol. Plant.* **15**, 473–497.
10. Catarina, C.S., Oltramari, A.C., Voltolini, J.A., Pedrotti, E.L., and Maraschin, M. (1999). *Ocotea odorifera* Mez. cell suspension cultures – an alternative approach for biomass and safrole production, in *Proceedings International Congress on Forestry*. EMBRAPA, Colombo, Parana, p. 381.
11. Morel, G.M. and Wetmore, R.H. (1951). Fern callus tissue culture. *Am. J. Bot.* **38**, 141–143.
12. Snyder, L.R., Kirkland, J.J., and Glajch, J.L. (1997). *Practical HPLC Method Development*. Wiley, New York, NY.
13. Oltramari, A.C., Wood, K.V., Bonham, C., Verpoorte, R., Caro, M.S.B., Viana, A.M., Pedrotti, E.L., Maraschin, R.P., and Maraschin, M. (2004). Safrole analysis by GC-MS of prototrophic (*Ocotea odorifera* (Vell.) Rohwer cell cultures. *Plant Cell Tissue Organ Cult.* **78**, 231–235.
14. Bax, A. and Davis, D.G. (1985). MLEV-17-based two-dimensional homonuclear magnetization transfer spectroscopy. *J. Magn. Reson.* **65**, 355–360.
15. Claridge, T.D.W. (1999). *High-Resolution NMR Techniques in Organic Chemistry*. Elsevier Science Ltd, Oxford, UK.

Chapter 29

The Production of 9-methoxycanthin-6-one from Callus Cultures of (*Eurycoma longifolia* Jack) Tongkat Ali

Mahmood Maziah and Noormi Rosli

Summary

Plant cell culture technology is potentially useful in producing high-valued secondary metabolites. *Eurycoma longifolia* root extracts are consumed as a health tonic but more popularly used as an aphrodisiac. Studies on the aphrodisiac properties and the possible compounds involved have been widely studied. There are many potentially useful compounds reported from the root extracts of *E. longifolia*. However, studies on the in vitro production of useful compounds from this plant have not been reported. This chapter will describe methods of callus induction and extraction of 9-methoxycanthin-6-one from *E. longifolia* Jack explants with emphasis on the tap and fibrous roots. This compound, known to have anti-tumour activity, is present in intact plant parts and in callus tissues of different explants.

Key words: Callus tissues, *Eurychoma Longifolia* Jack, 9-methoxycanthin-6-one, Explants, Roots

1. Introduction

Plants produce a variety of economically important secondary metabolites, some of which are used in health care, food, flavor, and cosmetics industries *(1)*. Others are used for the production of agrochemicals, pharmaceuticals, fragrances, colorings, and biopesticides *(2, 3)*. Natural production of beneficial bioactive compounds from plants generally involves extraction from the wild or cultivated plants. This method is often tedious, costly and results in inconsistent yields. In addition, the compounds of interest could be available only seasonally. Plant cell culture techniques can overcome some of these problems. This approach has the advantages over the conventional agriculture production as

S. Mohan Jain and Praveen K. Saxena (eds.), *Methods in Molecular Biology, Protocols for* In Vitro *Cultures and Secondary Metabolite Analysis of Aromatic and Medicinal Plants*, *vol. 547*
© Humana Press, a part of Springer Science+Business Media, LLC 2009
DOI: 10.1007/978-1-60327-287-2_29

it is independent on geographical seasonal variations, the continuous supply of products, uniform quality and yield is assured. There is an increasing number of reports on using plant tissues and cells to produce a wide range of different secondary compounds such as betaxanthins from *Beta vulgaris* callus cultures *(4)*, flavonoids from in vitro cultures of *Scutellaria baicalensis(5)*, rosmarinic acid from *Satureja hortensis* L. callus cultures *(6)*, and saponins from shoots, callus, suspension and root cultures of *Primula veris* L. *(7)*. Extracts from the roots of *E. longifolia* Jack, also locally known as 'Tongkat Ali,' are reputed to increase male virility and has gained notoriety as a male aphrodisiac *(8)*. Pharmacologic evaluation of the various compounds isolated from *E. longifolia* Jack showed that they also possess antimalarial *(8, 9)*, pesticidal *(10)*, and cytotoxic *(11)* properties. The presence of 9-methoxycanthin-6-one in plant parts of *E. longifolia* Jack has been reported by several researchers *(12–15)*. However, its production in callus cultures has not been reported. This compound is an alkaloid and was reported to have anti-tumor properties *(8, 12, 16)*. Because this is the first report on the 9-methoxycanthin-6-one production in callus cultures the focus of the study is to identify the basic requirements for callus culture of *E. longifolia* Jack such as types of basal media, carbon sources, auxins, pH and amino acids on the production of 9-methoxycanthin-6-one.

2. Materials

2.1. Induction and Establishment of Callus Cultures

1. Different plant parts. Leaves, cotyledons, leaf, rachises, stems, petioles, zygotic embryos, tap roots and fibrous roots obtained from five-year old plants grown in 50 L pots.
2. *Basal media.*: Murashige and Skoog (MS) (**Table 1**) *(17)* medium. The inorganic salts used as major and trace elements should be dissolved in double deionized water or highly pure grade water. Major elements are prepared individually. Trace elements can be combined with the exception of Ferric Na ethylene diamine tetraacetic acid (EDTA). Vitamins can be combined. Stocks of vitamins and organic supplements are to be stored in the refrigerator. It is important that all solutions are properly labeled (*see* **Note 1**).
3. Auxins: Dichlorophenoxyacetic acid (2,4-D) (Sigma Aldrich; Steinheim, Germany), 3, 6-Dichloro-*o*-aniscic acid (Dicamba) (Duchefa-Haarlem, The Netherlands), 4-Amino-3, 5, 6-trichlorpicolinic acid (Picloram) (Duchefa-Haarlem), naphthaleneacetic acid (NAA) (Sigma Aldrich), indoleacetic acid (IAA) (Sigma Aldrich): Prepare individual stock solutions

Table 1
Murashige and Skoog Medium

Major salts	Concentration (mg/L)
KNO_3	1900
NH_4NO_3	1650
$MgSO_4{\cdot}7H_2O$	370
$CaCl_2{\cdot}2H_2O$	440
KH_2PO_4	170
Minor salts	Concentration (mg/L)
H_3BO_3	6.20
$MnSO_4{\cdot}4H_2O$	22.30
$ZnSO_4{\cdot}7H_2O$	8.60
$Na_2MoO_4{\cdot}2H_2O$	0.25
$CuSO_4{\cdot}5H_2O$	0.025
$CoCl_2{\cdot}6H_2O$	0.025
KI	0.83
FeNaEDTA	65.00

(100 μM). 2,4-D and NAA can be initially dissolved in 2–5 mL ethanol. Then gradually add high purity grade water. Heat slightly for 2,4-D. Adjust the pH to 5.0 for NAA. IAA is dissolved in minimal volume of 1N KOH. Add water and store in dark bottle in the refrigerator Dissolve Picloram in 2–5 mL 0.2 N KOH and add water, then adjust the pH to 5.0 (*see* **Note 2**).

4. Gelrite (Phytotechnology Laboratories; USA).
5. Flat bottom vials (25 × 95 mm H) ((Phytotechnology Laboratories).
6. 150-mL Erlenmeyer flasks.

2.2. Analysis of 9-methoxycanthin-6-one

2.2.1. Thin Layer Chromatography

1. Thin layer chromatography (TLC) aluminium sheets coated with Silica gel 60 F_{254} (Merck catalog no. 1.05554).
2. 366-nm Ultraviolet lamp. 3. Chloroform ($CHCl_3$) (Merck; Germany): Methanol (CH_3 OH) (Sigma Aldrich).

2.2.2. High Performance Liquid Chromatography (HPLC)

1. High-performance liquid chromatography (HPLC) grade acetonitrile.
2. HPLC grade water.

3. Methods

3.1. Induction of Callus from Different Explants

Grow *E. longifolia* Jack plants in 50-L pots in the experimental field. Water the plants daily and supply with commercial fertilizers, 21N:21P:21K. Excise the *E. longifolia* explants from 5-yr old plants (**Fig. 1**).

3.1.1. Plant Preparation

Wash them thoroughly with detergent and rinse under running tap water for 30 min. Add two drops of Tween-20 as wetting agent into 15% *(v/v)* commercial sodium hypochlorite (Clorox) solution for 20 min, and final rinsing with sterile water. Cut sterilized explants in one cm pieces and transfer into vials (8.4 cm × 2.4 cm) each containing different concentrations of different auxins.

3.1.2. Media Preparation

Prepare MS basal media (*see* **Note 1**) containing 3% sucrose, 2.5% gelrite, and different concentrations 0, 1.0, 2.0, 3.0, 4.0 and 5.0 mg/L of auxins: 2,4-D, dicamba, picloram, NAA, and IAA

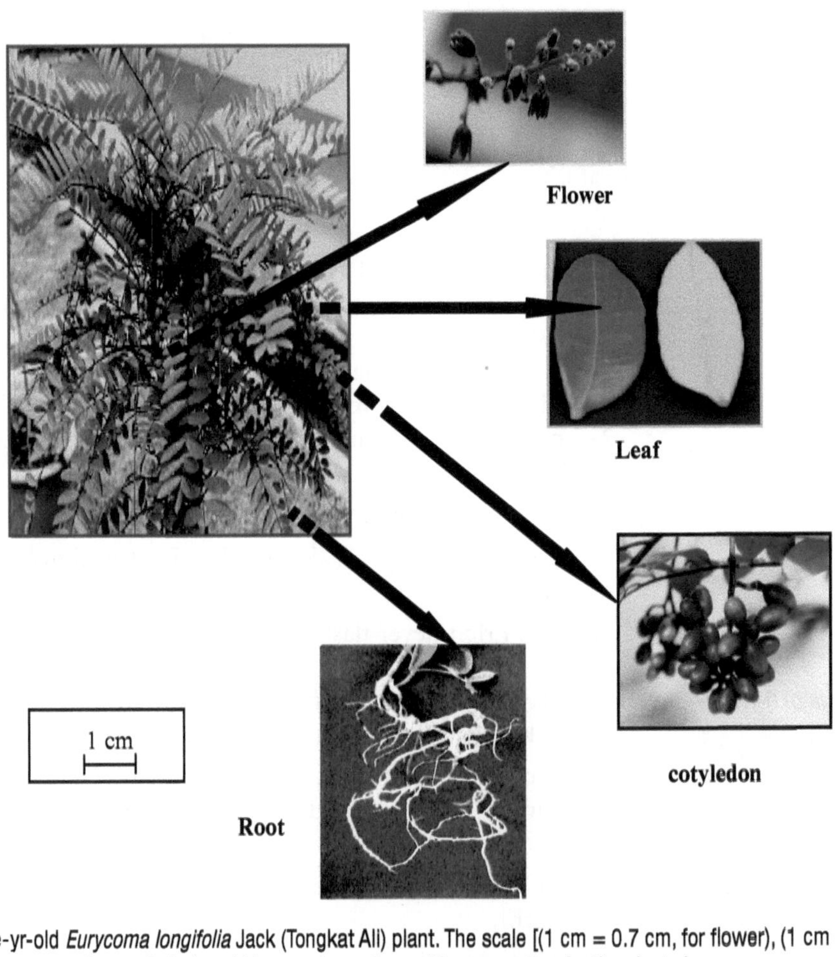

Fig. 1. Five-yr-old *Eurycoma longifolia* Jack (Tongkat Ali) plant. The scale [(1 cm = 0.7 cm, for flower), (1 cm = 4 cm, for leaf), (1 cm = 1 cm, for cotyledon) and (1 cm = 1 cm, for root)] representing for the plant above.

Fig. 2. Three-wk-old *Eurycoma longifolia* Jack callus after five passages. The scale (1 cm = 2 cm).

(see **Note 2**). Adjust the pH of the media to 5.7 prior to adding 2.5% gelrite and autoclave at 106 kPa, 121°C for 15 min. Avoid adding auxin in control. Incubate cultures at 25 ± 2°C in the dark and observe daily to determine the time of callus formation. Carefully aseptically excise callus tissues and transfer onto the fresh medium.

3.1.3. Culture and Maintenance of Callus Tissues

Multiply callus cultures onto fresh MS supplemented with 2.0 mg/L 2,4-D at 3-wek intervals. Callus cultures are well established after five passages and maintain them in the same medium (**Fig. 2**). Three-wk-old cultures are used for all the subsequent studies.

3.2. Effect of Different Auxins and Media Composition on 9-Methoxycanthin-6-One Production

Add five different auxin concentrations in MS medium prior autoclaving to determine their effects on 9-methoxycanthin-6-one production. Use different strength of basal media (full, half, quarter), MS, B_5 *(18)* (**Table 2**), SH *(18)* (**Table 3**) and WH *(19)* (**Table 4**). Harvest callus tissues after five weeks of culture. In all cases B_5 vitamins were used for all the basal media examined.

3.3. Effect of Different Carbon Sources on Callus Growth and 9-Methoxycanthin-6-One Production

Prepare the media MS medium containing 0.25% gelrite, 2.0 mg/L 2,4-D and different concentrations (0–5% *[w/v]*). of carbon sources such as sucrose, glucose, fructose, sorbitol and mannitol. Harvest callus after 5 wk of culture and analyze 9-methoxycanthin-6-one content (*see* **Note 3**).

3.4. Effect of Different Initial pH Values on Callus Growth and 9-Methoxycanthin-6-One Production

Prepare MS medium supplemented with 2.5% gelrite, 3.0% sucrose with the initial pH values adjusted at room temperature prior to autoclaving to 5.0, 5.5, 6.0, 6.5, 7.0, 7.5, and 8.0 using either 0.01N NaOH or 0.01 N HCl. Harvest callus tissue star-ting from the first week of culture and continue until the seventh week to determine callus biomass and 9-methoxycanthin-6-one production.

Table 2
Gamborg (B5) Medium

Major salts	Concentration (mg/L)
KNO_3	2,500
$(NH_4)_2SO_4$	134
$MgSO_4{\cdot}7H_2O$	250
$CaCl_2{\cdot}2H_2O$	150
$NaH_2PO_4{\cdot}H_2O$	150
Minor salts	Concentration (mg/L)
H_3BO_3	3.0
$MnSO_4{\cdot}H_2O$	10.0
$ZnSO_4{\cdot}7H_2O$	2.0
$Na_2MoO_4{\cdot}2H_2O$	0.25
$CuSO_4{\cdot}5H_2O$	0.025
$CoCl_2{\cdot}6H_2O$	0.025
KI	0.75
FeNaEDTA	65.00
Vitamins	Concentration (mg/L)
Nicotinic acid	1.0
Thiamine HCl	10.0
Pyridoxine HCl	1.0
Myo-inositol	100.0

Table 3
Schenck and Hildebrandt (SH)

Major salts	Concentration (mg/L)
KNO_3	2500
$MgSO_4{\cdot}7H_2O$	400
$CaCl_2{\cdot}2H_2O$	200
$NaH_2PO_4{\cdot}H_2O$	300
Minor salts	Concentration (mg/L)
H_3BO_3	5
$MnSO_4{\cdot}H_2O$	10

(continued)

Table 3
(continued)

Major salts	Concentration (mg/L)
$ZnSO_4 \cdot 7H_2O$	1.0
$Na_2MoO_4 \cdot 2H_2O$	0.1
$CuSO_4 \cdot 5H_2O$	0.02
$CoCl_2 \cdot 6H_2O$	0.1
KI	1.0
FeNaEDTA	40

Table 4
White's Medium(WH)

Major salts	Concentration (mg/L)
$Ca(NO_3)_2 \cdot 4H_2O$	200
KCl	65
$MgSO_4 \cdot 4H_2O$	360
Na_2SO_4	200
KNO_3	80
Na_2PO_4	16.5
Minor salts	Concentration (mg/L)
H_3BO_3	1.5
$MnSO_4 \cdot H_2O$	5.04
$ZnSO_4 \cdot 7H_2O$	2.67
MoO_3	0.001
$CuSO_4 \cdot 5H_2O$	0.01
$CoCl_2 \cdot 6H_2O$	0.1
KI	0.75
$FeSO_4 \cdot 7H_2O$	2.5

3.5. Effect of Different Phenylalanine, Tryptophan, Tyrosine and on Callus Growth and 9-Methoxycanthin-6-One Production

The MS medium is prepared that contains 3.0% sucrose and 2.5% gelrite. Transfer callus onto media containing different types and concentrations of amino acids; DL-tryptophan, L-phenylalanine and L-tryrosine and culture for 5 wk. Prepare stock solutions of amino acids in distilled water and filter-sterilize with a 0.2-μm polyethersulfone membrane. Add different amino acid concentrations into sterile 10 mL MS medium supplemented with 3% *(w/v)* sucrose (*see* **Note 4**).

3.6. Preparation of 9-Methoxycanthin-6-One Standard

Dissolve 5.0 μg/mL *(w/v)* 9-methoxycanthin-6-one in absolute methanol. Prepare the standard curve by preparing different dilutions of 9-methoxycanthin-6-one. Filter each of them through a Waters Sep-Pak Classic Cartridge prior to HPLC analysis.

3.7. Extraction of 9-Methoxycanthin-6-One from Callus Tissues

Harvest the callus tissues and dry in an oven at 45°C for 48 h or when the dry weight is constant. Weigh 1 g of dried samples, ground it with pestle and mortar before adding 20 mL solvent 4:1 (CH_3OH:$CHCl_3$). Filter the homogenate through four layers of miracloth, and centrifuge at 12,000*g* for 10 min at 4°C. Filter further the extracts using Waters Sep-Pak Classic Cartridge for HPLC analysis.

3.8. Thin Layer Chromatography

Spot 1.5 μL standard and samples on a TLC plate (*see* **Note 5**). Develop a plate with solvent $CHCl_3$:CH_3OH (2:8). 9-methoxycanthin-6-one is detected in fluorescence emission under irradiation of UV lamp (366 nm) (**Fig. 3**).

3.9. High Performance Liquid Chromatography

Perform chromatography on Agilent 1100 Series comprising of a manual injector, a diode array detector, a quaternary pump and vacuum degasser with external 20-μL sample loop. Perform the analyses with Novapack C18 steel cartridge column i.d. 3.9 × 150 mm. (Waters Associates; cat. no. No. 36975). The particle size is 4.0 μm. The mobile phase consists of a mixture of acetonitrile and distilled deionised water 40:60, HPLC grade, acidified with

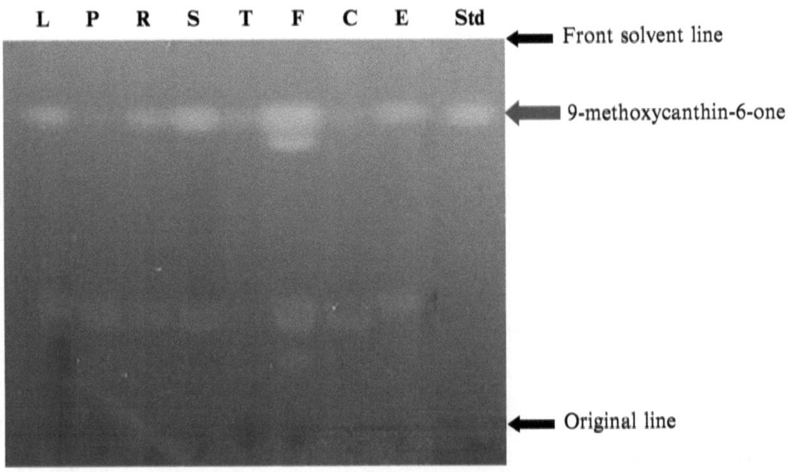

Fig. 3. TLC of 9-methoxycanthin-6-one on aluminium sheets coated with Silica gel 60 F_{254} and developed using chloroform and methanol solvent (Callus cultures). The quantity spotted to the actire TLC plate was 1.5 μl each. *L*, Leaf; *P*, Petiole; *R*, Rachis; *S*, Stem; *T*, Tap root; *F*, Fibrous root; *C*, Cotyledon; *E*, Embryo; *Std*, Standard 9- methoxycanthin- 6-one. The R_f value of 9-methoxycanthin-6-one was 0.83.

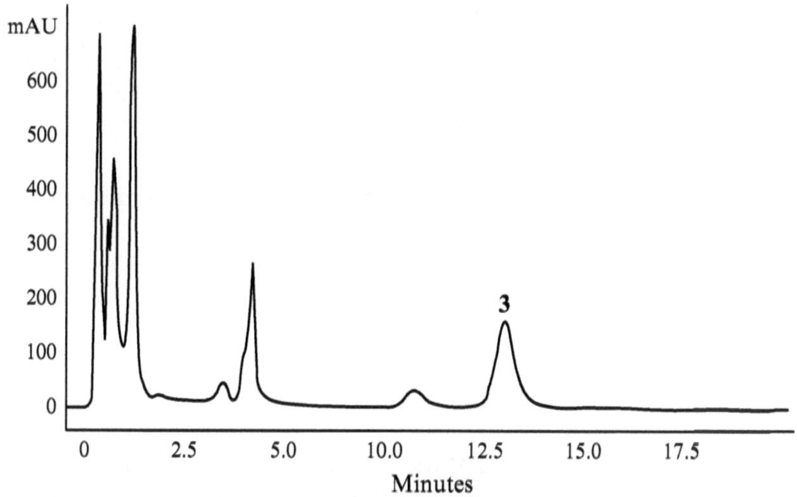

Fig. 4. HPLC profile showing 9-methoxycanthin-6-one peak (peak no. 3).

trifluroacetic acid to pH 2.5. The flow rate is 1.0 mL/min at room temperature. 9-methoxycanthin-6-one can be detected at 272 nm. The amount used for injection is 1.5 μL. Repeat three times. The concentration of the compound is determined by comparing the retention time values of the standard. **Figure 4** shows the chromatographic of HPLC profile. The retention time of 9-methoxycanthin-6-one is 11.40 min.

4. Notes

1. Concentrations of stocks generally prepared as 10X or 100X of the desired strength. FeNa EDTA is prepared separately to prevent precipitation Ferric Na EDTA and kept in dark bottle. Store each macro solution separately at room temperature, and similarly store micronutrient solutions in the refrigerator. Each major and minor element stock solution is stored preferably in glass containers.
2. The auxins are dissolved in 1N NaOH or KOH before diluting with double-distilled water and kept in the refrigerator.
3. The carbon sources used are: D-Sucrose (molecular weight [MW] 342.30), D(+) glucose monohydrate (MW198.17), D(−) fructose (MW180.6), D-Sorbitol (MW 182.20), and D-mannitol (MW182.17).
4. Three types of amino acids are used, DL-tryptophan (MW: 204.2), L-phenylalanine (MW: 165.2), and L-tryrosine (MW: 181.2).

Prepare stock solutions of amino acids in distilled water and filter-sterilize with a 0.2-μm polyethersulfone membrane. Use stock solutions to prepare a series of v ofvamino acid concentrations, 0, 1×10^{-4}, 1×10^{-3}, 1×10^{-2}, 1×10^{1}, 1.0, 1×10^{1}, 1×10^{2}, 1×10^{3}, 1×10^{4} μM.

5. The TLC plates used were aluminum plates coated with Silica gel 60 $F_{254.}$

Acknowledgments

The 9-methoxycanthin-6-one standard was kindly provided by Professor Chan Kit Lam, School of Pharmaceutical Science, University of Science Malaysia (USM), Penang. This project was supported by the Ministry of Science and Innovation. The authors would like to acknowledge the technical assistance of Mr. M. Hakiman.

References

1. Pascal, G. and Johan, M. (2002). Transcription factors: tools to engineer the production of pharmacologically active plant metabolites. *Trends in Pharmacological Sciences*, **23**: 563–569.
2. Ramachandra, S. and Ravishankar, G. (2002). Plant cell cultures: Chemical factories of secondary metabolites. *Biotechnology Advances*, **20**: 101–153.3.
3. Verpoorte, R. and Memelink, J. (2003). Engineering secondary metabolite production in plants. *Plant Biotechnology*, **13**: 181–187.
4. Trejo-Tapia, G., Balcazar-Aguilar, J.B., Martinez-Bonfil, B., Salcedo-Morales, G., Jarmillo-Flores, M., Arenas-Ocampo, M.L. and Jiménez-Aparicio, A. (2008). Effect of screening and subculture on the production of betaxanthins in *Beta vulgaris* L. var. "Dark Detroit" callus cultures. *Innovative Food Science & Emerging Technologies*, **9(1)**: 32–36.
5. Martin J. and Dusek J. (2007). Flavonoid accumulation in *Scutellaria baicalensis* Georgii *in vitro* cultures upon treatment with sodium cinnamate. *Ceska a Slovenska Farmacie*, **56**(6): 280–283.
6. Tepe, B. and Sokmen, A. (2007). Production and optimization of rosmirinic acid by *Stureja hortensis* L. callus cultures. *Natural Product Research*, **21**(13): 1133–1144.
7. Okrslar, V., Plaper, I., Kovac,M., Erjavac, A., Obermajer, T., Rebec, A., Ravnikar, M. and Zel, J. (2007). Saponins from tissue culture of Primula veris L. *In Vitro Cellular & Developmental Biology – Plant*, **43**(6): 644–651.
8. Kuo, P.C., Shi, L.S., Damu, A.G., Su, C.R., Huang, C.H., Ke, C.H., Wu, J.B., Lin, A.J., Bastow, K.F., Lee, K.H. and Wu, T.S. (2003). Cytotoxic and antimalarial β-carboline alkaloids from the roots of *Eurycoma longifolia*. *Journal of Natural Products*, **66**: 1324–1327.
9. Ping-Chung, K., Amooru, G.D., Kuo-Hsiung, L. and Tian-Shung, W. (2004). Cytotoxic and antimalarial constituents from the roots of *Eurycoma longifolia Bioorganic and Medicinal Chemistry*, **12**: 537–544.
10. Jankovic, T., Dijana, K. Savikin-Fodulovic, K., Menkovic, N. and Grubisic, D. (2002). Promoting and anti-parasitic activities of the quassinoids from *Eurycoma longifolia*, a medicinal plant in Southeast Asia. *Journal of Ethnopharmacology*, **82**: 55–58.
11. Morita, H., Kishi, E., Takeya, K., Itokawa, H. and Itaka, Y. (1993). Highly oxygenated quassinoids from *Eurycoma longifolia*. *Phytochemistry*, **33**(3): 691–696.
12. Jiwajinda, S., Santisopasri, V., Murakami, A., Kawanaka, M., Kawanaka, H., Gasquet, M., Eilas, R., Balansard, G. and Ohigashi, H.

(2002). *In vitro* anti-tumor promoting and anti-parasitic activities of the quassinoids from *Eurycoma longifolia*, a medicinal plant in Southeast Asia. *Journal of Ethnopharmacology*, **82**: 55–58.

13. Mitsunaga, K., Koike, K., Tanaka, T., Ohkawa, Y., Kobayashi, Y., Sawaguchi, T. and Ohmoto, T. (1994). Canthin-6-one alkaloids from *Eurycoma longifolia*. *Phytochemistry*, **35**: 799–802.
14. Ang., H.H., Hitotsuyanagi, Y., Fukaya, H. and Takeya, K. (2002). Quassinoids from *Eurycoma longifolia*. *Phytochemistry*, **59**: 833–837.
15. Choo, C.Y. and Chan, K.L. (2002). High performance liquid chromatography analysis of canthinone alkaloids from *Eurycoma longifolia*. *Planta Medica*, **68**: 382–384.
16. Kardono, L.B.S., Angerhofer, C.K., Tsauri, S., Padmawinata, K., Pezzuto, J.M. and Kinghorn, A.D. (1991). Cytotoxic and antimalarial constituents of the roots of *Eurycoma longifolia*. *Journal of Natural Product*, **54**(5): 1360–1367.
17. Murashige, T. and Skoog, F. (1962). A revised medium fro rapid growth and bioassays with tobacco tissue culture. *Plant Physiology*, **15**: 473–497.
18. Gamborg, O.L., Miller, R.A. and Ojima, K. (1968). Nutrient requirement of suspension cultures of soybean root cells. *Experimental Cell Research*, **50**: 151–158.
19. Schenk, R.V. and Hildebrandt, A.C. (1972). Medium and techniques for induction and growth of monocotyledonous and dicotyledonous plant cell cultures. *Canadian Journal of Botany*, **50**: 199–204.
20. White, P.R. (1963). An examination and correction of plant tissue culture basal medium formulations. *Plant Cell, Tissue and Organ Culture*, **28**: 147–150.

Part V

Altered Gravity and Biotechnology

Chapter 30

Plant Secondary Metabolism in Altered Gravity

Lindsey K. Tuominen, Lanfang H. Levine, and Mary E. Musgrave

Summary

Plans by the space program to use plants for food supply and environmental regeneration have led to an examination of how plants grow in microgravity. Because secondary metabolic compounds are so important in determining the nutritional and flavor characteristics of plants—as well as making plants more resistant to biotic and abiotic stresses—their responses to altered gravity are now being studied. These experiments are technically challenging because temperature, humidity, atmospheric composition, light, and water status must be maintained around the plant while simultaneously altering the *g*-load, either in the free-fall of orbital spacecraft or on a centrifuge rotor. In general, plants have shown increased accumulation of small secondary metabolites in microgravity ($<10^{-3}g$), while these have decreased in hypergravity (>1-g). Gravity-related changes in the plant environment as well as mechanical loading effects account for these responses.

Key words: Hypergravity, Microgravity, Bioregenerative life support systems, Secondary metabolites, Mechanical stress, Lignin, Glucosinolates, Isoflavonoids

1. Introduction

The goal of long-term human colonization of space necessitates the development of effective bioregenerative life support systems (BLSS). With the high expense of transporting goods into space and the high mental toll of leaving behind the familiar, life-filled surroundings in which humans evolved, BLSS can provide both economic and psychologic benefits to space pioneers. Such plant-based BLSS act as mini-ecosystems by providing food and oxygen, scrubbing out carbon dioxide, and filtering of wastewater. Currently these life support functions

S. Mohan Jain and Praveen K. Saxena (eds.), *Methods in Molecular Biology, Protocols for* In Vitro *Cultures and Secondary Metabolite Analysis of Aromatic and Medicinal Plants, vol. 547*
© Humana Press, a part of Springer Science+Business Media, LLC 2009
DOI: 10.1007/978-1-60327-287-2_30

are accomplished in space by physicochemical and storage means *(1)*. In this context, plant secondary metabolites are of great interest to the space program, not only because of their important roles in plant fitness, thus BLSS's sustainability, but also because of their culinary and nutraceutical values to human diet *(2)*.

Of the environmental factors that are changed for plants growing in spaceflight environments, none is more difficult to study than altered gravity. For near-term space exploration, only three levels of non-Earth normal gravity are projected to pertain to BLSS plants: microgravity ($<10^{-3}g$) as in free falling, orbiting spacecraft; lunar gravity (1/6 of that on Earth); and Martian gravity (1/3 of that on Earth). Therefore, a prerequisite for BLSS to be functional during a long-term mission in any of these gravity scenarios is its sustainability. The plants must be able to grow from seed to seed in successive generations under altered gravity conditions. After four decades of plant growth in microgravity, this requirement has now been achieved for a number of model species (*see* **ref.** *3* for review), and interest is now turning to the nutritional and nutraceutical properties of these space-grown plants *(4)*.

Gravity's effects on plant secondary metabolism would influence the success of BLSS in three ways. First, because astronauts are inevitably exposed to a high level of ionizing radiation, resulting in oxidative stresses and carcinogenesis *(5)*, a diet high in antioxidants *(6)* or anticarcinogenic compounds would be advantageous. Second, in space, where foods reportedly seem more bland *(7)*, it may be desirable to have levels of flavor and fragrance compounds even higher than in those grown on Earth. Finally, plant health, often maintained by virtue of plant secondary metabolites, is of pivotal importance for the sustainability of BLSS.

Plants appear to be more susceptible to disease when grown in microgravity. Ryba-White et al. *(8)* found microgravity-grown soybean seedlings to be more susceptible to attack by *Phytophthora sojae*. Bishop et al. *(9)* have noted that wheat plants grown in microgravity can be damaged by a species of *Neotyphodium*, a fungus that does not usually attack wheat, suggesting that stresses in the spaceflight environment may leave plants more open to opportunistic pathogens. Although there are redundant defense strategies in plants, biosynthesis of secondary metabolites such as lignins, isoflavonoids, phytoalexins, alkaloids, or glucosinolates play an important role in providing physical and chemical barriers against both obligate and opportunistic pathogens. Therefore, increased susceptibility of plants to disease in the microgravity environment may be linked to altered secondary metabolism.

2. Effects of Gravity on Secondary Metabolism

Well-controlled studies regarding the effect of gravity on small molecular secondary metabolites in plants are very scarce. Plants must be grown in hardware that can be mounted either in a pressurized sector of an orbital platform (for microgravity) or on a large-diameter rotor (for hypergravity). There is a small body of literature on the effects of "simulated microgravity" on bacterial and plant cell culture secondary metabolite production *(10–12)*; however, these studies utilize rotating-wall bioreactors in a 1*g* environment, a method that apparently only superficially simulates microgravity *(13)*. It has been especially challenging to study plant metabolism in microgravity because plant secondary metabolites are extremely sensitive to biological and environmental changes, and preservation (e.g. in-flight freezing of experimental tissue) in space became available only recently *(4)*.

Levine et al. *(14)* investigated the effect of microgravity on isoflavonoids in etiolated soybean (*Glycine max* (L.) Merr.) seedlings. A 6-d space shuttle-based microgravity exposure resulted in enhanced levels of isoflavone glycosides (daidzin, 6″-*O*-malonyl-7-*O*-glucosyl daidzein, genistin, 6″-*O*-malonyl-7-*O*-glucosyl genistein) in hypocotyls and root tissues, but concomitantly reduced levels of the same class of compounds in cotyledons. The average pool size of these compounds per seedling was not changed. This result, coupled with the fact that mRNA expression for two key enzymes (chalcone synthase and chalcone reductase) involved in isoflavonoid biosynthesis in hypocotyls and roots was very low in contrast to their high level of expression in cotyledons, suggests an altered rate of isoflavonoid translocation from the cotyledons to other organs under microgravity environments.

During shuttle flight STS-89, secondary metabolites, such as β-carotene, neoxanthin, violaxanthin, and lutein, were investigated in an aquatic plant Ceratophyllum (*Ceratophyllum demersum* L.) grown in liquid-filled hardware modules. There was no significant difference in the concentration of these metabolites between 9-d space flight- and Earth-grown materials *(15)*. As a result of ever present buoyant forces (and therefore different evolutionary selective pressures), free-floating aquatic plants may be less affected by gravity than their terrestrial counterparts. They should therefore be less susceptible to many direct and/or indirect microgravity-associated effects.

More recently, glucosinolate synthesis in *Brassica rapa* L. was studied by Musgrave et al. *(4)*. The genus *Brassica* is rich in diversity, providing nutritious vegetables such as broccoli and cauliflower, along with greens such as Chinese cabbage and even oil seeds (i.e., canola: *Brassica napus* L. and *B. rapa*). The genus is typified by the presence of glucosinolates in all parts of the plant, providing protection from generalist insect herbivores as well as characteristic taste and aftertaste attributes, such as pungent and burning sensations, for human diet. It was found that 3-butenyl glucosinolate concentration in space-grown *Brassica* stems was 75% greater than that in stems developed on Earth at 1*g*. Interestingly, increased gravity (hypergravity) had the opposite effect on glucosinolates *(16)*. In general, glucosinolate concentrations were lower in plants grown under hypergravity compared to the 1*g* control. Glucosinolate concentrations were significantly lower in 2*g* developed root and stem, but not in leaf tissue, compared with 1*g* controls. Increasing to 4*g* force resulted in a further decline in glucosinolate concentrations in all organs (root, stem and leaf) examined. These datasets present a snapshot of glucosinolate content across the gravity continuum **(Fig. 1)**. In this case, gravity clearly has a negative impact on glucosinolate biosynthesis.

Brassica rapa is also notable because palatability studies have been conducted on material grown in microgravity. In a series of experiments conducted over a 10-yr period on the Mir space station, a piece of hardware (Svet greenhouse) that was open to the cabin atmosphere was used to grow a variety of plants suitable for consumption as fresh greens *(17)*. In taste trials of four leaf vegetables produced in microgravity in 2000, Russian cosmonauts indicated a preference for red mustard (*B. juncea* L., cv. Red Giant) and mizuna (*B. rapa* L. var. *nipposinica* (Bailey) S. Kitamura) over two other varieties of *B. rapa (17)*. Given the changes found in the glucosinolate content of *B. rapa* cv. "Astroplants"

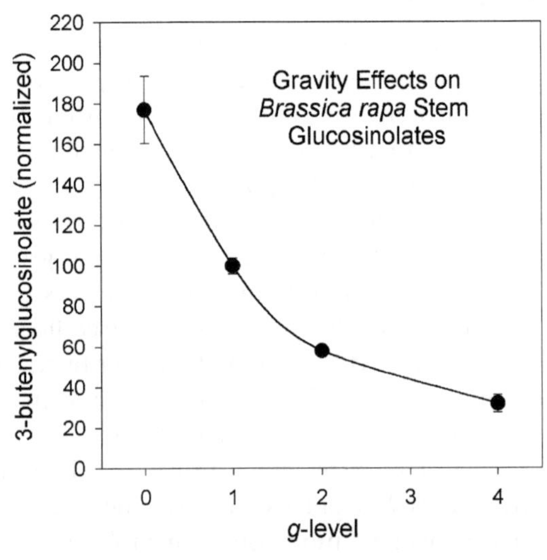

Fig. 1. Mean concentrations of 3-butenylglucosinolate in stems of *Brassica rapa* L. grown in microgravity on the International Space Station *(4)* or on a large diameter centrifuge providing acceleration equivalent to 2-*g* and 4-*g (16)*. Error bars are standard errors, for $n = 3$ (hypergravity) or 4 (microgravity). Values normalized against the 1-*g* controls (=100) for each experiment *(65)*.

across the gravity continuum **(Fig. 1)**, it would be interesting to determine if similar changes occur in these very closely related greens, and how this might influence taste preferences. Glucosinolate profiling in sprouts has suggested that red mustard and mizuna both contain higher total levels of these compounds and have higher anti-cancer potential than the other two varieties of *B. rapa (18)*, so it seems likely that increased glucosinolate levels produced in microgravity environments would be beneficial from both taste and health perspectives.

3. Hypotheses Regarding *g*-Effects on Secondary Metabolism (Fig. 2)

3.1. Role of the Gaseous Environment

The lack of buoyancy-driven convection in microgravity influences gas bioavailability to plants growing in orbital spacecraft (for review, *see* **ref.** *19)*. Closed (nonventilated) growth systems consistently failed to allow plants to complete the reproductive process in microgravity *(3)*. In closed chambers in microgravity, photosynthesis becomes limited by the diffusion rate of carbon dioxide *(19)*. Open growth systems or those with active air circulation have already been successfully employed in growing *Arabidopsis thaliana* (L.) Heynh. and *B. rapa* from seed to seed during spaceflight (for summary, *see* **ref.** *3)*, initially suggesting that this problem might be resolved with relative ease. Indeed, wheat (*Triticum aestivum* L.) plants grown on the International Space Station in the Biomass Production System, which has active ventilation, showed similar growth rates and canopy-level carbon exchange rates to plants grown under similar conditions on earth *(20)*.

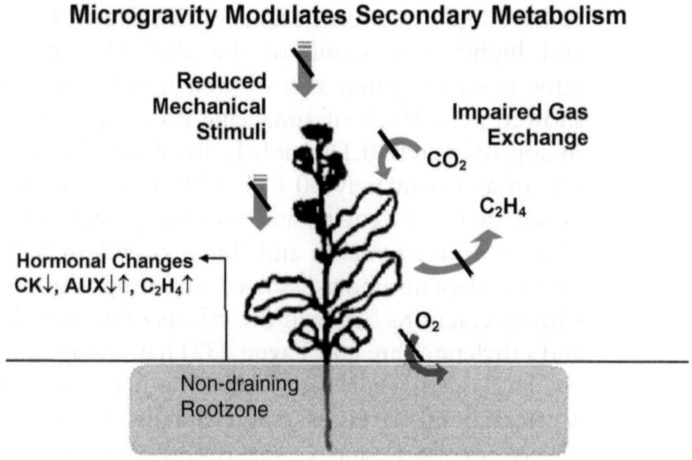

Fig. 2. Plant secondary metabolism is expected to change in altered gravity because of *g*-effects on its modulators: metabolic gases, hormonal activity, and mechanical stimuli. Microgravity effects are shown.

Whereas bulk air movement around plants appears to be part of the solution to this problem, gas exchange may continue to be limited by diffusion rate within plant organs (i.e., microenvironments). This could potentially limit both respiratory and photosynthetic capacity in particular regions of the plant even if canopy-level measurements appear to be normal *(19)*. For example, Musgrave et al. *(21)* hypothesized that the reduced seed size observed in microgravity-grown *B. rapa* is likely due to reduced oxygen and carbon dioxide exchange within the siliques during seed development.

The gaseous environment is known to play a role in secondary metabolite levels. With global climate change research coming to the forefront of ecology, most of the work has been centered on the effect of increased concentrations of carbon dioxide. Bryant et al. *(22)* have proposed that levels of carbon-based secondary metabolites, such as flavonoids, are positively correlated with increases in carbon:nitrogen ratio or negatively correlated with nitrogen concentration. Thus, Lambers *(23)* predicted increased levels of these secondary metabolites as atmospheric carbon dioxide levels increased, and experimental work suggests this hypothesis is applicable to soluble phenolics *(24)* and other carbon-based secondary metabolites as well *(25)*, although Idso and Idso *(26)* suggest that an adequate soil nutrient supply may be necessary for such increases to occur.

In considering gas diffusion rate rather than gas concentration *(27)* as the factor changed by gravity, both carbon dioxide and oxygen availability to plant tissues will be altered concomitantly. Several researchers have studied the effects of carbon dioxide concentration and aeration (a typical substitute for oxygen concentration) on secondary metabolite production in liquid cell cultures, although the two were not altered concomitantly. Pan et al. *(28)* found that the production of a taxane diterpene in *Taxus chinensis* Rehder cultures was increased slightly when treated with 2% CO_2. Bioreactors that recirculated exhaust gases from liquid cultures of *Catharanthus roseus* G. Don were shown to have higher concentrations of dissolved CO_2 in the medium, and higher production of the alkaloid ajmalicine *(29)*. The same research group also showed that *C. roseus* liquid cultures showed peak levels of ajmalicine production at a concentration of approximately 0.1 mmol/L dissolved CO_2 and at an air flow rate of approximately 60 L/h *(30)*. Linden et al. *(31)* have also noted that cell growth and secondary metabolite productivity of *Artemisia annua* L. and *Taxus cuspidata* Siebold and Zucc. cultures depend upon dissolved gas composition in the medium, with interactions between the effects of carbon dioxide, oxygen, and ethylene. Tate and Payne *(32)* have proposed that subambient levels of oxygen may reduce both the growth and secondary metabolite levels of cultured cells, although their work on *Daucus carota* L. and *C. roseus* suggested that the effect of carbon dioxide concentration is negligible. Tisserat et al. *(33)* have conducted one of the few studies on whole plants, examining levels of thymol and piperitenone oxide, respectively, in *Thymus vulgaris* L. (thyme) and *Mentha* spp. (mint). Interestingly, concentration of both metabolites was highest at 21% oxygen and significantly reduced at 5% oxygen. Thymol concentration was reduced at 32% and 42% oxygen, while piperitenone oxide concentration was reduced at 10% oxygen. The lack of significant differences in thymol concentration at 21% and 10% oxygen and in piperitenone oxide at 21%, 32%, and 42% oxygen suggests that the phenylpropanoid and isoprenoid biosynthetic pathways respond differently to altered oxygen availability.

3.2. Collateral Effects Resulting from Changes in Plant Hormones in Altered Gravity

Cytokinins are one class of hormones that may play a role in certain changes seen during spaceflight. Nan et al. *(34)* demonstrated that levels of the cytokinin isopentenyl adenosine were significantly lowered 48 h after exposure to acute water deficit stress, or to both water deficit and waterlogging in different zones, in roots of a cultivar of *T. aestivum* bred for growth during spaceflight. Under spaceflight conditions, both waterlogging and water deficit may occur, even simultaneously in the same plant *(35)*, because of changes in fluid dynamics in microgravity. Under water deficit or concurrent waterlogging and water deficit conditions used in the experiments of Nan et al. *(34)*, reduced cytokinin levels induce changes in morphology that lead to increased water and oxygen harvesting abilities *(36)*. Indeed, Levine et al. *(14)* have observed that etiolated soybean seedlings grown in spaceflight had significantly more lateral roots and significantly longer primary roots than ground control counterparts.

Indirect evidence suggests that auxin levels may be altered during spaceflight as well. Hoson et al. *(37)* have shown that elongation growth and cell wall extensibility of rice (*Oryza sativa* L.) coleoptiles were increased under spaceflight conditions, and cell elongation via cell wall loosening is a classic effect of increased auxin levels. Stresses associated with spaceflight could also cause changes in auxin levels. Nan et al. *(34)* demonstrated that the occurrence of acute water deficit stress in either the entire root system or a portion thereof resulted in significantly lower levels of indole acetic acid (IAA) in the entire root zone. Temporary reductions were also noted in leaves, whereas waterlogging stress resulted in a dramatic increase in leaf IAA levels that lasted less than 12 h *(34)*.

Cytokinins and auxins are known to alter production of secondary metabolites. Deikman and Hammer *(38)* demonstrated that levels of mRNAs for the enzymes phenylalanine ammonia-lyase (PAL1), chalcone synthase (CHS), chalcone isomerase (CHI), and dihydroflavonol reductase (DFR) all increased transiently in *Arabidopsis* plants spray-treated with the cytokinin 6-benzylaminopurine (BAP). All four of these enzymes are implicated in anthocyanin synthesis, and the first three are more generally associated with the synthesis of plant phenolic compounds including lignins, flavonoids, and other phenylpropanoids *(39)*. Rudat and Göring *(40)* have demonstrated an increase in betacyanin-producing cells in cultures of *Chenopodium album* L. when kinetin levels were increased in the medium. The isoflavonoid daidzein is accumulated in liquid cultures of *G. max* treated with kinetin or zeatin *(41, 42)*, and zeatin can increase the levels of ajmalicine in *C. roseus* suspension cell cultures *(43)*. Girod and Zryd *(44)* showed that the ratio of auxin to cytokinin was an important factor regulating the betaxanthin and betacyanin synthesis in red beet (*Beta vulgaris* L.) tissue cultures.

Hairy root cultures of *Hyoscyamus muticus* L., treated with auxins IAA or naphthalene acetic acid (NAA) showed significant increases in tropane alkaloid accumulation compared to untreated cultures *(45)*. Van der Plas et al. *(46)* found that NAA or IAA promoted the production of anthraquinones in *Morinda citrifolia* L. cell cultures while auxins such as 2,4-D abolished anthraquinone production in favor of increased primary metabolism. Similarly, Rudat and Göring *(40)* found that 2,4-D completely inhibited betacyanin synthesis, and Imbault et al. *(43)* and Canel et al. *(47)* have noted that the hormone abolishes alkaloid accumulation in suspension cultures of *C. roseus.* NAA-like auxins do not strictly increase all secondary metabolites. *Tagetes patula* L. hairy root cultures have shown reduced levels of thiophenes after IAA treatment-induced development of lateral roots *(48)*, and *C. roseus* cell cultures had reduced alkaloid levels in a medium treated with NAA *(47)*.

3.3. Mechanical Stress Effects

As reviewed by Klaus *(49)*, the gravitational constant (G) is neither a force nor an acceleration *per se* but rather a physical constant used to dimensionally derive the force (F_{12}) resulting from the attraction by one mass (m_1) on another mass (m_2) a distance (r) away ($F_{12} = (Gm_1m_2)r^{-2}$. In a sense, the influence of gravity is exerted via the mechanical signal, weight. The response to this force differs depending upon plant species, tissue type and the biochemical process under question. It is well known that wood formation is positively correlated with the weight of the crown in trees and other physical stresses such as touch, bending and wind *(50)*. Ko et al. *(51)* have demonstrated that the development of secondary xylem is correlated with the height of the plant and the weight (2.5 g) placed on the top of the immature inflorescence stem, suggesting that the weight carried by the stem is a primary signal for the induction of cambium differentiation. Furthermore, secondary metabolism related genes such as cinnamoyl coenzyme A (CoA) reductase and cytochrome P450, as well as the expression of a calcium-binding protein TCH2 (At5g37770), were upregulated by weight loading. TCH2 is also strongly induced by mechanical stimuli such as touch and wind *(52)*.

Microgravity generally reduces lignin formation in plants (**Table 1**), along with the activity of PAL, presumably because of

Table 1
Microgravity and Hypergravity Effects on Secondary Metabolism in Plants and Liquid Cultures.

Compound/enzyme	Change (from 1-*g*)	*g*	Exposure duration	Species	Tissue type	Reference
Plants						
Lignin	↔	Micro-*g*	10 d	*Triticum aestivum*	Seedling	*(53)*
Lignin	↓[c]	Micro-*g*	8 d	*Vigna radiata*	Seedling	*(54)*
PAL	↓	Micro-*g*	8 d	*Pinus* sp.	Seedling	*(54)*
Isoflavonoid	↔[d]	Micro-*g*	6 d	*Glycine max*	Seedling[a]	*(14)*
Glucosinolate	↑[b]	Micro-*g*	16 d	*Brassica rapa*	Stem	*(4)*
Lignin	↑	300*g*	24 h	*Arabidopsis thaliana*	Inflorescence	*(55)*
PAL	↑	300*g*	3 d	*Triticum aestivum*	Seedling	*(56)*
Glucosinolate	↓	2*g*, 4*g*	16 d	*Brassica rapa*	Stem	*(16)*
Liquid cultures; aquatics						
PAL	↔	Micro-*g*	10-d	*Nicotiana tabacum*	Protoplasts	*(57)*
Lutein, etc.	↔	Micro-*g*	9-d	*Ceratophyllum demersum*	Aquatic	*(15)*
Taxane	↑	3*g*, 24*g*	14-d	*Taxus cuspidata*	Cell culture	*(10)*

[a] Seedlings were dark-grown in closed chambers
[b–d] *Arrows* indicate an increase (↑), decrease (↓), or no change (↔) from 1-*g* controls.

the plant's weightlessness. Interestingly, reaction wood (characterized on Earth by increased deposition of lignin in opposition to the *g*-vector) formation was induced in Douglas fir (*Pseudotsuga menziesii* (Mirbel) Franco) upon receiving a compressing force exerted by mechanically bending the stem 45° away from its vertical alignment *(58)*. This demonstrates that increased lignification can proceed in microgravity when weight is replaced by another mechanical stimulus.

Hypergravity has been shown to inhibit shoot elongation, cell wall extensibility, and the plant's upright growth form in *Vigna*, *Arabidopsis* and *Brassica* *(59–61*, respectively). Inhibitors of stretch-activated mechanosensitive ion channels such as lanthanum and gadolinium ions eliminated the inhibitory effect of hypergravity on shoot elongation in azuki bean (*Vigna angularis* (Willd.) Ohwi and Ohashi) epicotyls *(59)* and *Arabidopsis* hypocotyls *(62)*. Gadolinium chloride also suppressed the promoting effect of prolonged hypergravity on lignification in the immature *Arabidopsis* inflorescence stem *(55)* as well as the effect of hypergravity (300*g*) on the decrease of extensibility of secondary wall that is largely dependent on the production of secondary metabolites (especially lignin content) *(60)*. These findings indicate a possible involvement of mechanoreceptors in the perception of hypergravity stimulus in plants *(55, 59, 60)*.

4. Summary and Conclusions

The effect of altered gravity on plant secondary metabolism has intrigued commercial entities because of the premise that if a plant develops in a microgravity environment, it might direct less of its metabolism toward lignin production, thereby permitting greater production of small molecular secondary metabolites *(1)*. **Table 1** evaluates this hypothesis by comparing how measured activities of the important enzyme PAL, the concentration of lignin, and the concentrations of various small molecular secondary metabolites change in microgravity and hypergravity. In microgravity, there was a trend toward increased secondary compound accumulation, with decreased PAL activity and lignin production. On the other hand, in hypergravity, secondary metabolites were reduced, while PAL activity and lignin formation increased. Changes in the supply of metabolic gases, nutrients, and phytohormones in micro- and hypergravity also modulate the production of secondary metabolites (especially for cells and plants growing in liquid culture), as has been the case for microbial cultures grown in spaceflight *(63, 64)*.

As this review has shown, work is just beginning in this area, but it promises to provide valuable information for support of the space program as well as for our basic understanding of gravity's interaction with plant secondary metabolism. By studying how plants grow in altered gravity, it may be possible to learn how to manipulate the same species on Earth to derive significant commercial benefits.

Acknowledgments

This research is supported by NASA grants NAG 10-329 and NNX07AT77G to MEM.

References

1. Ferl, R., Wheeler, R., Levine, H.G. and Paul, A. (2002). Plants in space. *Curr. Opin. Plant Biol.* **5**, 258–263.
2. Wink, M. (1999). Functions of plant secondary metabolites and their exploitation in biotechnology. Annual Plant Review, Vol. 3. Sheffield Academic, England.
3. Musgrave, M.E. (2002). Seeds in space. *Seed Sci. Res.* **12**, 1–16.
4. Musgrave, M.E., Kuang, A., Tuominen, L.K., Levine, L.H. and Morrow, R.C. (2005). Seed storage reserves and glucosinolates in Brassica rapa L. grown on the International Space Station. *J. Am. Soc. Hort. Sci.* **130**, 848–856.
5. Pence, B.C. and Yang, T.C. (2000). Antioxidants: radiation and stress. In: *Nutrition in Space-Flight and Weightlessness Models* (Lane, H.W. and Schoeller, D.A., eds.) CRC Press, Boca Raton, FL, pp. 233–251.
6. Fang, Y., Yang, S. and Wu, G. (2002). Free radicals, antioxidants, and nutrition. *Nutrition* **18**, 872–879.
7. Olabi, A.A., Lawless, H.T., Hunter, J.B., Levitsky, D.A. and Halpern, B.P. (2002). The effect of microgravity and space flight on the chemical senses. *J. Food Sci.* **67**(2), 468–478.
8. Ryba-White, M., Nedukha, O., Hilaire, E., Guikema, J.A., Kordyum, E. and Leach, J.E. (2001). Growth in microgravity increases susceptibility of soybean to a fungal pathogen. *Plant Cell Physiol.* **42**, 657–664.
9. Bishop, D.L., Levine, H.G., Kropp, B.R. and Anderson, A.J. (1997). Seedborne fungal contamination: consequences in space-grown wheat. *Phytopathology* **87**, 1125–1133.
10. Durzan, D.J. (1999). Gravisensing, apoptosis, and drug recovery in Taxus cell suspensions. *Gravit. Space Biol. Bull.* **12**, 47–55.
11. Demain, A.L. and Fang, A. (2001). Secondary metabolism in simulated microgravity. *Chem. Rec.* **1**, 333–346.
12. Gao, Q., Fang, A., Pierson, D.L., Mishra, S.K. and Demain, A.L. (2001). Shear stress enhances microcin B17 production in a rotating wall bioreactor, but ethanol stress does not. *Appl. Environ. Microbiol.* **56**, 384–387.
13. Moore, R. (1990). Comparative effectiveness of a clinostat and a slow-turning lateral vessel at mimicking the ultrastructural effects of microgravity in plant cells. *Ann. Bot.* **66**, 541–549.
14. Levine, L.H., Levine, H.G., Stryjewski, E.C., Prima, V. and Piastuch, W.C. (2001). Effect of spaceflight on isoflavonoid accumulation in etiolated soybean seedlings. *J. Gravit. Physiol.* **8**, 21–27.
15. Voeste, D., Levine, L.H., Levine, H.G. and Blüm, V. (2003). Pigment composition and concentrations within the plant (*Ceratophyllum demersum* L.) component of the STS-89 C.E.B.A.S. Mini-Module spaceflight experiment. *Adv. Space Res.* **31**(1), 211–214.
16. Darnell, R.D., Levine, L.H., Bisbee, P.A., Allen, J. and Musgrave, M.E. (2007). Glucosinolate production in hypergravity in *Brassica rapa*. *Gravit. Space Biol.* **21**, 33.
17. Levinskikh, M.A., Sytchev, V.N., Podolsky, I.G. and Bingham, G.E. (2001). Pioneering space experiments aimed at obtaining plant biomass to supplement crew food rations. *Gravit. Space Biol. Bull.* **15**(1), 53.
18. O'Hare, T.J., Wong, L.S. and Force, L.E. (2007). Glucosinolate composition and anti-cancer potential of seed-sprouts from horticultural members of the Brassicaceae. In: *Proceedings of the First International Symposium on Human*

Health Effects of Fruits and Vegetables (Desjardins, Y., ed.) **ISHS** Acta Horticulturae, Belgium, **744**, 181–187.

19. Porterfield, D.M. (2002). The biophysical limitations in physiological transport and exchange in plants grown in microgravity. *J. Plant Growth Regul.* **21**, 177–190.
20. Stutte, G.W., Monje, O., Goins, G.C. and Tripathy, B.C. (2005). Microgravity effects on thylakoid, single leaf, and whole canopy photosynthesis of dwarf wheat. *Planta* **223**, 46–56.
21. Musgrave, M.E., Kuang, A., Xiao, Y., Stout, S.C., Bingham, G.E., Briarty, L.G., Levinskikh, M.A., Sychev, V.N. and Podolski, I.G. (2000). Gravity independence of seed-to-seed cycling in *Brassica rapa*. *Planta* **210**, 400–406.
22. Bryant, J.P., Chapin, F.S. III and Klein, D.R. (1983). Carbon/nutrient balance of boreal plants in relation to vertebrate herbivory. *Oikos* **40**, 357–368.
23. Lambers, H. (1993). Rising CO_2, secondary plant metabolism, plant–herbivore interactions and litter decomposition. Theoretical considerations. *Vegetation* **104/105**, 263–271.
24. Penuelas, J. and Estiarte, M. (1998). Can elevated CO_2 affect secondary metabolism and ecosystem function? *Trends Ecol. Evol.* **13**, 20–24.
25. Lavola, A. and Julkunen-Tiitto, R. (1994). The effect of elevated carbon dioxide and fertilization on primary and secondary metabolites in birch, *Betula pendula* (Roth). *Oecologia* **99**, 315–321.
26. Idso, S.B. and Idso, K.E. (2001). Effects of atmospheric CO_2 enrichment on plant constituents related to animal and human health. *Environ. Exp. Bot.* **45**, 179–199.
27. Tuominen, L.K. and Musgrave, M.E. (2006). Tissue culture in synthetic atmospheres: diffusion rate effects on cytokinin-induced callus growth and isoflavonoid production in soybean (*Glycine max* (L.) Merr. cv. Acme) *Plant Growth Regul.* **49**, 167–175.
28. Pan, Z., Wang, H. and Zhong, J. (2000). Scale-up study on suspension cultures of *Taxus chinensis* cells for production of taxane diterpene. *Enzyme Microb. Technol.* **27**, 714–723.
29. Schlatmann, J.E., Nuutila, A.M., van Gulik, W.M., ten Hoopen, H.J.G., Verpoorte, R. and Heijnen, J.J. (1993). Scaleup of ajmalicine production by plant cell cultures of *Catharanthus roseus*. *Biotech. Bioeng.* **41**, 253–262.
30. Schlatmann, J.E., Moreno, P.R.H., Vinke, J.L., ten Hoopen, H.J.G., Verpoorte, R. and Heijnen, J.J. (1997). Gaseous metabolites and the ajmalicine production rate in high density cell cultures of *Catharanthus roseus*. *Enzyme Microb. Technol.* **20**, 107–115.
31. Linden, J.C., Mirjalili, N., Haigh, J.R. and Sun, X. (1995). An examination of dissolved gas effects on growth and secondary metabolism of *Taxus* and *Artemisia* cell cultures. *Abstr. Pap. Am. Chem. Soc.* **209**, 082.
32. Tate, J.L. and Payne, G.E. (1991). Plant cell growth under different levels of oxygen and carbon dioxide. *Plant Cell Rep.* **10**, 22–25.
33. Tisserat, B., Vaughn, S.F. and Silman, R. (2002). Influence of modified oxygen and carbon dioxide atmospheres on mint and thyme plant growth, morphogenesis, and secondary metabolism in vitro. *Plant Cell Rep.* **20**, 912–916.
34. Nan, R., Carman, J.G. and Salisbury, F.B. (2002). Water stress, CO_2 and photoperiod influence hormone levels in wheat. *J. Plant Physiol.* **159**, 307–312.
35. Campbell, W.F., Salisbury, F.B., Bugbee, B., Klassen, S., Naegle, E., Strickland, D.T., Bingham, G.E., Levinskikh, M., Iljina, G.M., Veselova, T.D., Sytchev, V.N., Podolsky, I., McManus, W.R., Bubenheim, D.L., Stieber, J. and Jahns, G. (2001). Comparative floral development of Mir-grown and ethylene-treated, earth-grown Super Dwarf wheat. *J. Plant Physiol.* **158**, 1051–1060.
36. Werner, T., Motyka, V., Strnad, M. and Schmülling, T. (2001). Regulation of plant growth by cytokinin. *Proc. Natl Acad. Sci. U. S. A.* **98**, 10487–10492.
37. Hoson, T., Soga, K., Mori, R., Saiki, M., Nakamura, Y., Wakabayashi, K. and Kamisaka, S. (2002). Stimulation of elongation growth and cell wall loosening in rice coleoptiles under microgravity conditions in space. *Plant Cell Physiol.* **43**, 1067–1071.
38. Deikman, J. and Hammer, P.E. (1995). Induction of anthocyanin accumulation by cytokinins in *Arabidopsis thaliana*. *Plant Physiol.* **108**, 47–57.
39. Taiz, L. and Zeiger, E. (1998). *Plant Physiology*, 2nd edn, Sunderland, MA: Sinauer Associates, Inc.
40. Rudat, A. and Göring, H. (1995). Induction of betacyanin formation in cell cultures of *Chenopodium album* under UV-light irradiation. *J. Exp. Bot.* **282**, 129–134.
41. Miller, C.O. (1969). Control of deoxyisoflavone synthesis in soybean tissue. *Planta* **87**, 26–35.

42. Miller, C.O. (1972). Modification of the cytokinin promotion of deoxyisoflavone synthesis in soybean tissue. *Plant Physiol.* **49**, 310–313.
43. Imbault, N., Thiersault, M., Dupéron, P., Benabdelmouna, A. and Doireau, P. (1996). Pravastatin: a tool for investigating the availability of mevalonate metabolites for primary and secondary metabolism in *Catharanthus roseus* cell suspensions. *Physiol. Plant.* **98**, 803–809.
44. Girod, P. and Zryd, J. (1991). Secondary metabolism in cultured red beet (*Beta vulgaris* L.) cells: differential regulation of betaxanthin and betacyanin biosynthesis. *Plant Cell Tiss. Organ Cult.* **25**, 1–12.
45. Vanhala, L., Eeva, M., Lapinjoki, S., Hiltunen, R. and Oksman-Caldentey, K. (1998). Effect of growth regulators on transformed root cultures of *Hyoscyamus muticus*. *J. Plant Physiol.* **153**, 475–481.
46. Van der Plas, L.H.W., Eijkelboom, C. and Hagendoorn, M.J.M. (1995). Relation between primary and secondary metabolism in plant cell suspensions: competition between secondary metabolite production and growth in a model system (*Morinda citrifolia*). *Plant Cell Tiss. Organ Cult.* **43**, 111–116.
47. Canel, C., Lopes-Cardoso, M.I., Whitmer, S., van der Fits, L., Pasquali, G., van der Heijden, R., Hoge, J.H.C. and Verpoorte, R. (1998). Effects of over-expression of strictosidine synthase and tryptophan decarboxylase on alkaloid production by cell cultures of *Catharanthus roseus*. *Planta* **205**, 414–419.
48. Arroo, R.R.J., Develi, A., Meijers, H., Vandewesterlo, E., Kemp, A.K., Croes, A.F. and Wullems, G.J. (1995). Effect of exogenous auxin on root morphology and secondary metabolism in *Tagetes patula* hairy root cultures. *Physiol. Plant.* **93**, 233–240.
49. Klaus, D.M. (2004). Gravitational influence on biomolecular engineering processes. *Gravit. Space Biol. Bull.* **17(2)**, 51–65.
50. Higuchi, T. (1997). *Biochemistry and molecular biology of wood*. Springer, New York, pp 263–290.
51. Ko, J.H., Han, K.H., Park, S. and Yang, J. (2004). Plant body weight-induced secondary growth in Arabidopsis and its transcription phenotype revealed by whole transcriptome profiling. *Plant Physiol.* **135**, 1069–1083.
52. Braam, J. (1992). Regulated expression of the calmodulin-related TCH genes in cultured *Arabidopisis cells*: induction by calcium and heat shock. *Proc. Natl Acad. Sci. U. S. A.* **99**, 3213–3216.
53. Levine, L.H., Heyenga, A.G., Levine, H.G., Choi, J., Davin, L.B., Krikorian, A.D. and Lewis, N.G. (2001). Cell-wall architecture and lignin composition of wheat developed in a microgravity environment. *Phytochemistry* **57**, 835–846.
54. Cowles, J.R., Scheld, H.W., LeMay, R. and Peterson, C. (1984). Experiments on plants grown in space: growth and lignification in seedlings exposed to eight days of microgravity. *Ann. Bot.* **54**(S3), 33–48.
55. Tamaoki, D., Karahara, I., Schreiber, L., Wakasugi, T., Yamada, K. and Kamisaka, S. (2006). Effects of hypergravity conditions on elongation growth and lignin formation in the inflorescence stem of *Arabidopsis thaliana* L. *J. Plant Res.* **119(2)**, 79–84.
56. Wakabayashi, K., Soga, K., Kamisaka, S. and Hoson, T. (2005). Changes in levels of cell wall constituents in wheat seedlings grown under continuous hypergravity conditions. Space life sciences: gravity-related effects on plants and spaceflight and man-made environments on biological systems. *Adv. Space Res.* **36**(7) 1292–1297, Special Issue 2005.
57. Hoffmann, E., Schonherr, K. and Hampp, R. (1996). Regeneration of plant cell protoplasts under microgravity: investigation of protein patterns by SDS–PAGE and immunoblotting. *Plant Cell Rep.* **15**, 914–919.
58. Kwon, M., Bedgar, D.L., Piastuch, W., Davin, L.B. and Lewis, N.G. (2001). Induced compression wood formation in Douglas fir (*Pseudotsuga menziesii*) in microgravity. *Phytochemistry* **57**, 847–857.
59. Soga, K., Wakabayashi, K., Kamisaka, S. and Hoson, T. (2005). Mechanoreceptors rather than sedimentable amyloplasts perceive the gravity signal in hypergravity-induced inhibition of root growth in azuki bean. *Funct. Plant Biol.* **32**, 175–179.
60. Nakabayashi, I., Karahara, I., Tamaoki, D., Masuda, K., Wakasugi, T., Yamada, K., Soga, K., Hoson, T. and Kamisaka, S. (2006). Hypergravity stimulus enhances primary xylem development and decreases mechanical properties of secondary cell walls in inflorescence stems of *Arabidopsis thaliana*. *Ann. Bot.* **97**, 1083–1090.
61. Allen, J., Kuang, A., Regan, M. and Musgrave, M.E. (2007). Gravity effects on growth form of *Brassica rapa* L. and *Arabidopsis thaliana* (L.) Heyhn. *Gravit. Space Biol.* **21**(1), 32.
62. Soga, K., Wakabayashi, K., Kamisaka, S. and Hoson, T. (2004). Graviperception in growth inhibition of plant shoots under hypergravity condition produced by centrifugation is independent

of that in gravitropism and may involve mechanoreceptors. *Planta* 218, 1054–1061.

63. Benoit, M.R., Li, W., Stodieck, L.S., Lam, K.S., Winther, C.L., Roane, T.M. and Klaus, D.M. (2006). Microbial antibiotic production aboard the International Space Station. *Appl. Microbiol. Biotech.* 70, 403–411.
64. Lam, K.S., Marnber, S.W., Pack, E.J., Forenza, S., Fernandes, P.B. and Klaus, D.M. (1998). The effects of space flight on the production of monorden by *Humicola fuscoatra* WC5157 in solid-state fermentation. *Appl. Microbiol. Biotech.* **49**, 579–583.
65. Allen, J., Bisbee, P.A., Darnell, R.L., Kuang, A., Levine, L.H., Musgrave, M.E. and van Loon, J.J.W.A. (2009). Gravity control of growth form in *Brassica rapa* and *Arabidopsis thaliana* (Brassicaceae): Consequences for secondary metabolism. *Am. J. Bot.* **96**, in press.

Chapter 31

The Role of Biotechnology in the Production of the Anticancer Compound Podophyllotoxin

Hemant Lata, Cassia S. Mizuno, and Rita M. Moraes

Summary

Podophyllotoxin is a plant-derived compound found in *Podophyllum* sp. that is used to produce semi-synthetic anticancer pharmaceuticals such as etoposide, teniposide, and etoposide phosphate. This chapter describes the role of biotechnology to produce podophyllotoxin and our attempts to domesticate *Podophyllum peltatum* L., also known as the American mayapple. The domestication research on mayapple included surveys of the natural population, identification of high yielding genotypes, propagation, cultivation, sustainable harvest procedures and the development of protocols for in vitro germplasm bank.

Key words: Podophyllum, Lignan, Secondary compounds, Medicinal crop

1. Introduction

The long-term objective on medicinal crops is the development of a production system specifically the sustainability of a renewable source of podophyllotoxin. A lignan that is biogenetically derived from the phenylpropanoid pathway is ubiquitously distributed among plant species with important roles in plant defense *(1, 2)*. Aryltetralin lignans are present in *Podophyllum* species but they are not restricted to this genus *(3)*. Among the lignans produced by *Podophyllum* species, (−)-podophyllotoxin (**Fig. 1**) is of great importance because of its biologic activity as a mitotic inhibitor *(4, 5)* and its therapeutic value as a starting compound of semi-synthetic chemotherapeutic drugs etoposide, teniposide, and etoposide phosphate *(6, 7)*.

S. Mohan Jain and Praveen K. Saxena (eds.), *Methods in Molecular Biology, Protocols for* In Vitro *Cultures and Secondary Metabolite Analysis of Aromatic and Medicinal Plants, vol. 547*
© Humana Press, a part of Springer Science+Business Media, LLC 2009
DOI: 10.1007/978-1-60327-287-2_31

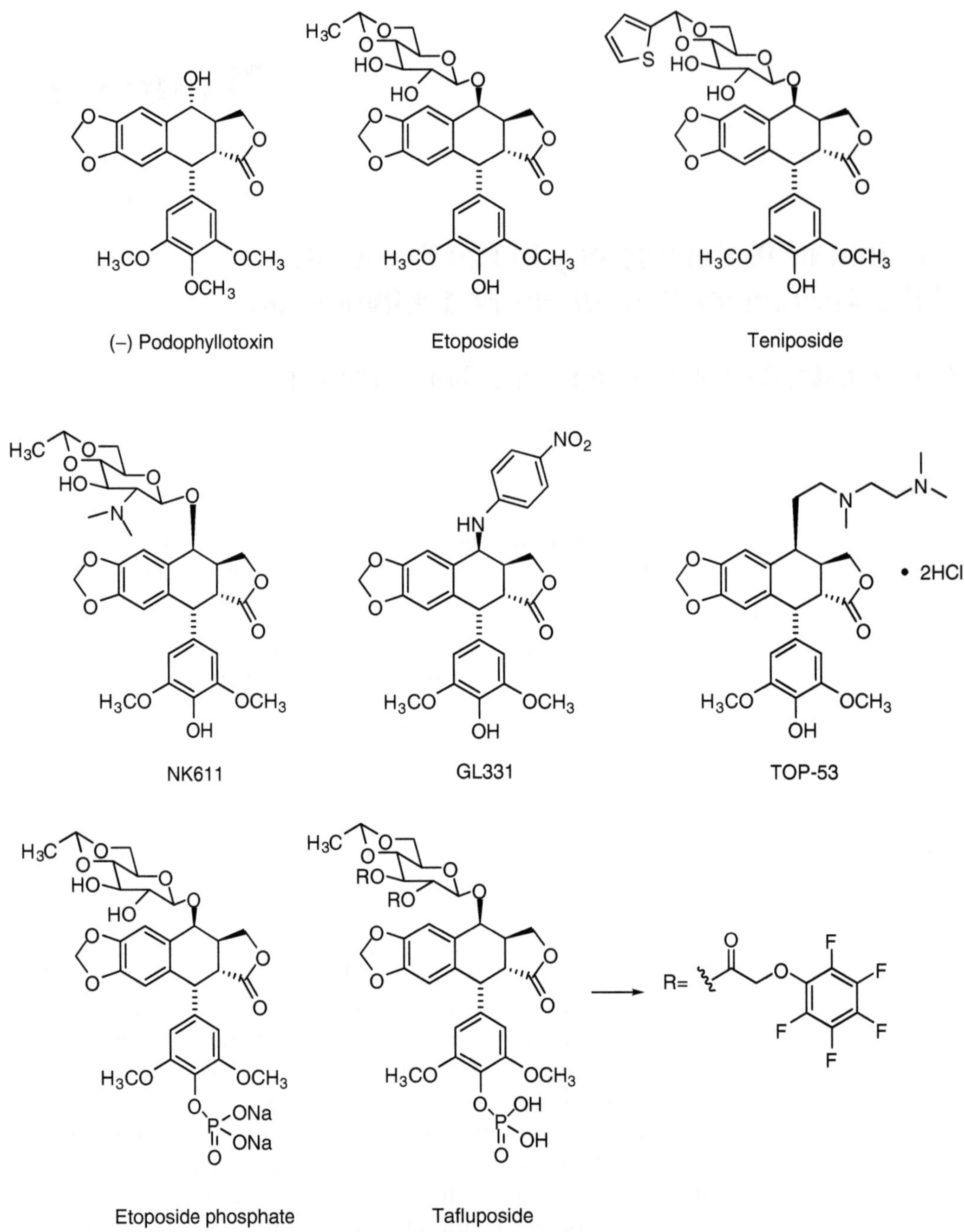

Fig. 1. Structures of (−)-podophyllotoxin and its derivatives.

The commercial source of podophyllotoxin is the Himalayan species *Podophyllum emodi* Wall. Alternative commercial sources are the *P. emodi* var. *chinense* Sprague, native to China, and *Podophyllum peltatum* L., commonly known as the mayapple, a North

American species. According to Stahelin and von Wartburg *(6)*, etoposide and teniposide are examples of pharmaceuticals developed from old folk-remedies. A current review on podophyllotoxin by Liu et al. *(8)* reports that species containing podophyllotoxin were widely used as folk-remedies by the traditional medicine of China, Japan, India, and the United States to treat several illness including gout, tuberculosis, syphilis, warts, and various tumors. In 1942, Kaplan *(9)* demonstrated that podophyllin, an ethanolic extract from the rhizome and roots of *P. peltatum*, was an effective treatment for genital warts (*Condylomata acuminatum*). Later King and Sullivan *(10)* reported that podophyllin and colchine had similar effects on treatment of warts. These works prompted interest in podophyllotoxin and related glucosides, thus leading researchers to dedicate extensive work on structure modifications to reduce gastrointestinal side effects and enhance anticancer activity. The work yielded two clinically approved antineoplastic drugs—etoposide and teniposide *(6)*.

The semi-synthetic drugs, etoposide and teniposide inhibit mitosis by forming a complex with DNA topoisomerase II *(11–13)*. According to Hartmann and Lipp *(14)*, etoposide was clinically approved for the treatments of small-cell lung cancer, non-small cell lung cancer, acute myelogenous leukemia, choriocariciona, advanced ovarian carcinoma, and testicular cancers. Teniposide approved therapy included central nervous system (CNS) tumors, malignant lymphoma, and bladder cancer. Etoposide continues to be tested in multi-drug regimes for better activity in treating several malignancies. Currently, etoposide in combination with doxorubicin, vincristine and cyclophosphamide has been clinically tested for the malignant rhabdoid tumor *(15)*.

Although, etoposide and teniposide are used in clinical oncology, they have deficiencies including poor water solubility, toxic effects, moderate potency, and drug resistance *(16)*. For the past 40 yr, research groups have been involved with other derivatizations to produce new analogs with superior pharmacological profiles and broader therapeutic uses. Several reviews were written on the synthesis and therapeutic properties of podophyllotoxin and related compounds *(7, 8, 17)*. Numerous derivatives varying epipodophyllotoxin basic structure have been proposed and synthesized. Among those are etoposide phosphate, NK611, GL331, TOP-53, and tafluposide and which were clinically tested.

Etoposide phosphate was designed by Bristol–Myers Squibb to overcome the limitations associated with the poor solubility of etoposide. It can be administered intravenously at higher doses and rapidly converted by phosphatase in the plasma to etoposide, thus constituting an improvement in the treatment *(16)*. Side effects and drug resistance led to other analogs that are in preclinical development, stages of phase I and II as listed below.

NK611 is a dimethylamino etoposide derivative produced by Nippon-Kayaku in 1986. Replacement of the hydroxyl group for a dimethylamino functionality at 2″-position on the sugar side chain of etoposide resulted in improved bioavailability. NK611 is 120-fold more soluble in water than the parent compound etoposide *(8)*. Ekimoto *(18)* reported that NK611 was more potent in topo II inhibition assay showing higher antitumor activity than its congener. NK611 is in Phase I clinical trials.

The *p*-aminoanilino derivative GL331 was synthesized by Lee et al. in 1992. Molecular modeling studies done by this group revealed that the C-4 position support bulk groups and that amino group are acceptable at that particular position *(19)*. Thus, the glycoside substitution of etoposide for an aniline group generated GL331. The amino group was selected for enhancing water solubility through salt formation. Huang et al. *(20)* reported that GL331 was 40 times more potent than etoposide in their cytotoxicity assay and it is effective in many multidrug-resistance cancer cell lines. This compound is in Phase IIa clinical trial against gastric carcinoma.

The 4β aminoalkyl derivative TOP-53 was discovered in 1993 by scientist at Taiho Pharmaceutical Co Ltd. *(21)*. The TOP-53 derivative replaces the glucopyranoside group with an aminoalkyl moiety that is directly connected to 4β position through C–C bond. Yoshida et al. *(22)* reported that the synthesis of TOP-53 was an attempt to improve interaction of compounds with intracellular molecules as phospholipids and improve cell penetration. TOP-53 is highly soluble in water. The TOP-53 inhibitory activity against topo II is two times more active than etoposide *(23)*. TOP-53 is in Phase II clinical trials.

Tafluposide (F 11782) is a lipophilic fluorinated etoposide phosphate derivative. The rational behind the F 11782 semi-synthesis was to develop a compound with improved lipophilicity in a search for an inhibitor of both topoisomerases. Furthermore, it was expected that a highly lipophilic molecule could improve cell penetration as well as the compound distribution in vivo *(24)*. Tafluposide is a potent inhibitor of topo I and II that acts synergistically with etoposide in producing DNA damage *(25)*.

The administration of podophyllotoxin derivatives causes complex physiologic reactions beyond inhibition of DNA topoisomerase and tubulin polymerization. A broad range of research on new biologic activities of epipodophyllotoxin analogs include reverse transcriptase inhibition and anti-HIV activity *(17, 26)*, immune modulator activity *(27)*, inhibition of 5-lipoxygenase, anti-rheumatic activity *(28)*, anti-psoriasis, insecticidal activity, and a plant growth inhibitor *(29–31)*. The therapeutic importance of these oncology drugs prompts synthetic chemists to devote their efforts to develop new routes to the total synthesis. The process has been accomplished but the yield is low as a result of the large number of steps involved *(7, 32)*.

Currently, the commercial source of podophyllotoxin is the Indian *Podophyllum* species, and it has acquired the status of an endangered species because of intense collection and lack of cultivation. To secure podophyllotoxin supply, we have examined the North American mayapple (*P. peltatum* L.) and its potential for podophyllotoxin production. The present chapter describes the biotechnological approaches that have being accomplished at the National Center for Natural Products Research (NCNPR), University of Mississippi and many other laboratories.

2. *Podophyllum* as the Source of Podophyllotoxin

The genus *Podophyllum* (Berberidaceae) has three species that can be commercially exploited as sources of podophyllotoxin. *P. emodi* Wall. (syn. *P. hexandrum* Royle) is a native plant to India and Nepal and is the commercial source of podophyllotoxin. *P. peltatum* L. (mayapple) grows in the United States whereas *P. emodi* var. *chinense* Sprague (syn, *Sinopodophyllum hexandrum* (Royle) T. S. Ying (bankari)) is from China. Extracts of dried rhizome of Bankakri and mayapple were used by Himalayans and the North American native populations as cathartics and cholagogues, respectively. In 1947, Hartwel and Shear (33) demonstrated that a single dose of resin was effective in reducing tumors, but severe abdominal pains were associated with the treatment. Extracts containing natural lignan glucosides from *Podophyllum* were tested to eliminate side effects and provide better pharmacologic results. The *Podophyllum* glucosides not only showed lower toxicity but also lower anticancer activity. These results led to the derivatization of podophyllotoxin and the development of etoposide and teniposide *(6)* (**Fig. 1**).

Podophyllotoxin and related lignans have also been described in other genera such as *Jeffersonia*, *Diphylleia* and *Dysosma* (Berberidaceae), *Catharanthus* (Apocynaceae), *Polygala* (Polygalaceae), *Anthriscus* (Apiaceae), *Linum* (Linaceae), *Hyptis* (Verbenaceae), *Teucrium*, *Nepeta* and *Thymus* (Labiaceae), *Thuja*, *Juniperus*, *Callitris*, and *Thujopsis* (Cupressaceae), *Cassia* (Fabaceae), *Haplophyllum* (Rutaceae), *Commiphora* (Burseraceae) and *Hernandia* (Hernandiaceae) *(3, 34–37)*.

After etoposide and teniposide discoveries, Meijer *(38)* estimated that the total demand in the United States for *P. peltatum* rhizomes and roots was 130 tons. The commercial interest turned to *P. emodi* when these rhizomes were found to contain more podophyllotoxin than the *P. peltatum (39)*.

Podophyllum emodi is a perennial rhizomatous herb found in the understory of subalpine forests of the Himalayas *(40)*. Due to podophyllotoxin demand in the international market for the past three decades, there has been a sharp decline in *P. emodi* populations and it has acquired endangered species status *(41)*. As of today, many Indian research institutions are making a great effort to rescue the species. The Department of Botany at the University of Delhi is collecting specimens for population biology and genetic diversity studies. The populations are being characterized at the cytogenetic level to assess the existing genetic diversity. Representative specimens are being transplanted to experimental plots in the Himalayan foothills *(40)*. Furthering the research approaches on replenishing *P. emodi*, researchers at the G.B. Pant Institute of Himalayan Environment and Development in Garhwal have successfully propagated the natural stocks using rhizome cuttings, viable seeds, and plants regenerated from embryogenic calli *(42)*.

The domestication of *P. peltatum*, as an alternative source of podophyllotoxin is another strategy to reduce the exploitation of *P. emodi* wild populations and secure a sustainable source of podophyllotoxin. For the past 10 yr, the domestication of *P. peltatum* has been our research aim at the NCNPR. A partial survey of *P. peltatum* natural populations resulted in the identification of several high yielding accessions *(43, 44)*. Our findings suggest that *P. peltatum* stores podophyllotoxin 4-*O*-β-D-glucopyranoside in leaf blades. Using a buffer extraction procedure, this compound is easily converted to podophyllotoxin *(45)* because of presence of a highly specific glucosidase *(46)*. The amount of podophyllotoxin 4-*O*-β-D-glucopyranoside varies from plant to plant, but some contain as much as 5% on a dry weight basis *(45)*. The buffer extraction of *P. peltatum* leaves yields podophyllotoxin in amounts similar to the ethanol extraction of *P. emodi* rhizomes and roots (40.0 mg/g on a dry weight basis) *(45, 47)*. Thus, the podophyllotoxin content in leaves of *P. peltatum* are equal to that of rhizomes of *P. emodi (43, 46)*, and the lignan profile in leaves of *P. peltatum* is of stable character that does not change over time *(48)*.

In continuing the effort to demonstrate the potential of *P. peltatum* as sustainable source of podophyllotoxin, a 2-yr-cultivation trial and a 4-yr-harvest time study were conducted. Results have shown that cultivation of *P. peltatum* is an effective production strategy *(49–51)*, leaves are renewable organs and source of podophyllotoxin, confirming that American mayapple has great potential to become a sustainable crop.

3. The American Maypple Podophyllotoxin Rich-Chemotypes

A partial survey to sample the natural population of *P. peltatum* was conducted in April of 1998, we visited six different states, and samples of 18 accessions were collected from 17 sites. Plants were separated by organ parts consisting of leaves, rhizomes, and roots. Ground tissues were extracted by the procedure described in the method of Canel et al. *(7)*, the *P. peltatum* accessions were classified in two chemotype groups: *(1)* podophyllotoxin-rich types with blades yielding between 85% and 94% podophyllotoxin of the total lignan content and *(2)* the peltatin-rich types with 65%–80% of the total lignan being α-peltatin *(43)*. Podophyllotoxin-rich accessions are of particular interest because biomass with high purity represents a significant economy in the process of purification. These rich accessions were proven to be stable chemotypes and confirmed as chemotypes by cultivating them in different growing conditions for three consecutive years *(48)*. Comparison of the different chemotypes showed an inverse relationship between podophyllotoxin content and α-peltatin content. This was also noticed during our partial survey of the natural range of *P. peltatum* in theUnited States and a GIS study mapping the Mississippi yielded high accessions and was later confirmed by cultivation studies *(43, 44, 50, 51)*.

4. In Vitro Studies for Production of Podophyllotoxin

Since total synthesis of podophyllotoxin is economically unfeasible for the foreseeable future, research attention has geared toward a sustainable and economic source of lignans. Our effort has been to search for high yielding plants of *P. peltatum*, propagation, conservation and cultivation. Nadeem et al. *(42)* have reported propagation and conservation studies of *P. emodi*, whereas other laboratories have dedicated their efforts on different biotechnology approaches to produce podophyllotoxin.

To achieve a sustainable bioproduction of podophyllotoxin has been the primary research goal of many laboratories, and many reviews have been written on in vitro procedures *(78, 36, 52–55)*. Kutney *(56)* has described the use of enzymes derived from plant cell cultures as reagents in biotransformation of suitable substrates to produce the end product, etoposide. Peroxidase of a *Nicotiana sylvestris* cell culture in a bioreactor catalyzed the oxidative cyclolization of a dibenzylbutanolide toward the cyclolignan derivative. In a similar way, cyclolignans have been obtained with different stereochemistries with cell lines of *P. peltatum*, *Catharanthus roseus*, and *Cassia didymobotrya (34)*, whereas Uden et al. *(57)* described a large-scale extraction of deoxypodophyllotoxin from the rhizome of *Anthriscus sylvestris* followed by its bioconversion into 5-methoxypodophyllotoxin β-D-glucosidase by cell cultures of *L. flavum*.

Calli and cell suspensions of different species of *Linum*, *Callitris*, and *Podophyllum (58–67)* have been reported as sources of aryltetralin lignans. According to Uden et al. *(60, 61)* bioproduction of aryltetralin lignans may be improved by feeding phenylpropanoid precursors to the media or by selecting elite root cultures. Cell suspensions of *Linum* spp. have been the model of intense studies to identify the enzymatic steps of the lignans pathway *(68–71)*. For Guillon et al. *(55)*, the knowledge gained on the limiting culture parameters of the metabolic pathway makes hairy root cultures a promising procedure for industrial production of secondary metabolites. Giri and Narasu *(52)* infected *Podophyllum emodi* with *Agrobacterium rhizogenes* for continuous production of *Podophyllum* spp. lignans. Using a similar procedure, Oostdam et al. *(72)* reported a 5- to 12-fold higher production of 6-methoxypodophyllotoxin than in *Linum* spp. cell suspension cultures.

In nature, plants produce secondary metabolites to defend themselves against attack by pathogens, and lignans play an important role in this type of plant defense *(1, 2)*. This class of compounds has antibiotic, antioxidant, and pesticidal properties which confer protection against pathogens. Plants elicit the same response when challenged by compounds of pathogenic origin or physical damage that simulates insect attack. Adding methyl jasmonate to cell cultures of *Forsythia intermedia*, Schmitt and Petersen *(73)* reported a 3- to 7-fold increase on the content of pinoresinol and matairesinol. A 2-fold increase in lignan production was noticed after the micropropagated plantlets were inoculated with arbuscular mycorrhiza during acclimatization *(74)*. Physical damage to the leaves of *P. peltatum* resulted in the accumulation of podophyllotoxin glucosides and aglucone at the wound site *(46, 75)*.

In the authors' laboratory, an in vitro propagation protocol to rapidly produce high-yielding *P. peltatum* plants was developed by growing rhizome tips on basal Murashige and Skoog (MS) medium containing sucrose and supplemented with benzyladenine and activated carbon. The podophyllotoxin content of in vitro rooted bud and plantlet cultures were similar to the content found in the wild *(76)*. According to Cousins and

Adelberg *(77)* commercial laboratories can supply starter-plants for field plantings generating uniform plant organs for direct production of active compounds. Micropropagation also allows for the creation and dissemination of large numbers of nursery plants without spreading pathogens across continents.

In addition to plant cell and organ cultures, Eyberger et al. *(78)* have identified *Phialocephala fortinii* a fungal endophyte isolated from *P. peltatum* rhizomes. Another fungal endophyte *Trametes hirsute* was also isolated from *Podophyllum.* Using high-performance liquid chromatography (HPLC), liquid chromatography mass spectrometry (LC-MS), LC/MS-MS, and nucleic magnetic resonance (NMR) Puri et al. *(79)* have shown that *T. hirsute* is capable to produce aryltetralin lignans.

5. In Vitro and *In Situ* Conservation of *P. peltatum* Germplasm

Conservation of medicinal genetic resources by in vitro techniques has become a reality and depends on procedures to reduce growth, avoiding frequent subcultures. The identification of *P. peltatum* elite genotypes led our effort in developing protocols to maintain the accessions in storage with maximum survival, minimum subculture changes, and without genetic changes. The aim is to maintain a repository for future development that will ensure a continuous supply with high quality plant material containing standardized amounts of lignans. Approaches to growth suppression using alginate encapsulation techniques and/or omission of some nutrients essential for normal growth were performed in an attempt to develop an efficient method for germplasm conservation at temperature conditions above freezing. Using synthetic seed technology, growth was minimal in all encapsulated buds stored on water/agar at the tested temperatures with maximum plant conversion after storage (**Table 1**), with no signs of shoot or root growth after 8 mo in storage. The cultures incubated at 10°C survived remarkably better than any other temperature with a 10-fold vigorous shoot proliferation (**Fig. 2**).

For *in situ P. peltatum* conservation we have established a GIS database (**Fig. 3**) that allowed us to map elite genotypes for future propagation and improvement *(44)*. This geodatabase stores phenotypic data such drug content, morphological characteristics and environmental factors that may influence growth and development of each accession.

Table 1
The Effect of Shoot Encapsulation on In Vitro Storage Conditions of *P. peltatum* explants

Media and support	Storage temperature (°C)	Re-growth[a] after storage (4 mo)	Plantlet conversion[b] (%)	Re-growth[a] after storage (8 months)	Plantlet conversion[b] (%)
Water + agar	5	13.0 ± 1.2 a	100	11.2 ± 1.7 ab	100
	10	15.8 ± 2.4 a	100	15.5 ± 1.9 a	100
	25	7.5 ± 1.0 c	100	5.0 ± 1.4 cd	100
2% Sucrose + agar	5	8.0 ± 2.4 bc	100	5.0 ± 3.0 cd	83
	10	12.0 ± 1.4 ab	100	11.7 ± 1.0 ab	100
	25	5.1 ± 1.5 cde	100	3.7 ± 3.1 cd	100
MS + agar	5	4.0 ± 2.0 cde	83	3.3 ± 2.8 cd	67
	10	6.5 ± 3.4 c	83	7.3 ± 4.0 bc	67
	25	1.7 ± 1.4 de	67	1.5 ± 1.5 d	50
MS + cotton fibers	5	2.0 ± 1.7 de	67	1.8 ± 1.4 d	67
	10	5.0 ± 3.1 cd	83	4.0 ± 3.2 cd	67
	25	1.2 ± 1.3 e	50	0.8 ± 1.3 d	33

Data represents the mean of three replicates with six explants for each treatment. Means followed by *same letter* do not differ statistically at $p = 0.05$ according the Tukey test

[a]Re-growth of encapsulated shoots is based on new shoots per explant induced in 30 d after the removal of the storage condition

[b]Plant conversion measures the number of encapsulated shoots that produced new buds in a 30-d period after storage

Fig. 2. In vitro germplasm storage of *P. peltatum*, (***a***) encapsulated axillary buds on water/agar; (***b***) sucrose/agar; (***c***) MS salts/agar; (***d***) MS salts/cotton wool; (***e***) shoot multiplication at 10°C after storage for 8 mo; (***f***) acclimatization in greenhouse.

6. Conclusion

In this chapter, an attempt has been made to elucidate the importance and uses of podophyllotoxin for its therapeutic applications with emphasis on in vitro technology as a possible alternative for the production of podophyllotoxin. We have therefore described our efforts to develop *P. peltatum* as a commercial source of podophyllotoxin and a specialty crop for small farmers. The chapter includes domestication, evaluation of wild crafted and in vitro established protocols to micropropagate and store elite germplasm, recent advancement of the knowledge in phytochemistry, and the regulation of secondary pathways as it pertains to podophyllotoxin production.

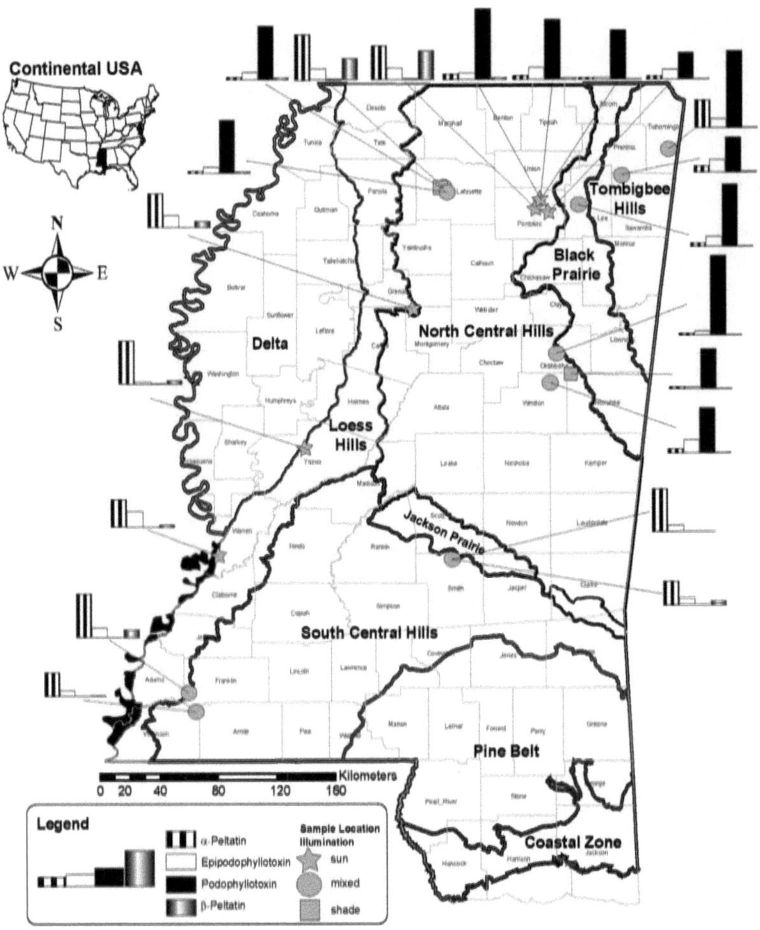

Fig. 3. *Podophyllum peltatum*'s geodatabase, *In situ* conservation of the genetic resources (Reprinted with permissiom from ref. *44*).

Acknowledgments

This work was partially funded by The University of Mississippi Office of Research and Sponsored Program and by USDA/ARS Specific Cooperative Research Agreement No. 58-6408-2-0009.

References

1. Fukuda, Y., Namiki, M. and Osawa, T. (1985). Studies on anti-oxidative substances in sesame seed. *Agric. Biol. Chem.* **49**, 301–306.
2. Figgitt, D.P., Denever, S.P., Dewick, P.M., Jackson, D.E. and Willians, P. (1989). Topoisomerase II: a potential target for novel antifungal agents. *Biochem. Biophys. Res. Commun.* **160**, 257–262.
3. Bedir, E., Khan, I. and Moraes R.M. (2002). Bioprospecting for podophyllotoxin. In: Janick, J., Whipkey, A. (Eds.) *Trends in New Crops and New Uses*, ASHS Press, Alexandria, pp. 545–549.
4. Loike, J.D. and Horwitz, S.B. (1976). Effects of podophyllotoxin and VP 16-213 on microtubule assembly *in vitro* and nucleoside transport in HeLa cells. *Biochemistry* **15**, 5435–5442.

5. Loike, J.D., Brewer, C.F., Sternlicht, H., Gensler, W.J. and Horwitz, S.B. (1978). Structure-activity study of the inhibition of microtubules assembly *in vitro* by podophyllotoxin and its congeners. Cancer Res. **38**, 2688–2693.
6. Stahelin, H.F. and Wartburg, A.V. von (1991). The chemical and biological route from podophyllotoxin glucoside to etoposide. *Cancer Res.* **51**, 5–15.
7. Canel, C., Moraes, R.M., Dayan, F.E. and Ferreira, D. (2000). Molecules of interest "podophyllotoxin". *Phytochemistry* **54**, 115–120.
8. Liu, Y.Q., Yang, L. and Tian, X. (2007). Podophyllotoxin: current perspectives. *Curr. Bioact. Compd.* **3**, 37–66.
9. Kaplan, I.W. (1942). *Condylomata acuminatum. New Orleans Med. Surg. J.* **94**, 388.
10. King, L.S. and Sullivan, M. (1946). The similarity of the effect of podophyllin and colchicine and their use in the treatment of condylomata acuminate. *Science* **104**, 244–245.
11. Loike, J.D. and Horwitz, S.B. (1976). Effect of VP 16-213 on the intracellular degradation of DNA in HeLa cells. *Biochemistry* **15**, 5443–5448.
12. Horwtiz, S.B. and Loike, J.D. (1977). A comparison of the mechanism of action of VP 16-213 and podophyllotoxin. *Lloydia* **40**, 82–89.
13. Minocha, A. and Long, B.H. (1984). Inhibition of the DNA catenation activity of type II topoisomerase by VP 16-213, VM – 26. *Biochem. Biophys. Res. Commun.* **122**, 165–170.
14. Hartmann, J.T. and Lipp, H.P. (2006). Camptothecin and podophyllotoxin derivatives. *Drug Saf.* **29**, 209–230.
15. Hosoi, H., Lehara, T., Tsuchiya, K., Misawa, A., Miyaji, M., Yagyu, S., Koizumi, M., Nishimura, T., Tokiwa, K. and Iwai, N. (2007). Continuous remission in an infant with chest wall malignant rhabdoid tumor after relapse. *J. Pediatr. Surg.* **42**, e9–e12.
16. Schacter, L. (1996). Etoposide phosphate: what, why, where, and how? *Semin. Oncol.* **6**(Suppl 13), 1–7.
17. Gordaliza, M., Garcia, P.A., Miguel del Corral, J.M., Castro, M.A. and Gomez-Zurita, M.A. (2004). Podophyllotoxin: distribution, source, application and new cytotoxic derivatives. *Toxicon* **44**, 441–459.
18. Ekimoto, H., Okamoto, K. and Maruyama, S. (1992). Preclinical characterization of NK611, a novel water-soluble etoposide analog. *Ann. Oncol.* **3**(Suppl 1), 98.
19. Lee, K.H. (1999). Novel antitumor agents from higher plants. *Med. Res. Rev.* **19**(6), 569–596.
20. Huang, T.S., Lee, C.C., Chao, Y., Shu, C.H., Chen, L.T., Chen, L.L., Chen, M.H., Yuan, C.C. and Whang-Peng, J. (1999). A novel podophyllotoxin-derived compound GL331 is more potent than its congener VP-16 in killing refractory cancer cells. *Pharm. Res.* **16**(7), 997–1002.
21. Terada, T., Fujimoto, K., Nomura, M.A., Yamashita, J., Wierzba, K., Yamazaki, R., Shibata, J., Sugimoto, Y. and Yamada, Y. (1993). Antitumor agents. 3. Synthesis and biological activity of 4β-alkyl derivatives containing hydroxy, amino, and amido groups of 4′-*O*-demethyl-4-desoxypodophyllotoxin as antitumor agents. *J. Med. Chem.* **36**(12), 1689–1699.
22. Yoshida, M., Kobunai, T., Aoyagi, K., Saito, H., Utsugi, T., Wierzba, K. and Yamada, Y. (2000). Specific distribution of TOP-53 to the lung and lung-localized tumor is determined by its interaction with phospholipids. *Clin. Cancer Res.* **6**(11), 4396–4401.
23. Utsugi, T., Shibata, H., Kumio, S., Aoyagi, K., Wierzba, K., Kobunai, T., Terada, T., Oh-hara, T., Tsuruo, T. and Yamada, Y. (1996). Antitumor activity of a novel podophyllotoxin derivative (TOP-53) against lung cancer and lung metastatic cancer. *Cancer Res.* **56**(12), 2809–2814.
24. Kruczynski, A., Etievant, C., Perrin, D., Imbert, T., Colpaert, F. and Hill, B.T. (2000). Preclinical antitumour activity of F 11782, a novel dual catalytic inhibitor of topoisomerases. *Br. J. Cancer* **83**(11), 1516–1524.
25. Jensen, L.H., Renodon-Corniere, A., Nitiss, K.C., Hill, B.T., Nitiss, J.L., Jensen, P.B. and Sehested, M. (2003). A dual mechanism of action of the anticancer agent F 11782 on human topoisomerase II. *Biochem. Pharm.* **66**(4), 623–631.
26. Chen, S.W., Wang, Y.H., Jin, Y., Tian, X., Zeng, Y.T., Lou, D.Q. and Tu, Y.Q. (2007). Synthesis and anti-HIV 1 activities of novel podophyllotoxin derivatives. *Bioorg. Med. Chem. Lett.* **17**, 2091–2095.
27. Pugh, N., Khan, I., Moraes, R.M. and Pasco, D. (2001). Podophyllotoxin lignans enhance IL-1 but suppress TNF – a mRNA expression in LPS – treated monocytes. *Immunopharmacol. Immunotoxicol.* **23**, 83–95.
28. Leander, K. and Rosen, B. (1988). Medicinal use for podophyllotoxin. *US Patent* **4** (788), 216.
29. Bohlin, L. and Rosen, B. (1996). Podophyllotoxin derivatives: drug discovery and development. *Drug Discover. Today* **1**, 343–351.

30. Di , X., Liu, Y., Liu, Y., Yu, X., Xiao, H., Tian, X. and Gao, R. (2007). Synthesis and insectidal activities of pyridine ring derivatives of podophyllotoxin. *Pestic. Biochem. Physiol.* **89**, 81–87.
31. Oliva, A., Moraes, R.M., Watson, S.B., Duke, S.O. and Dayan, F.E. (2002). Aryltetralin lignans inhibit plant growth by affecting the formation of mitotic microtubular organizing centers. *Pestic. Biochem. Physiol.* **72**, 45–54.
32. Bush, E.J. and Jones, D.W. (1995). Asymmetric total synthesis of (–)-podophyllotoxin. *J. Chem. Soc. Perkin Trans.* **1**, 151–155.
33. Hartwel, J.L. and Shear, M.J. (1947). Chemotherapy of cancer: class of compound under investigation and the active components of podophyllin *Cancer Res.* 7, 716.
34. Fransen, M.C.R. and Walton, M.J. (1999). Biotransformations. In: Walton, M.J., Brawn, D.E. (Eds.), *Chemicals from Plants, Perspectives on Plant Secondary Products*, Imperial College Press, London, pp. 277–325.
35. Gordaliza, M., Castro, M.A, Miguel del Corral, J.M. and San Feliciano, A. (2000). Antitumor properties of podophyllotoxin and related compounds. *Curr. Pharm. Des.* **6**, 1811–1839.
36. Petersen, M. and Alfermann, A.W. (2001). The production of cytotoxic lignans by plant cell cultures. *Appl. Microbiol. Biotechnol.* **55**, 135–142.
37. Puricelli, L., Innocenti, G., Piacente, S., Caniato, R., Filippini, R. and Capelletti, E.M. (2002) Production of lignans by Haplophyllum patavinum *in vivo* and *in vitro*. *Heterocycles* **56**, 607–612.
38. Meijer, W. (1974). Podophyllum peltatum – mayapple a potential new cash-crop plant of Eastern North America. *Econ. Bot.* **28**, 68–72.
39. Jackson, D.E. and Dewick, P.M. (1984). Aryltetralin lignans from *Podophyllum hexandrum* and *Podophyllum peltatum*. *Phytochemistry* **23**(5), 1147–1152.
40. Bhadula, S.K., Singh, A., Lata, H., Kuniyal, C.P. and Purohit, A.N. (1996). Genetic resources of *Podophyllum hexandrum* Royle, an endangered medicinal species from Garhwal Himalaya, India. *Int. Plant. Gen. Resour. Newslett.* **106**, 26–29.
41. Foster, S. (1993). Medicinal plant conservation and genetic resources: examples from the temperate northern hemisphere. *Acta Hortic.* **330**, 67–73.
42. Nadeem, M., Palni, L.M.S., Purohit, A.N., Pandey, H. and Nandi, S.K. (2000). Propagation and conservation of *Podophyllum hexandrum* Royle: an important medicinal herb. *Biol. Conserv.* **92**, 121–129.
43. Moraes, R.M., Burandt, C. Jr., Ganzera, M., Li, X., Khan, I. and Canel, C. (2000). The American mayapple revisited – *Podophyllum peltatum* – still a potential cash crop? *Econ. Bot.* **54**, 471–476.
44. Moraes, R.M., Momm, H.G., Silva, B., Maddox, V., Easson, G.L., Lata, H. and Ferreira, D. (2005). Geographic information system method for assessing chemo-diversity in medicinal plants. *Planta Med.* **71**, 1157–1164.
45. Canel, C., Dayan, F.E., Ganzera, M., Khan, I.A., Rimando, A., Burandt, C.L. and Moraes, R.M. (2001). High yield of podophyllotoxin from leaves of *Podophyllum peltatum* by *in situ* conversion of podophyllotoxin 4-O-β-D-glucopyranoside. *Planta Med.* **67**, 97–99.
46. Dayan, F., Kuhajek, J., Canel, C., Watson, S. and Moraes R. (2003). *Podophyllum peltatum* possesses a β-glucosidase with high substrate specificity for the aryltetralin lignan podophyllotoxin. *Biochim. Biophys. Acta* **1645**(1–2), 157–163
47. Jackson, D.E. and Dewick, P.M. (1985). Tumor-inhibitory aryltetralin lignans from *Podophyllum pleianthum*. *Phytochemistry* **24**, 2407–2409.
48. Moraes, R.M., Bedir, E., Barrett, H., Burandt, C., Canel, C. and Khan, I.A. (2002). Evaluation of *Podophyllum peltatum* accessions for podophyllotoxin production. *Planta Med.* **68**, 341–344.
49. Maqbool, M., Cushman, K.E. and Moraes, R.M. (2004). Mayapple: a review of the literature from a horticultural perspective. Annual Report of the North Mississippi Research and Extension Center. *Miss. Agric. For. Expt Sta. Info. Bull.* **375**, 313–319.
50. Cushman, K.E., Maqbool, M., Bedir, E., Lata, H., Khan, I.A. and Moraes, R.M. (2005). Growth and lignin content of American mayapple under increasing levels of shade. *Hortic. Sci.* **40**, 60–63.
51. Cushman, K.E., Moraes, R.M., Gerard, P.D., Bedir, E., Silva, B. and Khan, I.A. (2006). Frequency and timing of leaf removal affect growth and podophyllotoxin content of *Podophyllum peltatum* in full sun. *Planta* **72**, 824–829.
52. Giri, A. and Narasu, M.L. (2000). Production of podophyllotoxin from *Podophyllum hexandrum*: a potential natural product for clinically useful anticancer drugs. *Cytotechnology* **34**, 17–26.
53. Fuss, E. (2003). Lignans in plant cell and organ cultures: an overview. *Phytochem. Rev.* **2**(3), 307–320.
54. Wink, M., Alfermann, W., Franke, R., Wetterauer, B., Distl, M., Windhovel, J., Krohn, O., Fuss, E., Garden, H., Mohagheghzadeh

Eildi, E. and Ripplinger, P. (2005). Sustainable bioproduction of phytochemicals by plant *in vitro* cultures: anticancer agents. *Plant Genet. Resour.* **3**, 90–100.

55. Guillon, S., Tremouillaux-Guiller, J., Pati, P.K., Rideau, M. and Gantet, P. (2006). Harnessing the potential of hairy roots: down of a new era. *Trends Biotech.* **24**(9), 403–409.
56. Kutney, J.P. (1999). Biotechnology and synthetic chemistry – routes to clinically important compounds. *Pure Appl. Chem.* 71, 1025–1032.
57. Uden, W. van, Bos, J.A., Boeke, G.M., Woerdenbag, H.J. and Pras, N. (1997). The large-scale isolation of deoxypodophyllotoxin from rhizomes of *Anthriscus sylvestris* followed by its bioconversion into 5-vethoxypodophyllotoxin β-D-glucoside by cell cultures of Linum flavum. *Nat. Prod.* **60**(4), 401–403.
58. Kadakade, P.G. (1982). Growth and podophyllotoxin production in callus tissues of *Podophyllum peltatum. Plant Sci. Lett.* **25**, 107–115.
59. Uden, W. van, Pras, N., Visser, J.F. and Malingre, T.M. (1989). Detection and identification of podophyllotoxin produced by cell cultures derived from *Podophyllum hexandrum* Royle. *Plant Cell Rep.* **8**(3), 165–168.
60. Uden, W. van, Pras, N. and Malingre, T.M. (1990). The accumulation of podophyllotoxin β-D-glucosidase by cell suspension cultures derived from the conifer *Callistris drummondii. Plant Cell Rep.* **9**, 257–260.
61. Uden, W. van, Pras, N., Homan, B. and Malingre, T.M. (1991). Improvement of the production of 5-methoxypodophyllotoxin using a new selected root culture of *Linum flavum* L. *Plant Cell Tiss. Org. Cult.* **27**, 115–121.
62. Konuklugil, B., Schmidt, R.J. and Alfermann, A.W. (1999). Accumulation of aryltetralin lactone lignans in cell suspension cultures of *Linum nidiflorum. Planta Med.* **65**, 587–588.
63. Chattopadhyay, S., Srivastava, A.K., Bhojwani, S.S. and Bisaria, V.S. (2001). Development of suspension culture of *Podophyllum hexandrum* for the production of podophyllotoxin. *Biotechnol. Lett.* **23**, 2063–2066.
64. Chattopadhyay, S., Srivastava, A.K. and Bisaria, V.S. (2002). Optimization of culture parameters for production of podophyllotoxin in suspension culture of *Podophyllum hexandrum. Appl. Biochem. Biotechnol.* **102/103**, 381–393.
65. Chattopadhyay, S., Srivastava, A.K., Bhojwani, S.S. and Bisaria, V.S. (2002). Production of podophyllotoxin by plant cell cultures of *Podophyllum hexandrum* in bioreactor. *J. Ferment. Bioeng.* **93**, 215–220.
66. Chattopadhyay, S., Bisaria, V.S. and Srivastava, A.K. (2003). Enhanced production of podophyllotoxin by *Podophyllum hexandrum* using in situ cell retention bioreactor. *Biotechnol. Prog.* **19**, 1026–1028.
67. Pandey, H., Nandi, S.K., Chandra, B., Nadeem, M. and Palni, L.M.S. (2001). GA(3) induced flowering in *Podophyllum hexandrum* Royle: a rare alpine medicinal herb. *Acta Physiol. Plant.* **23**, 467–474.
68. Xia, Z.Q., Costa, M.A., Proctor, J., Davin, L.B. and Lewis, N.G. (2000). Dirigent-mediated podophyllotoxin biosynthesis in *Linum flavum* and *Podophyllum peltatum. Phytochemistry* **55**, 537–549.
69. Molog, G., Empt, U., Kuhlmann, S., Uden, W. van, Pras, N., Alfermann, A. and Petersen, M. (2001). Deoxypodophyllotoxin 6-hydroxylase, a cytochrome P450 monooxygenase from cell cultures of *Linum flavum* involved in the biosynthesis of cytotoxic lignans. *Planta* **214**(2), 288–294.
70. Hemmati, S., Schmidt, T. and Fuss, E. (2007). (+)-Pinoresinol/(−)-lariciresinol reductase from *Linum perenne* Himmelszelt involved in the biosynthesis of justicidin B. *FEBS Lett.* **581**, 603–610.
71. Hemmati , S., Schneider, B., Schmidt, T.J., Federolf, K., Alfermann, A.W. and Fuss, E. (2007). Justicidin B 7-hydroxylase, a cytochrome P450 monooxygenase from cell cultures of *Linum perenne* Himmelszelt involved in the biosynthesis of diphyllin. *Phytochemistry* **68**, 22–24.
72. Oostdam, A., Mol, J.N.M. and van der Plas, L.H.W. (1993). Establishment of hairy root cultures of *Linum flavum* producing the lignan 5-methoxypodophyllotoxin. *Plant Cell Rep.* **12**, 474–477.
73. Schmitt, J. and Petersen, M. (2002). Influence of methyl jasmonate and coniferyl alcohol on pinoresinol and matairesinol accumulation in a *Forsythia intermedia* suspension culture. *Plant Cell Rep.* **20**, 885–889.
74. Moraes, R.M., de Andrade, Z., Bedir, E., Dayan, F., Lata, H. and Khan, I. (2004). Arbuscular mycorrhiza improves acclimatization and increases lignan content of micropropagated mayapple (*Podophyllum peltatum* L.). *Plant Sci.* **166**, 23–29.
75. Bedir, E., Tellez, M., Lata, H., Khan, I., Cushman, K.E. and Moraes, R.M. (2006). Post-harvest and scale up extraction of American mayapple leaves for podophyllotoxin production. *Ind. Crops Prod.* **24**, 3–7.
76. Moraes-Cedeira, R.M., Burandt, C.L. Jr, Bastos, J.K., Nanayakkara, N.P.D. and McChesney, J.D. (1998) *In vitro* propagation of *Podophyllum peltatum. Planta Med.* **64**, 42–46.

77. Cousins, M.M. and Adelberg, J.W. (2007). *In vitro* plant and organ culture of medicinal and nutraceutical species in laboratory and industrial scales. *ISHS Acta Hortic.* **756**, 95–102.

78. Eyberger, A.L., Dondapati, R. and Porter, J.R. (2006). Endophyte fungal isolates from *Podophyllum peltatum* produces podophyllotoxin. *J. Nat. Prod.* **69**(8), 1121–1124.

79. Puri, S.C., Nazir, A., Chawla, A., Arora, R., Riyaz-ul-Hasan, S., Amna, T., Ahmed, B., Verma, V., Singh, S., Sagar, R., Sharma, A., Kumar, R., Sharma, R.K. and Qazi, G.N. (2005). The endophytic fungus *Trametes hirsuta* as a novel alternative source of podophyllotoxin and related aryl tetralin lignans. *J. Biotechnol.* **122**, 494–510.

Index

V

X

MIX
Papier aus verantwortungsvollen Quellen
Paper from responsible sources
FSC® C105338

If you have any concerns about our products,
you can contact us on
ProductSafety@springernature.com

In case Publisher is established outside the EU,
the EU authorized representative is:
Springer Nature Customer Service Center GmbH
Europaplatz 3, 69115 Heidelberg, Germany

Printed by Libri Plureos GmbH
in Hamburg, Germany